Place Matters

The editors and the publisher gratefully acknowledge the generous support provided by Ecotrust, NSF grant EIA-0113519, NOAA NESDIS through the Cooperative Institute for Ocean Satellite Studies, and an anonymous donor.

ecotrust

ECOTRUST is a non-profit conservation organization dedicated to strengthening communities and the environment from Alaska to California—the rich, beautiful place called Salmon Nation, where rivers flow through the largest temperate rain forest in the world and where humans first settled in North America. Based in Portland, Oregon, Ecotrust works within key sectors of this landscape: forest, fisheries, food and farms, and with native communities.

Place Matters

Geospatial Tools for Marine Science, Conservation, and Management in the Pacific Northwest

edited by Dawn J. Wright and Astrid J. Scholz

foreword by Sylvia A. Earle

Oregon State University Press
Corvallis

The paper in this book meets the guidelines for permanence and durability of the Committee on Production Guidelines for Book Longevity of the Council on Library Resources and the minimum requirements of the American National Standard for Permanence of Paper for Printed Library Materials Z39.48-1984.

Library of Congress Cataloging-in-Publication Data

Place matters : geospatial tools for marine science, conservation, and management in the Pacific Northwest / edited by Dawn J. Wright and Astrid J. Scholz.— 1st ed.

p. cm.

Includes index.

ISBN 0-87071-057-5 (pbk. : alk. paper)

1. Oceanography—Geographic information systems—Northwest, Pacific. 2. Marine sciences—Geographic information systems—Northwest, Pacific. 3. Marine ecology—Geographic information systems—Northwest, Pacific. 4. Fishery management—Geographic information systems—Northwest, Pacific. 5. Marine biological diversity conservation—Geographic information systems--Northwest, Pacific. I. Wright, Dawn J., 1961- II. Scholz, Astrid J.

GC38.5.P53 2005

333.91′6416′09795—dc22

2004026512

 First edition 2005
Printed in China

Oregon State University Press
500 Kerr Administration
Corvallis OR 97331
541-737-3166 • fax 541-737-3170
http://oregonstate.edu/dept/press

Contents

Foreword

Sylvia A. Earle

A map is the greatest of all epic poems. Its lines and colors show the realization of great dreams.

Gilbert Grosvenor, Editor, National Geographic, 1903 –1954

Two centuries ago, much of the northwestern American continent and most of the adjacent sea were unknown to all but the relatively small number of people whose ancestors had made their way across unexplored, uncharted wilderness ages before. Knowledge of what lay beyond the nearshore waters was earned the hard way, as scraps of information from ships' logs were pieced together to gain insight into currents, the location of islands and reefs, and the configuration of the shoreline. Ocean depths were determined by dropping weighted lines over the side, recording their length, and on a chart placing numbers that were somewhere in the vicinity of their actual location. Notes on vegetation and occurrence of wildlife were based on trudging the terrain and having face-to-face encounters with creatures along the way. Despite the daunting difficulties, explorers Meriwether Lewis and William Clark were inspired by the dream of mapping North America, and they succeeded in developing the first comprehensive image of hills and valleys, mountains and rivers that included much of the Pacific Northwest.

Imagine the amazement of Lewis and Clark if they could see this book of wondrously detailed information about places they struggled to comprehend, one data point at a time.

Early in the twentieth century, U.S. President Theodore Roosevelt took a personal interest in making sure that large tracts of the Pacific Northwest were protected as national parks, a safeguard for the nation's natural, historic and cultural heritage against rapid, often destructive exploitation. Similar actions were taken in Canada, resulting in great expanses of land and wildlife remaining in a relatively natural state. In a few decades, rivers were diverted and dammed, forests leveled, and cities developed, in a geological moment forever altering the distillation of thousands of millennia.

Imagine Roosevelt's astonishment if he could see changes over land and seascapes since his time, and how precious the protected areas have become, now engulfed by largely tamed terrain.

Throughout the 1900s, swift changes were also occurring in the sea, both through natural shifts in the shoreline and in the depths beyond, and through human activities. Offshore, fishing fleets ploughed the ocean floor with trawls and sophisticated new technologies were being applied to find, catch and market fish and other ocean wildlife on an unprecedented scale. Even so, from the surface much seemed unchanged. In 1951, Rachel Carson wrote in *The Sea Around Us*, her celebrated tribute to the ocean, that man " . . . has returned to his mother sea only on her own terms. He cannot control or change the ocean as, in his brief tenancy on earth, he has subdued and plundered the continents."

At that time people would not have been familiar with terms such as "geospatial" or "integrated database" and it would take time for them to grasp the capabilities of modern computers, cameras, satellites and sophisticated sensors, but they probably would be quick to comprehend the profound value of these technologies.

> *Imagine Carson's delight if she could read this book and discover how much has been discovered about the ocean in half a century – and her dismay about the enormity of what has been lost.*

In 50 years, more has been learned about the ocean than during all preceding history, largely owing to new technologies our predecessors could barely imagine. At last, even on cloudy days and foggy nights, with global positioning systems, there is no excuse for not knowing where in the world you are, land or sea. Satellites sweep the Earth making precise navigation routinely possible for one and all. Comprehensive knowledge of sea surface temperature and currents, biological productivity, wind speed and direction, salinity and even the configuration of the seafloor thousands of feet under the ocean's surface are gathered by satellites hundreds of miles in the sky.

Most wondrous of all, through geographic information systems (GIS) and related technologies, it is now possible to see our place in the world with new eyes, and to comprehend previously elusive connections among the ocean, atmosphere, land and living things. Linkages are revealed, causes and effects determined, and changes over time evaluated using computer analysis coupled with liberal doses of good, common sense. While less than 5% of the ocean below the surface has been seen directly, let alone explored, we nonetheless now comprehend that the ocean is the great blue engine that drives the way the world works, harbors 97% of the water, shapes climate, weather, planetary chemistry and temperature, generates most of the oxygen in the atmosphere, absorbs much of the carbon dioxide, and provides living space for about 97% of life on Earth.

We also now know that the ocean is not limitless in its capacity to accept noxious wastes, nor can it forever yield endless quantities of wildlife. In 50 years, populations of tuna, swordfish, sharks and other large fish have been reduced by ninety percent. More than a hundred dead zones have developed in coastal regions, worldwide. Coral reefs, kelp forests, seagrass meadows, natural coastal marshes and wetland ecosystems have sharply declined, and even deepwater systems, hosting ancient sponges, corals, and long-lived fish and other sea creatures, are feeling the harsh bite of fishing gear deployed from hundreds or thousands of feet above.

Even in the relatively untrammeled Pacific Northwest, pollution from upstream sources is inexorably altering the chemistry of downstream waters. The land and its living fabric of forests, waters, and wildlife, and the surrounding, living ocean that have made possible the prosperity of people for thousands of years, are now at risk. For the first time, enough data exist to be able see where we have come from, and realistically project where we might be going if we do this thing or that. We can, with good reason, wonder and worry what the consequences to the ocean, to the land, to wildlife, and to ourselves will be if we continue "business as usual."

> *Imagine the wrath of generations to come if, knowing what we know, we choose to allow the last 10% of the big fish to be consumed, or choose not to act to protect what remains of the healthy coral reefs, shallow and deep; the kelp stands, and the great blue heart of the living ocean beyond the coasts.*

Better yet, imagine the joy of those who look at this book a century or two from now and recognize the wisdom, the strength, and the power of the knowledge articulated here.

The key is to not only admire the good, thoughtful scientific analyses embodied here, but to heed the underlying messages. Johann Wolfgang von Goethe observed many years ago that "It is simply not enough to understand, but to act." It is not too late to halt the present avalanche of troubling trends and secure an enduring place for ourselves within the natural systems that sustain us. We can, if we will.

Place matters. . . .

Preface

The technologically sophisticated, spatially integrated database information management and display system known as the geographic information system or GIS, has been heralded as "the telescope, the microscope, the computer, and the Xerox machine of regional analysis and synthesis of spatial data" (Ron Abler, Executive Director, Association of American Geographers, 1988). Marine GIS has been touted as an application holding great potential for mapping and interpreting the ocean environment in unprecedented detail, from the seafloor to the sea surface, and from the nearshore, to the shoreline, the estuary and landward. This book is about how marine GIS is contributing to our understanding of the shores and ocean of the Pacific Northwest (i.e., northern California, Oregon, Washington, British Columbia, Yukon Territory, and Alaska, with some treatment given to southern California as well). With access to a plethora of data made easier by the Internet and the World Wide Web, and with improved, cheaper computer hardware and software, GIS is now functionally available to large segments of the population. This not only increases the utility of GIS for basic science and exploration of the marine environment, but also creates new opportunities for participatory management and cooperative research in the marine realm. The Pacific Northwest is emerging as a veritable hotbed of marine GIS research and development, with practitioners expanding the increasingly crucial technical role of GIS for ocean research, and its application to a variety of ocean science, policy and management issues.

This book is based on a special symposium that was held at the Annual Meeting and Science Innovation Exposition of the American Association for the Advancement of Science (AAAS), February 2004 in Seattle, Washington. For the first time, this prominent meeting, which attracts more science and science policy leaders than any other single meeting in the United States, sponsored a marine science track entitled "Living Oceans and Coastlines," consisting of twelve special symposia on research frontiers in coral reef studies, fisheries, marine mammals and marine birds, aquaculture, ecosystem-based marine management, design and assessment of marine protected areas, and general marine policy. The symposium "Place Matters: Geospatial Tools for Marine Science, Conservation, and Management" was one of two that specifically addressed issues of geospatial data and analysis in the oceans. In the symposium, scientists and practitioners from academic institutions, government agencies, and environmental organizations showcased how they are using marine GIS to handle and exploit present and future data streams from observatories, experiments, numerical models, and other sources, yielding insights into oceanographic, ecological, and socioeconomic conditions of the marine environment in the Pacific Northwest. The authors of chapters in this book were either participants in the AAAS symposium, or were invited to contribute their work on closely related projects elsewhere along the West Coast in order to round out the geographic range and policy applications of the selections combined here. The book is divided into three major sections: the first section is conceptual, laying out selected methods and models for conservation-based marine GIS; the second section describes working examples of marine GIS tools and large-scale implementations; while the third section focuses on the use of GIS by various environmental advocacy and local citizens' organizations. An epilogue includes a brief summary of consensus findings, but, more importantly, covers some of the most pressing theoretical challenges in marine and coastal GIS, and provides some insight into future trends in data access and exchange, representation, and modeling of marine data.

With the emergence of area-based management of fisheries and marine ecosystems, and the importance and prominence of marine and other environmental issues in the Pacific Northwest, this book should be of particular interest to a variety of scientific and lay readers. The accompanying Web site (http://www.ecotrust.org/placematters) includes some of the actual GIS tools, maps, and datasets that were developed or used by chapter authors, the Power Point files of the presentations given at the AAAS session, and some additional Web-based resources. All of these should be particularly helpful for use in high school or university classrooms, or professional workshops. GIS databases and tools are indeed on the leading edge of science, and help integrate various scientific disciplines and data sources to address management questions. With the ability to overlay biological, geological, and socioeconomic information, GIS tools bring together an interdisciplinary scientific audience, as well as managers, conservationists and other constituencies interested in the end products. Both in substance and practice, the use of GIS is pushing the community of researchers and practitioners towards increased collaboration, infrastructure and technical innovations, and creative meshing in hard-hitting policy arenas.

This is an exciting time, as scientists, managers and conservation organizations—often in collaboration with each other—are making advances in the way that data are collected, documented, used, shared, and saved. We hope that this book, and the accompanying materials on the companion Web site (http://www.ecotrust.org/placematters), help to either show the way, or to add inspiration and ideas to current efforts.

We wish to thank two anonymous referees for their careful and thoughtful reviews of all the chapters, which significantly improved the book's manuscript, as well as the support and patience of the editorial staff at Oregon State University Press.

A final note on nomenclature: Usage of terms and units of measurement are kept consistent within each chapter, but may vary across chapters. Typically, authors working in more academic settings use metric units, while researchers working in management contexts use the imperial units of nautical charts and agency datasets. Similarly, people working in fishing communities tend to adopt the term "fishermen" used by both fishing men and women to refer to themselves, rather than "fisher."

Dawn J. Wright, Oregon State University
Astrid J. Scholz, Ecotrust

PART I

Pushing the Spatial Envelope: Selected Models and Methods of Marine GIS

Chapter 1

Biogeographic Assessments of NOAA National Marine Sanctuaries:

The Integration of Ecology and GIS to Aid in Marine Management Boundary Delineation and Assessment

Mark E. Monaco, Matt S. Kendall, Jamison L. Higgins, Charles E. Alexander, and Mitchell S. Tartt

Abstract

The mission of the National Marine Sanctuary Program (NMSP), of the National Oceanic and Atmospheric Administration (NOAA), is to serve as the trustee for a system of marine protected areas to conserve, protect, and enhance biodiversity. To assist in accomplishing this mission, the NMSP has developed a partnership with NOAA's National Centers for Coastal Ocean Science (NCCOS) to conduct biogeographic assessments of marine resources within and adjacent to the marine waters of all national marine sanctuaries over the next five years. NCCOS's Biogeography Program is leading the joint effort to define species distribution patterns and map associated habitats. Biogeography provides a framework to integrate species distributions and life history data with information on the habitats of the region to characterize marine resources in a sanctuary. The biogeographical data are integrated in a geographic information system (GIS) to enable visualization of species' spatial and temporal patterns, and to predict changes in abundance that may result from a

Mark E. Monaco, Matt S. Kendall, and **Jamison L. Higgins**
NOAA/NOS National Centers for Coastal Ocean Science, Center for Coastal Monitoring and Assessment, Biogeography Program, 1305 East West Highway, Silver Spring, MD 20910
Corresponding author: mark.monaco@noaa.gov
matt.kendall@noaa.gov
jamie.higgins@noaa.gov

Charles E. Alexander and **Mitchell S. Tartt**
NOAA/NOS National Marine Sanctuaries Program, 1305 East West Highway, Silver Spring, MD 20910
charles.alexander@noaa.gov
mitchell.tart@noaa.gov

variety of natural and anthropogenic perturbations or management strategies. For example, the biogeographic assessment of three central/northern California sanctuaries was used to delineate "hot spots" based on community metrics (e.g., biodiversity). In addition, accurate and highly resolved digital benthic habitat maps have been developed for Gray's Reef National Marine Sanctuary to define species habitat utilization patterns to identify areas for special protection. Plans are to conduct assessments in all national marine sanctuaries over the next several years through utilization of the biogeographic process.

Background

The mission of the National Marine Sanctuary Program (NMSP), of the National Oceanic and Atmospheric Administration (NOAA), is to serve as the trustee for a system of marine protected areas (MPAs), to conserve, protect, and enhance their biodiversity, ecological integrity, and cultural legacy. Since 1972, 13 national marine sanctuaries, representing a wide variety of ocean environments, have been established, each with management goals tuned to their unique diversity (an additional site is proposed for the Northwestern Hawaiian Island Coral Reef Ecological Reserve). These goals may include restoration, monitoring, protecting healthy areas, and public education and outreach programs to generate understanding about the NMSP's role as a coastal steward. While some human activities in these marine protected areas are regulated or prohibited, achieving compatibility between conservation objectives and among multiple uses such as research, monitoring, commercial, and recreational activities is central to the design of the sanctuary system. Monitoring and managing this range of goals and activities requires an approach that integrates human use with ecology and geography. The science of biogeography can be used to support these complex management challenges, and to ensure successful stewardship of sanctuary resources.

Since their establishment, many of the sanctuaries have witnessed increased pressure on marine resources from natural and anthropogenic phenomena, including climatic variation and degradation of habitats. In order for the NMSP to increase management capabilities, it is imperative that the spatial and temporal distributions of biota and habitats within sanctuaries be delineated. Thus, the NMSP has developed a partnership with NOAA's National Centers for Coastal Ocean Science (NCCOS) to conduct biogeographic assessments of marine resources within and adjacent to the marine waters of all national marine sanctuaries over the next five years (Kendall and Monaco, 2003). The study of biogeography provides a framework to integrate species distributions and life history data with information on the habitats of the region to characterize marine resources in a sanctuary. When the biogeographic data are integrated into a geographic information system (GIS), it enables users to visualize spatial and temporal distributions, and to conduct ecological forecasts of potential changes in species distributions that may result from a variety of natural and anthropogenic perturbations. In addition, based on specific ecological metrics (e.g., species diversity), biologically significant areas can be delineated.

What is Biogeography?

To understand how the NMSP can benefit from implementing biogeographic approaches to management, it is necessary to first define biogeography. A general definition of biogeography is: the study of spatial and temporal distributions of organisms and their habitats, and the historical and biological factors that produced them (Cox and Moor, 1993). The complexity of products from biogeographic analysis range from simple distribution maps for a single species or a particular habitat, to more complex products that combine these simple data layers to create maps of biodiversity or habitat diversity (Nelson and Monaco, 2000; Kendall et al., 2001). More commonly, however, biogeographic products are even more complex and integrate several biological and physical parameters at once. For example, biogeographic analyses often integrate environmental variables with knowledge of organisms' habitat affinities and physiological limitations (Rubec et al., 1999; Brown et al., 2000). This can result in a type of biological "weather map," which is used to express the probability of encountering an organism in a given area (Monaco et al., 1998; Monaco et al., 2001). Other analyses may integrate the distribution of multiple species with some anthropogenic influence to determine how the biotic community is being affected (Christensen et al., 1997; Livingston et al., 2000). Still other products may focus on the changes in distribution of organisms or habitats through time. For example, the displacement of a native species by exotics can be analyzed in time series to predict future impacts and develop containment strategies. Even the response of organisms to climate change and sea-level rise can be predicted and used to examine longer-term management scenarios. These integrated products provide information not only about where specific organisms and habitats may be found, but also about why they are present only in those locations and how their distribution will change through time.

Biogeographic studies may range in spatial scale from continental to individual watersheds, estuaries, or even smaller areas. A single study may encompass several spatial scales to understand the distances over which ecosystem components interact, or they may focus on a single scale to identify the specific details of a particular organism's distribution or other system component. Biogeographic studies may also include a range of time scales which may resolve changes in habitat or species distribution on daily, monthly, seasonally, or even on much longer scales. The scale and resolution of biogeographic studies is often subject to logistical, financial, and technological limitations. Financial and personnel allocations determine the amount of time that can be spent in the field collecting fine-scale information. As a result, the final extent and detail of the biogeographic data used in any study is typically customized to carefully balance logistical constraints while meeting the specific issues and objectives relevant to that locality.

Application of Biogeographic Concepts to Aid Sanctuary Management

The system of national marine sanctuaries encompasses a mosaic of seascape components, oceanographic conditions, and geomorphological features that include coral reefs, kelp forests, whale migration corridors, deep-sea canyons, and even underwater archeological sites. They range in spatial extent from 0.65 km^2 in Fagatele Bay, American Samoa, to over 13,727 km^2 in Monterey Bay, California. Despite vastly different management objectives and seascape features, all sanctuaries are responsible for managing spatial resources and activities. To properly manage these resources, sanctuary staff requires a thorough understanding of resource distribution relative to sanctuary boundaries.

Biogeographic analysis is an ideal tool for sanctuary managers to utilize for the conservation of biodiversity and ecosystem integrity across the spectrum of spatial and time scales that these issues encompass. Even basic biogeographic data layers are lacking in many sanctuaries, such as a simple inventory of organisms and habitat that the sanctuary was designed to protect. Completing a biogeographic assessment of the distribution of such resources within and across sanctuary boundaries is critical for placing them into their wider ecological context and to understanding how the ecosystem composition changes through time. This baseline information is central to all management decisions and is the foundation for more sophisticated analyses. Coupling data layers on the distribution of animals and their habitats with data on human and natural threats provides a powerful analytical tool for managers of those resources. Using the biogeographic approach, managers can explore the potential changes in resource distribution that result from alternative management practices. Scenarios of interest may include a better awareness of possible effects of an oil spill on sanctuary resources, understanding the impact of marine zoning within national marine sanctuaries, or the alteration of current MPA boundaries or regulations governing the resources of a given sanctuary and the surrounding areas (Fig. 1.1: see page 53). The ability to conduct ecological forecasts across the range of spatial and temporal scales encountered by sanctuary managers is a valuable asset when answering inquiries regarding the expected impact of proposed regulations, conveying the trade-offs in resource use among multiple interest groups, and in generating public support during the often contentious period of public comment that precedes changes in management practices.

A Focus on Diverse Management Needs

The goal of this five-year effort is to bring a biogeographically based approach to the management of natural resources within the national marine sanctuaries. More specifically, this effort will be incorporated into the management plan review process for each applicable national marine sanctuary (Table 1.1). Significant areas of this review process where biogeographic analysis can support information needs and decision-making include: environmental characterizations, boundary evaluations, zoning, and threat assessment.

To successfully establish a biogeographic approach to management in the national marine sanctuaries, two objectives must be addressed. First, an assessment of current information holdings and a prioritized list of data needs must be completed for each sanctuary. Second, for this assessment to have lasting benefits, the ability to repeatedly collect, manipulate, and analyze biogeographic data must be established at each site. This will be achieved by conducting a collaborative biogeographic study, bringing together the NMSP and NCCOS staff to address a specific management issue at each sanctuary.

Representative Biogeographic Assessment

The following text describes representative results from implementation of the biogeographic approach that can be applied to each applicable national marine sanctuary (Table 1.1). The initial biogeographic

Table 1.1. Status of biogeographic studies at NOAA National Marine Sanctuaries.

NMS Site	*Biogeographic Activity*	*Status*
Monterey Bay	Biogeographic Characterization	Completed
Gulf of Farallones	Biogeographic Characterization	Completed
Cordell Bank	Biogeographic Characterization	Completed
Gray's Reef	Benthic Habitat Mapping	Completed
Fagatele Bay	Benthic Habitat Mapping	Underway
NW Hawaiian Is. CR Ecosystem Reserve	Biogeographic Characterization	Underway
Channel Islands	Boundary Alternative Assessment	Underway, Year 2
Stellwagen Bank	Biogeographic Data for Management Plan Review	Underway, Year 2
Olympic Coast		Underway
Florida Keys		TBD
Flower Garden Banks		TBD
Hawaiian Islands Humpback Whale		TBD
Monitor NMS		TBD
Thunder Bay		TBD

assessment outlined in the five-year NCCOS/NMSP plan was implemented in the spring of 2001 to conduct a 24-month investigation to assess biogeographic patterns of selected marine species found within and adjacent to the boundaries of three contiguous West Coast national marine sanctuaries (Fig. 1.2: see page 53; NOAA National Ocean Service, 2002). These sanctuaries—Monterey Bay, Gulf of the Farallones, and Cordell Bank—are conducting a joint review to update sanctuary management plans. To support the management plan review process, the Biogeography Program is leading a partnership effort to conduct a robust analytical assessment to define important biological areas and time periods within the region. Phase I of this project was recently completed and provided data, analytical results, and descriptions of ecosystems and their linkages; it also identified data gaps, and suggested future activities now underway in Phase II (NOAA National Centers for Coastal Ocean Science, 2003).

Phase I of this effort was a biogeographic assessment of existing data on the distribution and abundance of marine fishes, marine birds, marine mammals and their associated habitats. The study did not attempt to define biogeographic patterns along the entire U.S. West Coast nor in nearshore environments (e.g., estuaries). Rather, the study area was restricted to the marine area from Point Arena in Mendocino County (38°54′32″ N, the northern bound) to Point Sal in northern Santa Barbara County (34°54′05″ N, the southern bound). This relatively large study area enabled the assessment to extend beyond the limits of current sanctuary boundaries to place study results in the context of northern/central California Coast biogeographic patterns. Results of this assessment are being used to assist the NMSP in addressing issues such as evaluating potential modification of sanctuary boundaries and changes in management strategies or administration, based on the principles of biogeography.

The biogeographic assessment was formulated around three closely integrated study components: (1) an Ecological Linkages Report, (2) biogeographic analyses, and (3) development of GIS data for incorporation into NMSP's Marine Information System (MarIS) (Figs. 1.1 and 1.3: see pages 53 and 54). The majority of results from the assessment were presented as a suite of GIS maps to visually display biogeographic patterns across the study area. The body of the document provided examples of the entire suite of digital map products found on a companion CD-ROM. The spatial data and additional information, such as digital species distribution maps and additional details on analytical methodologies, were also presented on the CD-ROM.

The overall project process, milestones, and associated time frame are shown in Fig. 1.3. The following text provides an overview of each of the study components and examples of associated biogeographic products to demonstrate the types of information that can be generated across the system of national marine sanctuaries.

Ideally, biogeographic assessments utilize significant amounts of data that have been collected over the entire spatial extent of the study area over a long time period. However, such a wealth of data is rarely available. In many instances, little information exists to accurately characterize the study area or associated living marine resources. This paucity of comprehensive data can limit the efficacy of biogeographic assessments, but additional analytical methods can be used to complement the assessment. In addition to analysis of databases, two additional tasks were used to conduct the assessment. First a synthesis of existing information was compiled and presented in an Ecological Linkages Report to provide background information on species, habitats, and a general ecological characterization of marine ecosystems and linkages within the study area. Second, species habitat suitability modeling was conducted for fishes to define potential species distributions based on known habitat affinities and physiological limitations (Brown et al., 2000).

In addition, a critical component of the assessment process was the extensive effort to have the data, analytical approaches, and results peer reviewed. Initial analytical results were presented to experts familiar with the marine ecosystem off northern/central California, as well as to the originators of the data sources, in an attempt to improve the analyses (NOAA National Ocean Service, 2002). The role of expert review and input was considerable, and the contributions made by experts significantly enhanced the assessment. Thus, the integration of the synthesis of ecological linkage information, statistical analyses, species habitat suitability modeling, and peer review resulted in this biogeographic assessment product.

Study Components

The Ecological Linkages Report provides the context to understand overall biogeographic product results, relative to the biogeography of the U.S. West Coast (Airamé et al., 2003). The bulk of the report describes ecosystems in the region, key species associated with these ecosystems, ecosystem status, and linkages among them. The report presents the latitudinal range distributions of species groups, such as invertebrates, fish, marine birds and marine mammals. These maps provide an overview of marine species' distributions along the entire west coast of North America by documenting the accepted northern and southern range endpoints of species that occur in all or part of this region. In addition, the report identifies gaps in current knowledge about regional ecosystems.

Biogeographic Analyses. This study component introduces the methods used to conduct the assessment and the results of the biogeographic analyses. This component of the assessment is the cornerstone of the overall biogeographic product to support the NMSP joint management plan review process. The data, analyses, and supporting information are linked using statistical and GIS tools to portray in space and time significant biological areas or "hot spots." The term "hot spot" is defined based on specific criteria or metrics (e.g., species diversity, high species abundance). The vast majority of the analytical results are displayed as a series of maps to visualize where the analyses identified biologically significant areas.

There are many different ways to analyze and organize biogeographic information; however, to efficiently support the management plan process, only a limited number of analytical options were invoked. These analyses were selected based on reviewers' comments on the Project's Interim Atlas Product (NOAA National Ocean Service, 2002), feedback from technical review meetings, and peer-review workshops. Thus, a very difficult step in the project was to select and rely on the most appropriate analyses to characterize the various components of the marine ecosystem that exist in the study area. The inclusion of the GIS-based products on the companion CD-ROM will enable NOAA staff, advisory councils, and research partners to query data and information relevant for questions and issues that are not specifically addressed in this product.

The first analyses focused on a suite of assemblage analyses to assess the biogeography of fishes and a few macro-invertebrates. Primary data included fisheries-independent data, such as those collected by researchers from the National Marine Fisheries Service (NMFS; Fig. 1.4: see page 54), and fisheries-dependent data, such as those collected by the California Department of Fish and Game for recreational fisheries. These datasets, although not spatially or temporally comprehensive, are the most robust datasets that exist for the entire region, and provide considerable information on the distribution of several hundred fish and invertebrate species.

Species Habitat Suitability Models. Due to limitations in the spatial and temporal extent of data and to complement the assemblage analyses of fishes, species habitat suitability index (HSI) models were developed (Brown et al., 2000). This was done primarily to accommodate the paucity of empirical data in nearshore areas and to target species of special significance to the sanctuaries. An extensive literature review of the life history characteristics of individual species resulted in information on species' habitat affinities that were converted into quantifiable habitat suitability index values (Monaco et al., 1998). The life history information and associated species habitat suitability index values are found on the CD-ROM. These derived values were input into an equation and used to predict potential distributions based on an affinity for the mosaic of bathymetry and bottom habitats found throughout the region (Fig. 1.5: see page 55). The species habitat suitability models were validated through statistical and spatial analyses, using fishery-independent survey data.

Marine Birds. The Biogeography Program contracted H.T. Harvey & Associates and R.G. Ford Consulting to define and assess biogeographic patterns of marine birds found within the study area. These experts used multivariate statistical methods and GIS to develop a series of maps that displayed seasonal marine bird distributions, estimated densities, and diversity (Fig. 1.6: see page 55). The results are reported as maps (i.e., hot spots) and associated data tables to visualize important locations and time periods for marine birds in the study area. Phase II of the assessment will present a robust technical report on the methods and results summarized in the Phase I map and tabular products.

Marine Mammals. The Biogeography Program contracted H.T. Harvey & Associates and R.G. Ford Consulting Co. to work with the project team and local marine mammal experts to identify biogeographic patterns and important areas and time periods for marine mammals occurring in the study area. In addition, NOAA/NMFS scientists provided marine mammal sightings data along the entire West Coast to aid in analyzing marine mammal biogeographic patterns relative to the study area. These experts used a GIS to develop a preliminary series of maps that show occurrence patterns and important areas and time periods for 13 marine mammals in the study area. Phase II of this assessment will incorporate additional data, include additional marine mammal species maps, and map selected community metrics. A robust technical report on this work is also planned.

Integration Analyses. Many possible combinations of the data layers could be integrated for the biogeographic assessment. In most instances, however, it was not appropriate to integrate all results across taxa. Therefore, to minimize confounding of results and to focus on the "protection of biodiversity" component of the NMSP mission, the integration of patterns in species diversity and density was utilized to define biologically significant areas across species groups. In addition, results of individual species habitat suitability models were integrated across species. Thus, an approach was developed to integrate individual species habitat suitability models into a single cumulative suitability metric indicating areas of high potential groundfish abundance. These results were coupled with fish and marine bird metrics to define a map of integrated biological hot spots (Fig. 1.7: see page 56). In an attempt to achieve the most explanatory and informative integration of the diversity and cumulative suitability results, analyses were conducted to detect recurring spatial patterns that were present among the multiple species groups. Thus, areas that showed significant biological concentrations, high species diversity, or usage by multiple species groups were delineated. These areas of significant biological importance contributed to defining and assessing biogeographic patterns within the study area (Fig. 1.7).

Concluding Comments

Spatially explicit biogeographic assessments provide a robust set of analytical results and GIS data to strengthen the sustainable management of marine resources within and adjacent to the national marine sanctuaries. A primary use of the biogeographic assessments will be to support the NMSP as it continues to conduct management plan reviews and updates for all national marine sanctuaries. In addition, the Biogeography Program will assist the NMSP in further analyses and presentations of the data and analytical results to address specific research and management questions. Plans are to continue to implement the biogeographic approach across all national marine sanctuaries as described in Kendall and Monaco (2003).

Acknowledgements

The authors wish to thank all of the partners who have provided data and reviewed the process and initial analytical results of the national marine sanctuary biogeographic assessments. The evolving projects would not be possible without the contributions of data, information, references, and efforts of many colleagues in federal, state, private sector, and non-governmental organizations. We especially thank our Biogeography Team members who have led or provided support in the development of assessment products.

References

Airamé, S., Gaines, S., and Caldow, C., 2003. Ecological Linkages: Marine and Estuarine Ecosystems of Central and Northern California, NOAA Technical Memorandum, Silver Spring, MD, NOAA, National Ocean Service, 164 pp.

Brown, S. K., Buja, K. R., Jury, S. H., Monaco, M. E., and Banner, A., 2000. Habitat suitability index models for eight fish and invertebrate species in Casco and Sheepscot Bays, Maine, *North American Journal of Fisheries Management*, 20: 408-35.

Christensen, J. D., Monaco, M. E., and Lowery, T. A., 1997. An index to assess the sensitivity of Gulf of Mexico species to changes in estuarine salinity regimes. *Gulf Research Reports*, 9:219-29.

Cox, C. B., and Moore, P. D., 1993. *Biogeography, an Ecological and Evolutionary Approach*, 5th edition, Oxford, England, Blackwell Scientific Publications, 326 pp.

Kendall, M. S., and Monaco, M. E., 2003. Biogeography of the National Marine Sanctuaries: A Partnership Between the NOS Biogeography Program and the National Marine Sanctuary Program, NOAA Technical Memorandum, NOAA. Silver Spring, MD, NOAA, 8 pp.

Kendall, M. S., Kruer, C. R., Buja, K. R., Christensen, J. D., Finkbiener, M., and Monaco, M. E., 2001. Methods Used to Map the Benthic Habitats of Puerto Rico and the US Virgin Islands, NOAA Technical Memorandum, NOS NCCOS CCMA 152 (On-Line), Available from U.S. National Oceanic and Atmospheric Administration. National Ocean Service. National Centers for Coastal Ocean Science. Biogeography Program, (CD-ROM).

Livingston, R. J., Lewis, F. G., Woodsum, G. C., Niu, X.–F., Galperin, B., Haung, W., Christensen, J. D., Monaco, M. E., Battista, T. A., Klein, C. J., Howell, R. L., IV, and Ray, G. L., 2000. Modeling oyster population response to variation in freshwater input, Estuarine, Coastal, and Shelf *Science*, 50:655-72.

Monaco, M. E., Christensen, J. D., and Rohmann, S. O., 2001. Mapping and monitoring of US coral reef ecosystems, *Earth System Monitor*, 12(1): 1-7.

Monaco, M. E., Weisberg, S. B., and Lowery, T. A., 1998 Summer habitat affinities of estuarine fish in U.S. mid-Atlantic coastal systems *Fisheries Management and Ecology* 5: 161-71.

Nelson, D. M., and Monaco, M. E., 2000. National Overview and Evolution of NOAA's Estuarine Living Marine Resources (ELMR) Program NOAA Technical Memorandum NOS NCCOS CCMA 144. Silver Spring, MD NOAA, NOS, Center for Coastal Monitoring and Assessment, 60 pp.

NOAA National Centers for Coastal Ocean Science (NCCOS), 2003. A Biogeographic Assessment off North/Central California: To Support the Joint Management Plan Review for Cordell Bank, Gulf of the Farallones, and Monterey Bay National Marine Sanctuaries: Phase 1-Marine Fishes, Birds and Mammals, NOAA Technical Memorandum, Silver Spring, MD. 145 pp.

NOAA National Ocean Service, 2002. NOAA's National Centers for Coastal Ocean Science and National Marine Sanctuary Program. Interim Product: A Biogeographic Assessment off North/Central California: To Support the Joint Management Plan Review for Cordell Bank, Gulf of the Farallones and Monterey Bay National Marine Sanctuaries, NOAA Technical Memorandum, Silver Spring, MD. 38 pp.

Rubec, P. J., Bexley, J. C. W., Norris, H., Coyne, M. S., Monaco, M. E., Smith, S. G., and Ault, J. S. 1999. Suitability modeling to delineate habitat essential to sustainable fisheries, American Fisheries Society Symposium, 22:108-33.

CHAPTER 2

Mapping Global Fisheries Patterns and Their Consequences

Reg Watson, Jackie Alder, Villy Christensen, and Daniel Pauly

Abstract

Despite increasing reports of fisheries collapses world wide, investigations of the effects of fishing on the global marine environment have been constrained by the paucity of fisheries landings data on suitable spatial scales. Working to overcome this, we have developed new databases and approaches that demonstrate basin-scale reductions in biomass and landings due to intensifying fishing effort, and equally disturbing, reductions in the size and trophic level of species landed. Starting with international, regional, and national datasets acquired from many sources, we have collated global datasets and mapped fisheries landings from 1950 to the present to a system of 30-min spatial cells. To facilitate this, we have also developed databases describing the global distribution of all fished species, as well as the fishing patterns/access rights of all fishing nations. Our methods effectively "reverse engineer" landing records to approximate the original catch patterns. The process includes disaggregation of records bundled as "miscellaneous fishes" and the "de-flagging" of reflagged fishing vessels. Our results have revealed rich evidence of dramatic change, which includes declines in catch, reductions in fish length, and a general reduction in the tropic level of landings. The analyses have also uncovered major problems in 'official' datasets, including significant over-reporting that has masked decades of decline. As we proceed, we remain committed to making our data available through our Web pages at www.seaaroundus.org.

Introduction

To say that the commercial fisheries of the world require careful management would be contested by few people, largely because we read daily of fisheries collapses, lost livelihoods, and shattered fishing

Reg Watson, Jackie Alder, Villy Christensen and **Daniel Pauly**
Fisheries Centre, University of British Columbia, 2259 Lower Mall, Vancouver, B.C. Canada V6T1Z4
Corresponding author: r.watson@fisheries.ubc.ca; fax: 604-822-8934

communities. To continue that managers require information about the impact of past fishing on commercial stocks and on the ecosystems that support them would also rarely be opposed. It would, however, surprise many to learn that the information available to do this is sadly often not up to the task. Some developed countries have comprehensive reporting systems, but many more countries do not. Beyond records of port landings, there is often scant information on where fish are actually caught, even as fisheries are pursued farther offshore following the demise of valuable inshore stocks.

Besides national fisheries statistical systems, there are a number of regional fishery management bodies. These are most highly developed in the North Atlantic, where most catch is reported from relatively small statistical areas. Elsewhere, as in the Central Pacific, reporting areas are enormous (Fig. 2.1: see page 56). Statistical reports from such areas do not allow the kind of analysis that managers often require. Indeed, in many coastal waters it is not possible to closely estimate catches at the small spatial scale often used in ecosystem models. Clearly more spatial precision is required.

The Sea Around Us Project (SAUP; www.seaaroundus.org) has a mandate to examine the impacts of fishing on the marine environment. This requires global fisheries data on a scale fine enough to use in ecosystem models. In short, it requires a new approach to working with existing fisheries data, one that uses all possible secondary data sources to help map fisheries landings with a precision not previously possible. Whereas original reporting areas may measure up to 48 million km^2 (such as FAO's Eastern Central Pacific area), our methods allocate landings to a system of spatial cells measuring only 1/2 degree latitude by 1/2 degree longitude—or averaging under 1,400 km^2 in area. This means we have more than 180,000 spatial cells for which we estimate fisheries landings—a formidable task, but worth the effort.

Fisheries managers also have limited information about the future of fisheries under varying potential management policies, which can impact on their willingness to change current approaches to management. Ecosystem approaches to fisheries management have been mandated by many agencies in recent years. Current developments in modeling software and improved knowledge of fisheries through the mapping work of the SAUP and others have enabled researchers to quantitatively explore the future fisheries of specified ecosystems. The improved understanding and role of scenarios also provides a qualitative framework for describing what could happen.

Scenarios are plausible, challenging, and relevant stories about how the future might play out. They widen our perspective on what the future can include and highlight key issues that might have been missed or dismissed. Using scenarios allows us to re-think the present and pursue changes so that we can influence the future. Using scenarios such as those being investigated by the Millennium Ecosystem

Assessment process (Millennium Ecosystem Assessment, 2003), in combination with quantitative ecosystem models, we describe the possible futures of fisheries using examples from the Gulf of Thailand, the Central North Pacific and the Benguela Current off the west coast of southern Africa.

Methods

Data Sources – Getting the Best Mix

For many parts of the world we need to rely on data provided by the Food and Agriculture Organization (FAO) of the United Nations (UN) which are supplied voluntarily by UN member states. These "official" data vary considerably in quality. They may be biased (Watson and Pauly, 2001) or just incomplete, and they do not include either illegal or discarded catch—they can more accurately be thought of as "reported landings." As the reporting areas for the global data are typically very extensive, it is advantageous to use regional datasets, which typically use smaller reporting areas, whenever possible. Such data exist, for example, for the Northeast Atlantic from the International Council for the Exploration of the Sea (ICES), for the Northwest Atlantic from the Northwest Atlantic Fisheries Organization (NAFO), and from regional UN bodies in the Mediterranean, and along the western coast of Africa. In theory there are other regional bodies that use more compact reporting areas than those used in the global FAO dataset, but in practice these are not readily available except from groups such as ours. Most regional bodies have data starting in the early 1970s, whereas FAO's global landings dataset starts in 1950. By 1950, large-scale fisheries were already well advanced in the North Sea. The FAO data therefore do not provide a glimpse of the unfished past. In addition, there are national datasets. Our project has had strong cooperation from many countries, and we are, as an example, currently using data from Canada for its east coast as these afford even more precise reporting areas than our initial starting point. Future versions will incorporate smaller scale datasets from many areas.

Allocation of Fisheries Landings

Most landings statistics contain clues, e.g., species-specific information, that make it possible to assign the records to a comparatively small portion of the whole nominal area. Species have known geographic distributions, or at the very least, they have preferences and some have very limited ranges. Knowing the range of the species reported is of great help in determining where the catch could have come from. You cannot catch a given species where is does not occur, and it is more likely to have come from the parts of its geographic distribution where it is most abundant or accessible than from the extremes of its range.

Similarly, we usually know the year the landing was reported, and which country reported landing it. Since the declaration of exclusive economic zones (EEZ) or fishing zones by most maritime states in the 1970s and '80s, access to the coastal waters of many countries has been limited. It is common for fishing arrangements to be required before other countries' fleets may access sovereign waters. While these arrangements vary according to the acceptance of maritime claims and the surveillance/enforcement capabilities of the countries involved, the threat of eventually attracting detection, fines, penalties, and seizures leads most fleets to accept some sort of arrangement. Using databases of such fishing arrangements, and observations of fishing activities by foreign fleets, it is often possible to further reduce the possible area from which reported landings were actually taken. As the coastal waters (EEZ) presently account for about 90% of global landings, it is very valuable to know who is fishing, and where, in order to increase the spatial precision of landing reports. If a reporting county may not, or does not, access the coastal waters of another country, then this area can be eliminated as a possible source for the reported catch.

Eliminating from the statistical areas reported in the official datasets both those areas that are outside the distribution of the animals caught and those where the reporting country may or simply does not fish greatly increases spatial precision (Watson et al., 2004). Furthermore, it is possible to create a gradient of likelihood in a system of global spatial cells (30 min longitude and latitude) based on the relative abundance of the reported taxon, which allows reported catch to be prorated amongst a collection of spatial cells. Cells where the fishing nation does not, or—for lack of a fishery access agreement—may not fish can be eliminated as possible locations for the reported catch. Each catch record is processed in turn, with the catch rate of each spatial cell adjusted accordingly. The resulting spatial database allows for queries stratified by year, country fishing, and/or taxon fished. Because cells also belong to collections representing major statistical areas, the exclusive economic zone of coastal states, large marine ecosystems, and other groupings, results can be presented for any of these aggregate categories.

There are several complications in using this method. First, it requires a spatial database showing the geographic distribution of all commercial species. We were able to build such a database by adapting and adding to existing efforts (Froese and Pauly, 2000; FAO, 2001). Second, and more challenging, was the construction of a database of fisheries access arrangements. This task had already been started by the FAO in their Farisis database (FAO, 1998), but required updating and the creation of a complementary database that records observations of countries fishing in the waters of other nations (many arrangements are not in the public domain for obvious commercial reasons). Such fishing arrangements are usually limited to certain species or gear types, and there may be quotas imposed on how much may be caught. These qualifications and limitations must also be considered in our process for spatially allocating landings.

Even with the best "cocktail" of data sources, the starting data are still problematic. Considerable uncertainty exists around the question of what the reported catch actually consists of, i.e., its taxonomic or biological composition. For example, if we are to use the geographic distribution of the animal to limit where catches of it originate, we must first identify it, and usually to at least the family level. All too often, considerable portions of landings reported by nations are simply described as "miscellaneous" or the equally unfortunate "nei" (not elsewhere included). Whether for sake of expediency, lack of resources, taxonomic problems, or other reasons, there is no useable biological identification for considerable segments of global fishery landings statistics. Countries like China and North Korea have the largest tonnage of landings provided with these vague labels, while some smaller tropical countries may have their entire national catch reported this way. FAO staff work hard to improve reporting statistics and even attempt to interpret and recover missing information, but unfortunately some datasets remain incomplete. Our approach has been to use the spatial and temporal distribution of landings by species, genus, and families to guide the proportioning of "miscellaneous" taxa into useable taxonomic groups.

Our method is further complicated by determining "who" (which country) is taking the catch. This is the issue of "reflagging" or the use of flags of convenience for fishing vessels. This practice is very prevalent, but fortunately the European Parliament (2001) has tried to identify the frequency and trends in this practice. Our approach entails recognizing which vessels are likely "reflagged," as this influences what coastal waters the vessel can access. For example, if a vessel from Spain has been reflagged to the country of Belize, then the landings reported by Belize may include landings from coastal waters where vessels from Spain have access, even if these same waters are not usually accessible by vessels from Belize.

Last, but not least, is the problem of illegal, unreported and unregulated (IUU) catches. This is a serious problem when catch data is supposed to be used in the analysis of the impacts of fishing on the marine environment. For example, much of the non-target species killed by trawlers is never reported. Aside from the mortality caused by the fishing gear on the bottom, sometimes as much as eight times the weight of animals is discarded as retained. (Alverson et al.,1994). Quotas, trip limits and commercial expediency encourage the discarding of less valuable, under-sized, or over-quota animals. Without unbiased observers, this catch is never documented. Our project is addressing these reporting problems and will be attempting to map various types of IUU catches on the same spatial scale as is used for those officially reported landings (Pitcher et al., 2002).

The SAUP process for allocating global landings to spatial cells is one of constant refinement. New, more detailed data are obtained to replace more general data. National datasets replace regional datasets that replace global ones. The allocation process itself uncovers errors in the reporting process—some notable examples of such data artifacts include species reported caught in areas where the animal does not occur, obvious reflagging of fishing vessels that mask the true fishing country, or even significant biases in the statistics themselves (Watson and Pauly, 2001). For the last four years, we have refined the method, included more detailed and comprehensive data sources, produced rules to correct misidentifications and reporting biases, and improved the databases used in the process. The results from our allocation process, aggregated by EEZ, large marine ecosystems (LME), and for areas of the high seas, are available on our Web site (www.seaaroundus.org). We have also made the distributions of the target species analyzed in the SAUP available for inspection and comment. Similarly, the contents of our databases that describe fishing access arrangements and observations of foreign fishing are available on-line. We are currently engaged in a process of contacting experts from maritime nations to inspect our results and offer suggestions and comments. This process will lead to further refinement and greater collaboration with the providers and users of fisheries data.

Ecosystem Modeling

Using Ecopath with Ecosim (EwE; Christensen and Pauly, 1992; Walters et al., 1997; Pauly et al., 2000; Walters et al., 2000; Christensen and Walters, 2004), we used the data compiled using the above methods to explore four scenarios developed as part of the international Millennium Ecosystem Assessment (Millennium Ecosystem Assessment, 2003). EwE is an ecological modeling software suite for personal computers, some components of which have been in development for nearly two decades. The approach is thoroughly documented in the scientific literature with over 100 ecosystems models developed to date (for a literature list, see www.ecopath.org). EwE uses two main components: Ecopath—a static, mass-balanced snapshot of the system, and Ecosim—a time-dynamic simulation module for policy exploration that is based on an Ecopath model.

Ecosim can be used in "gaming" mode, where the user can explore policy options. This is achieved by "sketching" fishing rates over time and examining the results (catches, economic performance indicators, biomass changes) for each sketch. In addition, formal optimization methods can be used to search for fishing policies that would maximize a particular policy goal or "objective function" for management. The objective function represents as a weighted sum of the four objectives:

economic, social, legal, and ecological. Assigning alternative weights to these components is a way to see how they conflict or tradeoff with one another in terms of policy choice.

The goal function for policy optimization is defined by the user in Ecosim, based on an evaluation of four weighted policy objectives:

- maximize fisheries rent;
- maximize social benefits or value of landings;
- maximize mandated rebuilding of species;
- maximize ecosystem structure or "health."

Maximizing profits is based on calculating profits as the value of the catch (catch * price, by species) less the cost of fishing (fixed + variable costs). Meeting this objective often results in phasing out all but the most profitable fleets, and the elimination of ecosystem groups competing with or preying on the more valuable target species. The derived fishing effort is often lower than the current, as profit may be reduced by lowering the effort.

Social benefits are calculated as number of jobs relative to the catch value, and are fleet specific. Therefore, social benefits are largely proportional to fishing effort. Optimizing effort often leads to even more extreme (with regards to overfishing) fishing scenarios than optimizing for profit.

Mandated rebuilding of species (or guilds) explores policies that focus on preserving or rebuilding the population of a given species in a given area. This corresponds to setting a threshold biomass (relative to the biomass in Ecopath) for the species or group, and optimizing towards the fleet effort structure that will most effectively ensure this objective. The outcomes of this policy option are case-specific.

Maximizing ecosystem structure (or "health") uses E.P. Odum's ecosystem "maturity" concept, where large, long-lived organisms dominate (Christensen, 1995). In this case, optimizing a group-specific biomass/production ratio as a measure of longevity often results in a reduction of fishing effort for all fleets, except those targeting species with low weighting factors.

Fisheries Scenarios

Four scenarios were harmonized with three ecosystem models as described in the section that follows,using EwE to investigate the future of fisheries to 2050.

Humanity uses scenarios to think about the future all of the time, much of it through narrative storytelling. In the last 50 years, however, the use of scenarios has expanded to include quantitative approaches. The storylines are developed to be plausible, challenging, and relevant descriptions of future events that may take place, and not necessarily what will take place. Scenarios also offer support for more informed and rational decision making in the present and in the future.

The use of scenarios began in the 1960s, but it was not until the 1970s that they were used to link natural resources sufficiency to population growth and consumption that led to its use today. In the fisheries sector, the use of scenarios began in the late 1980s. Since then, there have been six studies that use scenarios with different foci to describe how fisheries may develop past 2010 (Table 2.1).

The Scenarios Working Group of the Millennium Assessment developed the scenarios used in this study over several months of consultation in 2003. The four scenarios are Global Orchestration, Order from Strength, Adapting Mosaics and Technogarden, and they describe the range of possible future directions policy-makers may take in the management of ecosystems. The scenarios differ by the specific conditions necessary for change, proposed rate of change, and the policies that are needed to facilitate change.

GLOBAL ORCHESTRATION

This scenario is about finding the right balance between ecological structure, economic rent, and social benefits. In this scenario, many regional fishing agreements as well as the Fish Stocks and Compliance agreements are strengthened and implemented, and perverse subsidies are reduced or eliminated. Areas where there is good governance will see improvements in selected fisheries, especially for those of high economic value and not necessarily those species integral to ecosystem stability and structure. Some fisheries with extensive ecosystem impacts are phased out and habitats closed to allow the area and stocks to recover. Depending on the degree of degradation, it may take decades to see improvement using these measures. In areas where climate change has a severe impact (e.g., Caribbean reef fisheries), improvements may not eventuate. Effort will be reduced in some fisheries through economic incentives (e.g., bycatch reduction devices), and marine protected areas (MPAs) will be used to improve fisheries habitats. Areas where high exploitation combined with habitat destruction continues will see local stock collapses.

ORDER FROM STRENGTH

In this scenario, fishing and environmental management agreements break down, and impacts such as climate change continue, and in some areas, intensify. Despite these breakdowns, distant-water fleets from rich nations continue to expand into waters of countries not as severely impacted from overfishing to secure fish food as well as fishmeal for national food production. Domestically, fisheries are sustainably managed. This approach does not necessarily maintain ecosystem functions and services. Since some nations will focus on export-driven, high-value fisheries, which are often low-trophic, short-lived

text continues on page 23

Table 2.1: Summary of harmonizing the storylines and EwE models

Global Orchestration

Gulf of Thailand

2000-2010	Optimize profits from shrimp and jobs (70/30), climate change moderate to high
2010-2030	Optimize profits, jobs and then ecosystems (50/30/20), climate change impact reducing
2030-2050	Optimize profits and ecosystem (biomass) (50/50)

North Benguela

2000-2010	Optimize profits and jobs (50/50) Climate change moderate
2010-2030	Optimize profits and jobs (30/70) climate change moderate to low, increase catch of fish for fish food
2030-2050	Optimize profits, jobs and ecosystems (50/20/30), increase the catch of small pelagics

Central North Pacific

2000-2010	Optimize profits from tuna and jobs (80/20) – climate change low
2010-2030	Optimize profits from tuna and jobs (70/30), climate change stable
2030-2050	Optimize profits and ecosystems (50/50) rebuilding of bigeye

Order from Strength

Gulf of Thailand

2000-2010	Optimize profits of the invertebrate fishery and jobs (50/50)
2010-2030	Optimization mix continues (50/50) but effort increasing since Thailand feels the effects of national EEZs and despite agreements it has no room to expand Distant Water Fleet (DWF) which is now concentrated in Gulf of Thailand.
2030-2050	Climate change has significant impact (high impact) and ecosystem severely destabilized; rebuilding stocks of demersal species continues with objective of optimizing jobs rather than profits.

North Benguela

2000-2010	Optimize profits and jobs (50/50) of high value fisheries; DWF increases effort (moderate to high of current species as EU pushes for food security & Africa's debt mounts.
2010-2030	Climate change starts low with build-up over this decade to medium impact. Rebuilding of biomass starts late in this period but there is still concern with maintaining jobs. (30/50/20)
2030-2050	Mix of profit and job optimization (60/40). Increased fishing effort with switch through time to fishmeal species for domestic and international aquaculture operations and also internal food security

table continues

Central North Pacific

2000-2010 Optimize profits from the tuna fishery as well as jobs (75/25); distant-water fishing effort remains stable since countries focused on national issues

2010-2030 Optimize profit and jobs (85/15). Japan returns to drift netting. DWF has moderate increase as US secures food and increases presence in Pacific for security.

2030-2050 Profit optimization not as important as jobs (60/40). Japan stops drift netting by 2040; DWF effort remains stable

Adapting Mosaics

Gulf of Thailand

2000-2010 Optimize profits of the invertebrate fishery and jobs (70/30)

2010-2030 Climate change starts in earnest (moderate to high impact), optimize for profits, shift to rebuilding stocks of demersal species starts

2030-2050 Climate change has significant impact (high impact) and ecosystem severely destabilized, rebuilding stocks of demersal species continues with objective of optimizing jobs rather than profits.

North Benguela

2000-2010 Optimize profits and jobs (40/60) and maintain food and fishmeal fisheries

2010-2030 Climate change starts low with build up over this decade to medium impact. Rebuilding of biomass starts late in this period but there is still concern with maintaining jobs. (30/50/20)

2030-2050 Climate change continues to high impact with some destabilization of the system, food security becomes an issue and therefore focus is on maximizing biomass for fish feed since it goes to aquaculture that ensures a stable supply of food. (0/100/0)

Central North Pacific

2000-2010 Optimize profits from the tuna fishery; turtle exploitation ceases

2010-2030 Climate change minimal if any impact, severe exploitation of bigeye until close to 2030 when stock rebuilding commences, at the same time shift to optimizing for jobs with profit (70/30)

2030-2050 Climate change has a low impact, bigeye rebuilding continues, optimize for ecosystem especially for top predators. International MPA to rebuild stocks. (50 /50)

Techno Garden

Gulf of Thailand

2000-2010	Optimize profit
2010-2030	Optimize pelagic catch (cost of fishing lower) followed by ecosystem optimization (since impacts can be engineered)
2030-2050	Optimize pelagic catch – by 2040 ecosystem irrelevant due to technology advances – profits maximized by using Gulf of Thailand to produce quality fishmeal for prawn aquaculture.

North Benguela

2000-2010	Optimize profit
2010-2030	Optimize profits while increasing pelagics (50/50) for fish food since technology makes aquaculture widespread and demand for fish meal up despite artificial feed improvements
2030-2050	Optimize profits from fish used in fishmeal. Basically supplies European demand for aquaculture.

Central North Pacific

2000-2010	Optimize profit
2010-2030	Optimize profit – but with costs lowered since technology improves. Possible to have more tuna caught younger for ranching (2015-2030)?
2030-2050	Optimize profits – but fish changes to species for fishmeal since technology cracks tuna hatchery technology.

invertebrate species, the result may be that large, long-lived species are eliminated from the system overall. These changed systems are vulnerable to severe events and therefore food and fishmeal supplies are highly variable. In areas without appropriate management systems, destructive fishing practices continue and stocks eventually decline along with inshore ecosystems. Towards 2050, developed nations reduce their net outflows of fish products to secure food supplies and social benefits. There is a significant reduction in effort starting with distant-water fleets, which are seen as threats to national food security. Areas are closed to fishing, where appropriate fisheries with low biomass production and destructive impacts, (e.g., long-living species) are phased out.

ADAPTING MOSAICS

This scenario includes a significant culture shift to maintaining the ecological structure of coastal and marine ecosystems that includes the recovery of long-lived, high-trophic-level species. In this scenario, there is considerable variation in the state of fisheries and the ecosystems as different management regimes, including variations on individual transferable quotas, community quotas, adaptive management, and community-based management, are tested. Global fishing agreements are largely ignored and regional fishing agreements decline in importance, but regional fisheries bodies are maintained for technical support for local and regional initiatives rather than for management advice. Initially, efforts are focused on coastal areas, but once it is realized that oceanic systems need to be included to ensure the recovery of long-lived species, ocean management is also embraced. Correspondingly, regional fisheries agreements and management bodies begin to coordinate information and learning exchanges.

TECHNOGARDEN

In this scenario, technology in the fisheries sector is primarily used to optimize economic returns and therefore the focus is on producing high-value, short-lived species, such as prawns, lobsters, squids, and salmon, through capture fisheries and aquaculture. The capture fisheries will not be purposely phased out because of the need to maintain the genetic resources should major failures take place in the aquaculture sector. Food provisioning by marine and coastal ecosystems is an important service and one which is well served by technology in addressing destructive fishing practices as well as a number of aquaculture development issues. The dominance of large corporations in the fisheries sector implies that ecosystem services other than food provisioning are considered to be secondary, except in a few isolated areas where other services such as tourism, biodiversity, and water regulation are given the same or higher priority.

This scenario also includes the development of ecologically sound aquaculture ventures in the coast and oceans. Given the importance of maintaining ecosystem structures, environmental concerns and uncertainty around genetically modified organisms (GMO), the expansion of aquaculture slows dramatically until these issues are resolved, which takes two to three decades. Technology plays an important role in developing fishmeal and fish oil replacements. Slow expansion only takes place under strict environmental and GMO policies.

Case Studies

GULF OF THAILAND

The Gulf of Thailand is located in the South China Sea. It is a shallow, tropical, coastal shelf system that has been heavily exploited since the 1960s. Prior to the early1960s, fishing in the area was primarily small scale with minimal impact on the ecosystem. However, a trawl fishery was introduced in 1963, and since then the area has been subjected to intense, steadily increasing fishing pressure (Pauly, 1979; Pauly and Chuenpagdee, 2003). The system has changed from a highly diverse ecosystem with a number of large, long-lived species (e.g., sharks and rays), to one that is now dominated by small, short-lived species that support a highly valued invertebrate fishery. Shrimp and squid caught primarily by trawl gear are economically the dominant fisheries in the

Gulf of Thailand. The bycatch of the trawl fishery is used for animal feed. The Gulf of Thailand model is well established and detailed in an FAO technical report (FAO/FISHCODE, 2001).

CENTRAL NORTH PACIFIC

The modeled area of the Central North Pacific is focused on epipelagic waters from 0°N to 40°N latitude and between 150°W to 130°E longitude (Cox et al., 2002).Tuna fishing is the major economic activity in the area, after tourism in the Hawaiian Islands. The tuna fishery is divided into deepwater, longline fisheries that target large-sized bigeye, yellowfin, and albacore tuna, and surface fleets that target all ages/sizes of skipjack tuna, small-sizes of bigeye, yellowfin, and albacore using a range of gear, including purse-seine, large-mesh gillnet (i.e., driftnet), small-mesh gillnet, handline, pole-and-line, and troll (Cox et al., 2002).Recent assessments of the tuna fisheries indicate that top predators, such as blue marlin (*Makaira* spp.) and swordfish (*Xiphias gladius*), declined since the 1950s, while their prey, small tunas, have increased. The Central North Pacific model is described in detail in Cox et al. (2002).

NORTH BENGUELA

The North Benguela Current is an upwelling system off the west coast of Southern Africa. This upwelling system is highly productive resulting in a rich living marine resource system that supports small, medium and large pelagic fisheries (Heymans et al., 2003). The system undergoes dramatic changes due to climatic and physical changes and therefore the marine life production can be quite variable. Sardine or anchovy used to be the dominant small pelagics; both species how however been at very low abundance for years as indicated by surveys in the late 1990s (Boyer and Hampton, 2001). The North Benguela ecosystem model is now used by the Namibian Fisheries Research Institute and is described in detail in Heymans et al. (2004).

Harmonizing of the scenarios and ecosystems models is summarized in Table 2.1. The landings, value of the landings and the diversity of the landings were used to investigate the differences between the various scenarios (Table 2.1) for each ecosystem.

Results and Discussion

Reporting Biases

By mapping landings into a global grid of cells it is possible to compare landings between locations while considering depth, latitude, primary productivity etc. In this way reporting anomalies become readily apparent. Systematic over-reporting from China was documented in this way (Watson and Pauly, 2001) and by estimating the expected landing levels it became apparent that this bias had maintained global landing totals for many years when actually landings had been declining. Since that time a more complete analysis of global reporting anomalies has been prepared for FAO.

Plans for future work include validation of our "top-down" approach to mapping fisheries catches by comparisons with more localized and intensive "bottom-up" efforts such as spatial models of the groundfish fisheries along the N.W. coast of the U.S (Scholz et al., this volume). Such comparisons are very instructive for those involved. For those working on global mapping it tests whether significant local spatial structures have been reconstructed by the rule-based disaggregation process from large-scale data. This could be very beneficial to modify future versions of the procedures involved. To those working on a smaller scale it helps examine whether there are spatial discontinuities when the boundaries of the mapped domain are reached, in other words, how representative are the local conclusions, and how significant are they in a broader spatial context.

Decades of Decline

Maps of landings for all years since 1950 can establish when spatial cells were first fished (where "first fished" is defined as the point in time when allocated landings first reached 10% of their all-time maximum). Mapping the decades when the maximum is reached over time, we can see a pattern of global expansion and decline (Fig. 2.2; see page 57). The parts of the North Atlantic where large-scale commercial fisheries started (for example the North Sea) were the first to decline, and in some cases they peaked as early as the 1960s (Watson et al., 2003). Generally areas to the south are either currently at their highest historical landings or in the case of areas nearer the Antarctic, there has been so little fishing history as to make this analysis impractical. The collapse of inshore stocks (Christensen et al., 2003) combined with the technical ability to fish deeper (Roberts, 2002) and further from ports have driven fisheries to expand. Our mapping method demonstrates that more and more of the deeper ocean depths are exposed to fishing each year (Pauly et al., 2003). While the largest offshore expansion has involved the tuna fleets, many other distant-water fleets fish thousands of kilometers from home ports for smaller pelagic fishes, most of which is reduced to meal and oil.

Reduction in the Average Length of Landings

If annual landings are combined with the maximum length of taxa landed it is possible to calculate the average length of the animals landed, and map the change in this revealing statistic. Most people are aware that historical accounts of fisheries cite larger animals as well as more abundant catches then are currently landed. The composition of landings, however, has also changed, often accompanied by a decline in the average length of animals landed. In some areas of the North

Atlantic where fisheries were well developed even before our data series began in the 1950s, there has been a decline of nearly one meter in the average length of animals reported (Fig. 2.3; see page 57). As we move to fishing smaller fishes and concentrate on small invertebrates like shrimp this trend is likely to accelerate.

Reduction in the Trophic Level of Landings

Similarly, it is possible to use biological databases such as FishBase (www.fishbase.org) to estimate the mean trophic level for each reported commercial species. If this information is combined with spatially allocated landings it is possible to produce a map showing the change in mean trophic level in commercial landings (Fig. 2.4; see page 57). In this way, we can see that the oldest commercial fisheries in the North Atlantic are also those where the trophic level of animals landed has been reduced the most (Pauly and Watson, 2003). This process has been show in bivariate graphs by Pauly et al. (1998), but it was never previously possible to map the changes.

Modeled Futures

Gulf of Thailand

The Gulf of Thailand landings vary between scenarios (Fig. 2.5). Overall landings are low in the Adapting Mosaic, which optimizes the value of the invertebrate fishery and jobs. Landings are maximized in the Order from Strength, which primarily optimizes the overall value of the fisheries and jobs, and Technogarden, which optimizes the value of small pelagic and lower trophic fish to support the aquaculture industry. The Global Orchestration also optimizes a mix of value and jobs initially, but after 2010 it changes to balancing value, jobs, which is reflected in the substantial decline in landings after 2010. In 2040 the focus is changed again to optimizing the ecosystem and the value of the fisheries.

The value of the landings is optimized in the Technogarden scenario due to the focus on supporting fisheries that are used in the fishmeal industry (Fig. 2.5b). The value of the landings in the Order from Strength scenario is similar to the Technogarden even though the management includes a focus on jobs. Some of the lowest landed values are in the Adapting Mosaic scenarios. Landing diversity initially differs substantially between the four scenarios with the Technogarden scenario having the highest diversity (Fig. 2.5c). By 2050, however, landing diversity declines below 2000 levels for all the scenarios as the fisheries are optimized for either invertebrate or small pelagic fisheries.

Central North Pacific

Changes in landing are substantially different initially between the four scenarios initially (Fig. 2.6). Between 2010 and 2030 landings for the Technogarden and Order from Strength scenarios are similar despite

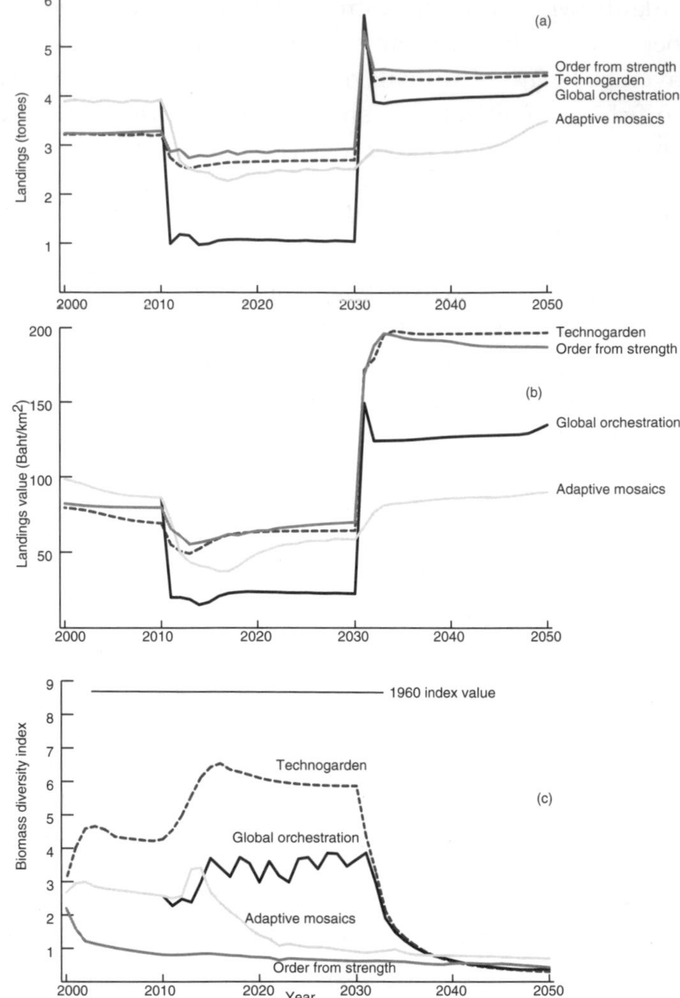

Figure 2.5. Future scenarios results for the Gulf of Thailand showing a) landings, b) value of landings and c) diversity index of landings.

optimizing value in the Technogarden scenario and a mix of value and jobs in the Order from Strength scenario. The two scenarios diverge even further after 2040. Landings in the Adapting Mosaics scenario, which initially focus on optimizing value from the tuna fisheries and later on rebuilding the bigeye stocks and optimizing jobs, remain relatively constant. Landings in the Global Orchestration scenario drop substantially after 2010 which corresponds with a shift from optimizing primarily value followed by jobs to a mix which also optimizes value but an increased emphasis on jobs.

Overall the value of the landings are highest in the Technogarden and Global Orchestration scenarios, which is as expected since the focus is on optimizing higher valued tuna fisheries (Fig. 2.6b). Although the landings in the Adapting Mosaic remained constant, the value of landing increases, while in the Order from Strength scenario the value of landings decrease. The diversity of the landings remains relatively constant for the Techno and Global Orchestration scenarios due to their focus on tuna fisheries (Fig. 2.6c). The Adapting Mosaic scenario yields

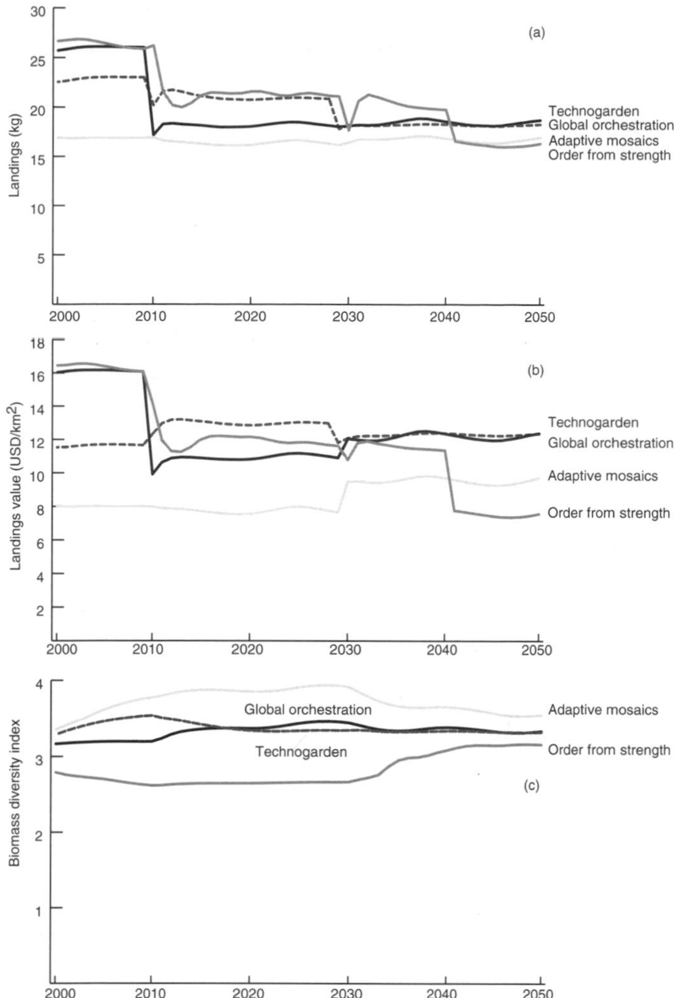

Figure 2.6. Future scenarios results for the Central North Pacific showing a) landings, b) value of landings and c) diversity index of landings.

the highest diversity of the landings. The diversity changes, however, as the focus in the Adapting Mosaic scenario changes from optimizing value to rebuilding bigeye stocks and optimizing the ecosystem. Diversity of the landings is lowest in the Order from Strength scenario, which is focused initially on optimizing value from tuna fisheries.

North Benguela

Until 2040, landings for the Technogarden and Global Orchestration scenarios follow the same trend (Fig. 2.7). These scenarios initially focus on optimizing value and then diverge in 2010, where management focus continues on optimizing value in the Technogarden scenario. In the Global Orchestration scenario, however, jobs are optimized as well as value. By 2040, the Technogarden scenario continues to optimize value but the Global Orchestration scenarios tries to balance value, jobs and the ecological values. Over the same time period, landings in the Mosaic scenario, which focuses on optimizing value and jobs initially and then shifts to rebuilding ecosystem, increases very slowly until

2040. After 2040, the focus changes to optimizing biomass and consequently landings increase. By 2050, however, landings in these three scenarios approach similar levels. The Order from Strength scenario landings are significantly less than the other scenarios due to the focus on optimizing value initially from the high value, distant-water fisheries. In 2010, the focus changes to optimizing jobs and later rebuilding the ecosystem, resulting in substantial increases in landing. This increase, however, is short-lived due to management changing to a mix of optimizing value and jobs.

The value of the landings also follows a similar trend to the landings, with differences due to fisheries that are optimized (Fig. 2.7b). If jobs are the focus, then the fisheries that are optimized may not yield as high a value as fisheries that employ fewer people but target high-valued species. Diversity of the landings follows a similar trend to value. By 2050, the diversity of the landings begin to approach the same level with the Technogarden and Global Orchestration scenarios yielding slightly higher levels than the other scenarios (Fig. 2.7c).

Figure 2.7. Future scenarios results for the North Benguela showing a) landings, b) value of landings and c) diversity index of landings.

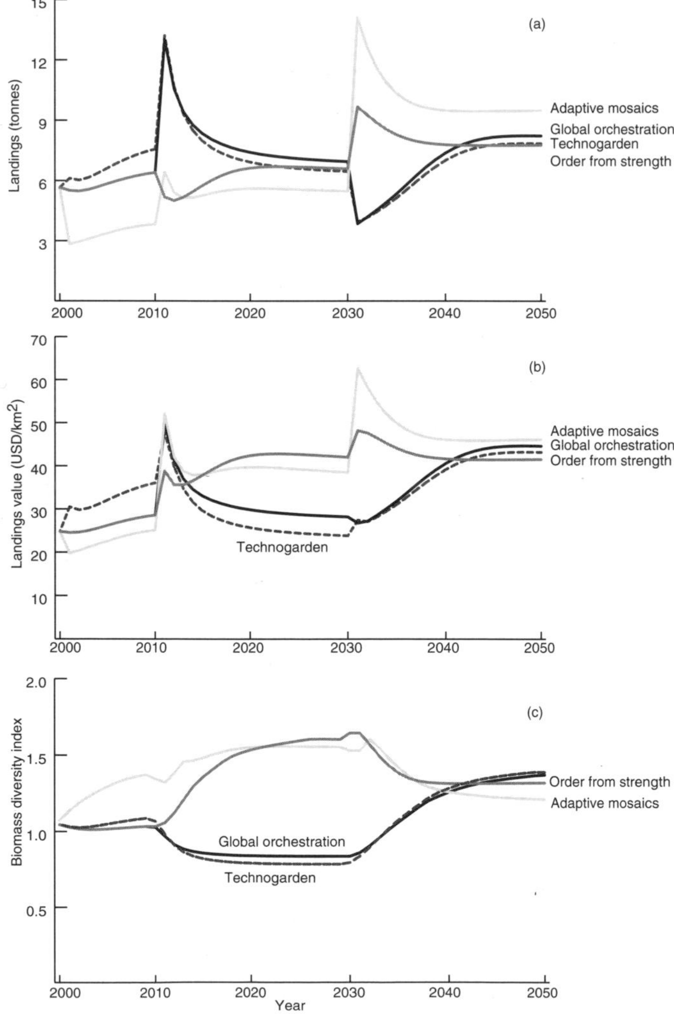

Conclusions

Conventional fisheries data do not provide spatial information on landings and their trends suitable to support either broad global analysis or fine-scale spatial ecosystem modeling. Using additional data, such as marine distributions, fishing access, and fishing patterns, considerable spatial precision can be achieved. By treating landing data this way, we can demonstrate worrying patterns present since the 1970s and before, notably reductions in landings, and a decrease in the mean size and trophic level of animals landed in some of the major fisheries of the world.

Landings data can be combined with new approaches in ecosystem modeling to examine the impact of future scenarios on marine resources, and on the people and industries that depend on them. Choices made in the management of marine resources will greatly affect the outcomes, but they will have different impacts on different places.

Our preliminary exploration of the future of fisheries indicates that it is not too late to reverse current trends in capture fisheries around the world. The future depends on where policy-makers chose to focus their interests: profits, jobs, or ecosystems.

In all three case studies discussed here, which represent very different ecosystems, testing the four future scenarios yielded varied outcomes. Policies that were totally focused on maximizing profits did not necessarily maintain diversity or support employment. Similarly, a policy that was focused on employment did not necessarily maximize profits or maintain ecosystems.

The diversity of the stocks exploited can be enhanced if the policy favors maximizing the ecosystem or rebuilding stocks. Diversity, however, is lost if the sole objective of management is to maintain or increase profits. Our results demonstrate that society is going to have to take a more active role in exploring the right balance or tradeoffs between profits, jobs, and ecosystems.

Acknowledgments

The authors would like to thank database providers, including Rainer Froese and the "FishBase team," CephBase (particularly Catriona Day), the World Conservation Monitoring Centre (Cambridge), and the Food and Agriculture Organization (Rome). We are grateful for the assistance of Fred Valdez, Wilf Swartz, and the other "Sea Around Us" project (SAUP) staff. We thank the editors and their referees for their valuable suggestions. The SAUP is an activity initiated and supported by the Pew Charitable Trusts of Philadelphia.

References

Alverson, D. L., Freeberg, M. H., Murawski, S. A., and Pope, J. G., 1994. A global assessment of fisheries by-catch and discards, FAO Fisheries Technical Paper 339, 233 pp.

Boyer, D. C. and Hampton, I., 2001. An overview of the living marine resources of Namibia, *South African Journal of Marine Science*, 23:5-35

Christensen, V., 1995. Ecosystem maturity - towards quantification, *Ecological Modelling*, 77:3-32.

Christensen, V., Guénette, S., Heymans, J. J., Walters, C. J., Watson, R., Zeller, D., and Pauly, D., 2003. Hundred-year decline of North Atlantic predatory fishes, *Fish and Fisheries*, 4: 1-24.

Christensen, V., and Pauly, D., 1992. Ecopath II - a software for balancing steady-state ecosystem models and calculating network characteristics, *Ecological Modelling*, 61:169-85.

Christensen, V., and Walters, C.J., 2004. Ecopath with Ecosim: methods, capabilities and limitations, *Ecological Modelling*, 172: 109-39.

Cox, S. P., Martell, S. J. D., Walters, C. J., Essington, T. E., Kitchell, J. F., Boggs, C., and Kaplan, I., 2002. Reconstructing ecosystem dynamics in the central Pacific Ocean, 1952-1998. 1. Estimating population biomass and recruitment of tunas and billfishes, *Canadian Journal of Fisheries and Aquatic Sciences*, 59:1724-35.

European Parliament, 2001. Revised Working Document 3 on the Role of Flags of Convenience in the Fisheries Sector, Committee on Fisheries, European Parliament, DT\452747EN.doc PE 309.162/REV, 5 pp.

Food and Agriculture Organization (FAO) of the United Nations, 1998. FAO's fisheries agreements register (FARISIS), Committee on Fisheries, 23rd Session, Rome, Italy, February 15-19, (COFI/99/Inf.9E), 4 pp.

FAO, 2001. Fisheries Department, Fishery Information, Data and Statistics Unit. FISHSTAT Plus, Universal software for fishery statistical time series, Version 2.3.

FAO/FISHCODE, 2001. Report of a Bio-economic Modelling Workshop and a Policy Dialogue Meeting on the Thai Demersal Fisheries in the Gulf of Thailand held at Hua Hin, Thailand, 31 May - 9 June 2000, FI: GCP/INT/648/NOR: Field Report F-16 (En), Rome, FAO, 104 pp.

Froese, R., and Pauly, D. (eds.), 2000. FishBase 2000: Concepts, Design and Data Sources, Los Baños, Philippines, distributed with 4 CD-ROMs; updates on www.fishbase.org.

Heymans, J. J., Shannon, L. T., and Jarre, A., 2004. The Northern Benguela Ecosystem: Changes over three decades (1970, 1980 and 1990). *Ecological Modelling* 172(2-4): 175-96.

Millennium Ecosystem Assessment, 2003. *Ecosystems and Human Well-being: A Framework for Assessment*, Washington, DC, Island Press, 245 pp.

Pauly, D. 1979. Theory and management of tropical multispecies stocks, ICLARM Studies and Reviews 1, Manila, 35 pp.

Pauly, D., Alder, J., Bennett, E., Christensen, V., Tyedmers, P., and Watson, R. 2003. World fisheries: the next 50 years, *Science* 302: 1359-61.

Pauly, D., Christensen, V., Dalsgaard, J., Froese, R., and Francisco T., Jr., 1998. Fishing down marine food webs, *Science* 279: 860-63.

Pauly, D., Christensen, V., and Walters, C., 2000. Ecopath, Ecosim, and Ecospace as tools for evaluating ecosystem impact of fisheries, *ICES Journal of Marine Science*, 57: 697–706.

Pauly, D., and Chuenpagdee, R., 2003. Development of fisheries in the Gulf of Thailand large marine ecosystem: Analysis of an unplanned experiment, in Hempel, G., and Sherman, K. (eds.), *Large Marine Ecosystems of the World: Change and Sustainability*, Amsterdam, Elsevier Science.

Pauly, D., and Watson, R. 2003. Counting the last fish. *Scientific American*, July 2003, 42-47.

Pitcher, T. J., Watson, R., Forrest, R., Valtasson, H., and Guénette, S., 2002. Estimating illegal and unreported catches from marine ecosystems: a basis for change, *Fish and Fisheries*, 3: 317-39.

Roberts, C. M., 2002. Deep impact: the rising toll of fishing in the deep sea, *Trends in Ecology and Evolution*, 17(5): 242-45.

Scholz, A. J., Mertens, M., and Steinback, C., this volume. The OCEAN Framework—Modeling the Linkages between Marine Ecology, Fishing Economy, and Coastal Communities, in Wright, D. J., and Scholz, A. J. (eds.), *Place Matters–Geospatial Tools for Marine Science, Conservation and Management in the Pacific Northwest*, Corvallis, OR, Oregon State University Press.

Walters, C., Christensen, V., and Pauly, D. 1997. Structuring dynamic models of exploited ecosystems from trophic mass-balance assessments, *Reviews in Fish Biology and Fisheries*, 7:139-72.

Walters, C., Pauly, D., and Christensen, V., 1998. Ecospace: prediction of mesoscale spatial patterns in trophic relationships of exploited ecosystems, with emphasis on the impacts of marine protected areas, *Ecosystems* 2: 539-54.

Walters, C., Pauly, D., Christensen, V., and Kitchell, J. F., 2000. Representing density dependent consequences of life history strategies in aquatic ecosystems: Ecosim II, *Ecosystems*, 3:70-83.

Watson, R., and Pauly. D., 2001. Systematic distortions in world fisheries catch trends, *Nature*, 414: 534-36.

Watson, R., Tyedmers, P., Kitchingman, A., and Pauly, D., 2003. What's left: the emerging shape of the global fisheries crisis, *Conservation in Practice*, 4(3): 20-21.

Watson, R., Kitchingman, A., Gelchu, A. and Pauly, D., 2004. Mapping global fisheries: sharpening our focus, *Fish and Fisheries* 5: 168-77.

Chapter 3

The Benefits and Pitfalls of Geographic Information Systems in Marine Benthic Habitat Mapping

H. Gary Greene, Joseph J. Bizzarro, Janet E. Tilden, Holly L. Lopez, and Mercedes D. Erdey

Abstract

The application of geographic information system (GIS) technology to the characterization of marine benthic habitats has greatly increased the speed and resolution of seafloor mapping efforts. GIS is a powerful tool for the visualization and imaging of seafloor characteristics and has also proven useful for the quantification of mapped substrate types, determination of slope inclination and rugosity, and other spatial analyses. With the use of GIS, geologists and digital cartographers can create marine benthic habitat maps to assist scientists and policy-makers in the management of commercial groundfish stocks and the designation of marine protected areas. However, without a complete understanding of mapping procedures and the technology used to obtain source data (e.g., multibeam swath bathymetric and backscatter imagery), maps and GIS products may be misinterpreted and used in ways that are inappropriate or misleading.

Introduction

The use of geographic information systems (GISs) has proven to be extremely effective in the compilation and presentation of maps of various types and scales. GIS technology (especially by ESRI®) is presently the tool of choice for the scientific community involved in the mapping of marine benthic habitats because of its flexibility and ease in adding, modifying, and analyzing data. However, a lack of proper understanding and documentation of the quality, manipulation, and limitations of source data and derivative habitat interpretations is leading to confusion and potentially inappropriate use of habitat maps presented

H. Gary Greene, Joseph J. Bizzarro, Janet E. Tilden, Holly L. Lopez, and **Mercedes D. Erdey**

Center for Habitat Studies, Moss Landing Marine Laboratories, 8272 Moss Landing Rd., Moss Landing, CA 95039

Corresponding author: greene@mlml.calstate.edu

in GIS. Though the use of GIS in seafloor mapping is still in its early stages of development, protocols must be established to more clearly identify data type, quality, interpretive processes, and authors of habitat interpretations (genealogy).

Marine benthic habitat maps are critical to state and federal fisheries agencies for the development of management and conservation policies and as a basis for habitat-related studies. These maps play a crucial role in the evaluation, extension, and selection of marine protected areas (MPAs) that are being established to conserve overexploited groundfish species (Yoklavich et al., 1997; O'Connell et al., 1998). The demand for these maps and related GIS products has led to a community-wide compilation and interpretation frenzy. In many cases, groups and agencies have rapidly incorporated, and possibly incompletely documented, GIS datasets that are being utilized by government, academic, industry and non-governmental organizations in mapping and monitoring marine benthic habitats and developing management plans for groundfish species.

Although GIS has facilitated a great increase in the quality and quantity of marine benthic habitat maps, in some cases users are unaware of its limitations. Even when Federal Geographic Data Committee (FGDC) compliant metadata was included, our map products have occasionally been misinterpreted and incorrectly used because our interpretive processes and/or the quality of source data was not fully understood. Given the widespread compilation and use of habitat maps and their importance in fisheries management, this could become a serious problem. The objective of this paper is therefore to briefly discuss the advantages, or benefits, and disadvantages, or pitfalls, encountered in using GIS in mapping marine benthic habitats. Possible solutions to the problems outlined herein are also suggested.

Discussion

Habitats: Definitions

The word "habitat" has been used in many ways and the concept has inconsistent connotations to scientists of different disciplines. The following basic definition is found in the *Merriam-Webster's Dictionary* (2004): "1.a. the place or environment where a plant or animal naturally or normally lives and grows. 2. the place where something is commonly found." The *Glossary of Geology* (Bates and Jackson,1980) provides an only slightly more specific definition: "the particular environment where an organism or species tends to live; a more locally circumscribed portion of the total environment." Essential Fish Habitat is defined in the Sustainable Fisheries Act (1996) as: "waters and substrate necessary for spawning, breeding, feeding or growth to maturity," which again is so generally descriptive as not to be very useful. Due to the vague nature of this verbiage, the definition of marine habitats by NOAA has been legally challenged and is in the process of being re-defined. None of these general descriptors is useful in characterizing marine benthic habitats, which are necessarily defined on a species-specific basis and may be highly variable for different populations or life stages.

In this paper, we consider a marine benthic habitat as a set of seafloor conditions that is commonly associated with a species or local population thereof. Subsets of the overall habitat of a species may be utilized differentially for foraging (subsistence), refuge, reproduction or rest. Physical (e.g., temperature, current speed and direction, depth), chemical (e.g., salinity, nutrients, minerals), geological (e.g., substrate type, seafloor morphology) and biological (e.g., species density, % cover of sessile or encrusting flora and fauna) parameters can be used to determine a species' habitat associations. These various datasets can be presented in GIS in both tabular (attribute) and visual form. Multiple layers can be overlaid to depict the various seafloor conditions in a coordinated fashion and used to interpret marine benthic habitats.

Since specific habitat associations for a species are not often known during the compilation and interpretation of seafloor data, it is not appropriate to describe interpretive maps of the seafloor as "habitat" maps. We therefore propose the term "potential habitat" to describe a set of distinct seafloor conditions that may be utilized differentially by a species. Once habitat associations are determined, they can be used to create maps that depict the actual distribution and abundance of a species in relation to its known habitat types.

Habitats: Characterization

There are two basic approaches to characterizing habitats. One is the top-down approach advocated by biologists and the other is the bottom-up approach characteristic to geologists. Biologists pioneered the description of habitats and developed habitat characterization schemes based on flora and fauna in the terrestrial and coastal environment (e.g., CEC 1997; FGDC 1997). These schemes typically describe forest, brush, and micro-vegetation from the crest of mountains to the intertidal zones, with substrate being the third or fourth descriptor. However, while flora and fauna change, substrate, or geology, may often be continuous from onshore to offshore.

A bottom-up classification scheme can link terrestrial and seafloor conditions in a continuous fashion, a process that is much more difficult to accomplish with biological parameters. In seafloor areas (such as the deep sea) where demersal fauna and flora are sparse or non-existent, biology may be absent or restricted to infauna. Organisms that are present in these regions are often difficult to identify or quantify. Conversely, seafloor conditions can be efficiently imaged geophysically and described geologically due to tremendous advances in remote sensing technology. As a result of these considerations, a geological bottom-up characterization of habitats seems more appropriate for deep-water (>~30 m) marine benthic regions.

A GIS-Compatible Classification Scheme for Potential Marine Benthic Habitats

We constructed a detailed, GIS-compatible classification scheme for the characterization of potential marine benthic habitats. Although the classification scheme is in flux, it is presented for reference purposes along with an explanation for its use (Appendices 3.1, 3.2). This scheme is generally based on geomorphological and physiographical scales, depth, seafloor induration (hardness), texture and sessile biology. Potential habitats are divided into four types based on size (scale) and depth: Mega-, Meso-, Macro- and Microhabitats after Greene et al. (1999). Mega-, meso-, and macrohabitats are typically interpreted from seafloor imagery (e.g., sidescan sonar, multibeam imagery) or geologic data. Imaging and characterization of microhabitats is typically more difficult and time-consuming and is usually best accomplished from in situ or video observations.

Data Sources and Map Construction

Many different types of data are being used to characterize potential and actual marine benthic habitats (Greene et al., 1995, 1999, 2000; Yoklavich et al., 1995, 2000; Auzende and Greene, 1999; Gordon et al., 2000; Todd et al., 2000; Kostylev et al., 2001). These data range from previously constructed seafloor geologic, geomorphic, geophysical, sediment, biologic and bathymetric maps to remotely collected seafloor imagery (e.g., single beam echosounder, multibeam bathymetric and backscatter, sidescan sonar, and seismic reflection profile, LIDAR, laser-line scan, and hyperspectral data) and in situ observational and photographic (video and still photo) data obtained with the use of submersibles, ROVs, camera sleds, or by free diving. As previously mentioned, these data are used for the interpretation of seafloor morphology and substrate types that can be represented as either potential habitats or actual habitats. Thematic maps depicting substrate types, benthic habitats, physiography, bathymetry, and morphology can be constructed from interpretation of remotely and in situ collected data. Various spatial analysis tools enable an interpreter using GIS to construct maps detailing seafloor complexity and seafloor slope, and to quantify substrate and morphologic features useful in determining critical habitat parameters. In the following section, we discuss some of the advantages and disadvantages of the contemporary practice of mapping potential or actual marine benthic habitats.

Advantages (Benefits)

GIS is an excellent tool for developing base maps and for layering various thematic datasets above and below a base map. Often in deep-water habitat mapping, the base map is a bathymetric map of some sort. Typical source data for base maps consist of either a bathymetric contour map or a multibeam bathymetric map, commonly presented as an artificial,

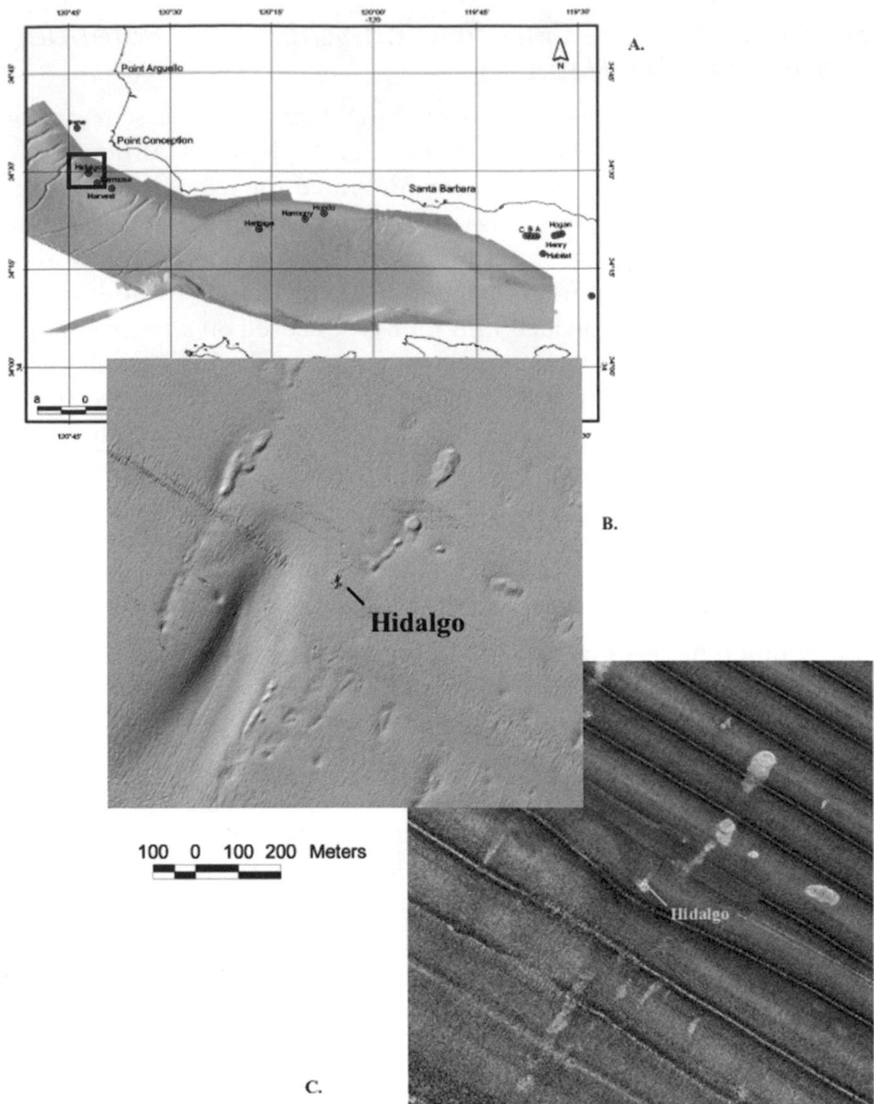

Figure 3.1. (a) Simrad EM 300 (30 kHz) color-shaded multibeam bathymetric image of seafloor beneath oil platforms, including Hidalgo, in the Santa Barbara region. Box shows location of (b) and (c). (b) Simrad EM 300 (30 kHz) multibeam artificial sun-illuminated bathymetric image of the seafloor around the Hidalgo oil platform. See (a) for location. (c) Backscatter image obtained with a Simrad EM 300 (30 kHz) system showing seafloor texture around the Hidalgo oil platform; bright areas are hard mounds, darker areas are unconsolidated sediment. Data courtesy of the Monterey Bay Aquarium Research Institute. See (a) for location.

sun-illuminated relief map (Fig. 3.1a, b), which is digitally constructed from x, y, z data that represent accurately positioned soundings. These types of maps are easily displayed using GIS. The next type of map, or overlay to the base map, is typically multibeam backscatter (Fig. 3.1c) or sidescan sonar (Fig. 3.2) imagery, which provides information about seafloor texture and substrate types. If geologic data and/or geologic maps are available, these data can then be incorporated into a GIS project as another layer. Many other datasets and maps can also be included and represented. Potential marine benthic habitats are interpreted from these multiple data layers and the ultimate interpretive map consists of polygons that have been attributed to distinct habitat types (Fig. 3.3; see page 58).

Ease of geo-referencing and incorporating maps and data from a variety of sources is a distinct advantage of GIS and facilitates the inclusion of both analog data, which can be scanned and digitized, and digital data. This allows for the utilization of historical data sources that may otherwise be overlooked. Once incorporated into a GIS project, these data can be layered and used collectively as a basis for habitat interpretations.

GIS is also a convenient tool for updating habitat maps. Because of the ease of inputting and layering geo-referenced data, habitat maps

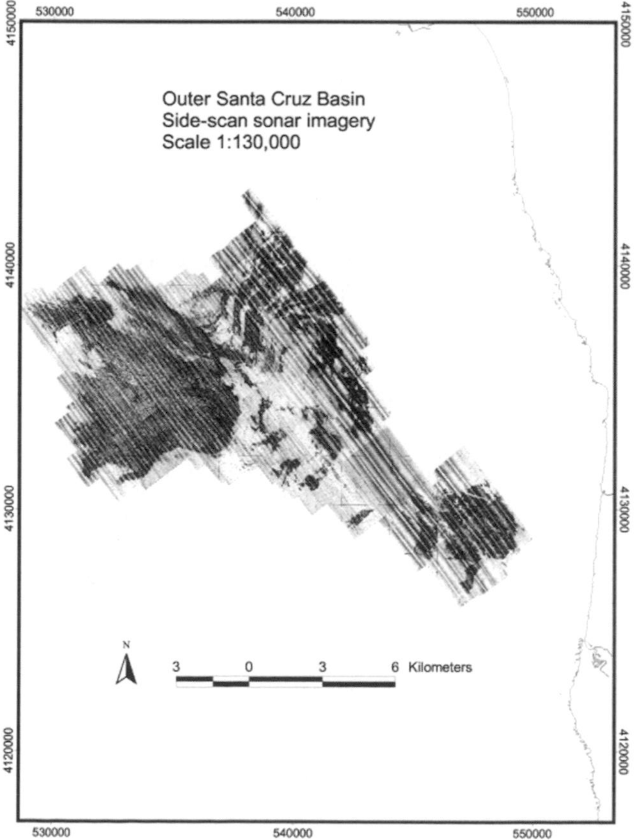

Figure 3.2. Example of a sidescan sonar mosaic. This dataset was collected on the continental shelf north of Santa Cruz, California. Dark areas are hard rock exposures; light areas are unconsolidated sediment (likely sand). Data courtesy of Delta Oceanographics, Inc. and Fugro-Pelagos, Inc.

can be readily updated once new data become available. This enables users to conduct time-series analyses that may be essential to monitoring studies. These studies are especially important in areas where dynamic seafloor processes occur and may temporally alter habitats (Fig. 3.4; see page 58).

Excellent quantification and spatial analysis tools are available in GIS programs. With these tools, polygon areas can be quantified and can be summed by habitat type to determine habitat-specific areas (Fig. 3.5; see page 59). Seafloor slope can also be calculated using x,y,z data typically collected with multibeam systems (Fig. 3.6; see page 59). Rugosity, based on neighborhood statistics, can also be calculated with these data. Maps derived from these analyses can be constructed and represented as thematic layers in a GIS project.

Disadvantages (Pitfalls)

Probably the most serious problem in the use of GIS for marine benthic habitat mapping is the lack of attention paid to the type and quality of data used to construct a habitat map and the incomplete documentation of the history of data collection, modification, interpretation, and genealogy. This type of information is included as metadata in either read-me files or, more recently, compiled in ArcCatalog®. However, metadata is often isolated from GIS map projects and may not be readily accessible or considered by the users. Even when it is easily incorporated (e.g., ArcGIS®) it is presented in a lengthy, written format. One of the main benefits and primary uses of GIS programs is to facilitate the visualization and incorporation of a wide variety of data sources into a project. Metadata (especially for data type and quality) should therefore be displayed in a similar format to increase utilization and comprehension by the user.

Without detailed knowledge of data type and the quality, it is difficult to assess the accuracy of derivative habitat maps. For example, a habitat map may have been constructed from a previously published offshore geologic map that was produced from the interpretation of seismic reflection profiles and seafloor sampling (Fig. 3.7; see page 60). Although closed polygons were constructed, their resolution would be such that their boundaries in most areas are only approximately located. This map could be merged with higher resolution maps constructed from state-of-the-art multibeam bathymetry and backscatter data (Fig. 3.8; see page 60) that would then exhibit seamless polygons synthesized from all datasets (Fig. 3.9; see page 125). However, without knowledge of these facts, a user may assume that the habitat map was created from sources of equal quality and therefore should be of uniform accuracy. It is easy to imagine the pitfalls of this thinking, which could adversely skew management regulations or other decision-making tasks.

A critical component in the creation of habitat maps is the scale at which the source data were interpreted. Without this information, it becomes very difficult to determine accuracy of habitat interpretations. Although it is possible to infer relative differences in data resolution by the differing sizes and shapes of delineated polygons (Fig. 3.9), the true differences are unobtainable from habitat maps or metadata.

Solutions

While metadata is useful for referencing technical information, it is not visually informative and therefore often overlooked or not well understood, especially by managers and scientists who may not have a technical understanding of GIS. Some of the most important information needed, such as data type, quality, and scale, are best presented visually rather than in a written format. We therefore propose that a data type and quality layer be developed that would correspond to all marine habitat maps presented in a GIS product. This layer would essentially be a map (Fig. 3.10; see page 125) that would exhibit area (with tracklines when appropriate), type, and quality of data used in the interpretation of marine benthic habitat maps. Information on data source, collection, associated publications, scale, and genealogy could be listed in the attribute table for this map layer and easily accessed for polygons or regions of interest. This type of metadata presentation would more efficiently and effectively serve GIS map users in evaluating accuracy and quality and determining data sources.

Conclusions

GISs provide excellent tools for the compilation and presentation of marine benthic habitat maps. They are especially valuable in exhibiting various thematic layers that can be used to compile and manipulate different and disparate datasets in a manner that allows for the construction of very comprehensive habitat and other thematic maps (Fig. 3.11; see page 126). However, the lack of a convenient protocol to clearly and illustratively convey information such as source data type, quality, scale, and genealogy hampers the ability of a user to assess the accuracy and quality of the resultant habitat maps. Inclusion in metadata is a necessary, but circuitous way of displaying this critical information. We, therefore, propose a protocol that consists of a distinct layer within a GIS project that exhibits and lists data type, quality, source, collection date, interpretation scale(s), genealogy and associated publications (bibliography) used in the construction of marine benthic habitat maps (Fig. 3.11). If adapted by the marine benthic habitat mapping community, a standard methodology would exist to better determine and understand the specific details of seafloor datasets used in the characterization and assessment of marine benthic habitats.

Acknowledgements

We wish to thank Lee Murai for his assistance in constructing the figures. Many of the habitat maps and other data presented in this paper are the result of grants from the Dickinson Foundation, SeaDoc Society (formerly the Marine Ecosystem Health Program) managed at UC Davis, NOAA Coastal Services Center (grant #NA17OC2646), NOAA National Sea Grant (grant #R/F-181A), NOAA National Marine Fisheries Service, Alaska Department of Fish & Game, and California Department of Fish and Game.

References

Auzende, J. M., and Greene, H. G. (eds.), 1999. Marine Benthic Habitats, Special Issue of *Oceanologica Acta*, 22(6), 726 pp.

Bates, R. L., and Jackson, J. A., 1980. *Glossary of Geology*, Falls Church, VA, American Geological Institute, 751 pp.

CEC (Commission for Environmental Cooperation), 1997. *Ecological regions of North America: toward a common perspective*, Montreal, Quebec, Canada, 71 pp. Map (1:12,500,000).

FGDC (Federal Geographic Data Committee), 1997. National Vegetation Classification and Information Standards Approved by FGDC, http://www.nbs.gov/fgdc.veg/standards/vegstd-pr.htm.

Greene, H. G., Yoklavich, M. M., Sullivan, D., and Cailliet, G. M., 1995. A geophysical approach to classifying marine benthic habitats: Monterey Bay as a model, in O'Connell, T., and Wakefield, W. (eds.), *Applications of Side-scan Sonar and Laser-line Systems in Fisheries Research*, Alaska Dept. Fish and Game Special Publication No. 9, 15-30.

Greene, H. G., Yoklavich, M. M., Starr, R., O'Connell, V. M., Wakefield, W. W., Sullivan, D. L. MacRea, J. E., and Cailliet, G. M., 1999. A classification scheme for deep-water seafloor habitats, *Oceanologica Acta*, 22(6): 663-78.

Greene, H. G., Yoklavich, M. M., O'Connell, V. M., Starr, R. M., Wakefield, W. W., Brylinsky, C. K., Bizzarro, J. J., and Cailliet, G. M., 2000. Mapping and classification of deep seafloor Habitats, ICES 2000 Annual Science Conference, Bruges, Belgium, ICES paper CM 2000/T:08.

Gordon, D. C., Jr., Kenchington, E. L. R., Gilkinson, K. D., McKeown, D. L., Steeves, G., Chin-Yee, M., Vass, W. P., Bentham, K., and Boudreau, P. R., 2000. Canadian imaging and sampling technology for studying marine benthic habitat and biological communities, ICES 2000 Annual Science Conference, Bruges, Belguim, ICES Paper CM 2000/T:07.

Kostylev, V. E., Todd, B. J., Fader, G. B. J., Courtney, R. C., Cameron, G. D. M., and Pickrill, R. A., 2001. Benthic habitat mapping on the Scotian Shelf based on multibeam bathymetry, surficial geology and seafloor photographs, *Marine Ecology Progress Series*, 219: 121-37.

O'Connell, V., Wakefield, W. W., and Greene, H. G., 1998. The use of a no-take marine reserve in the eastern Gulf of Alaska to protect essential fish habitat, in Yoklavich, M. M. (ed.), Marine Harvest Refugia for West Coast Rockfish: A Workshop, NOAA Technical Memorandum. NOAA-TM-NMFSC-255: 125-32.

Merriam-Webster. 2004. *Merriam-Webster's Online Dictionary*. http://www.m-w.com/dictionary.htm.

Public Law 104-297. October 11, 1996. Sustainable Fisheries Act.

Todd, B. J., Kostylev, V. E., Fader, G. B. J., et al., 2000. New approaches to benthic habitat mapping integrating multibeam bathymetry and backscatter, surficial geology and seafloor photographs: A case study from the Scotian Shelf, Atlantic Canada, ICES 2000 Annual Science Conference, Bruges, Belgium, ICES Paper CM 2000/T:16.

Vedder, J. G, Greene, H. G., Clarke, S. H., and Kennedy, M. P. 1986. Geologic map of the mid-southern California continental margin. in Greene, H. G., and Kennedy, M. P. (eds). *California Continental Margin Marine Geologic Map Series. Area 2, Sheet 1.* California Division of Mines and Geology, 1:250,000 scale.

Yoklavich, M. M., Cailliet, G. M., Greene, H. G., and Sullivan, D., 1995. Interpretation of side-scan sonar records for rockfish habitat analysis: Examples from Monterey Bay, in O'Connell, T., and Wakefield, W. (eds.). *Applications of Side-scan Sonar and Laser-line Systems in Fisheries Research,* Alaska Dept. Fish and Game Special Publication No. 9, 11-14.

Yoklavich, M., Starr, R., Steger, J., Greene, H. G., Schwing, F., and Malzone, C., 1997. Mapping benthic habitats and ocean currents in the vicinity of central California's Big Creek Ecological Reserve, NOAA Technical Memorandum, NMFS, NOAA-TM-NMFS-SWFSC-245, 52 pp.

Yoklavich, M. M., Greene, H. G., Cailliet, G. M., Sullivan, D. E., Lea, R. N., and Love, M. S., 2000. Habitat association of deep-water rockfishes in a submarine canyon: an example of a natural refuge, *Fisheries Bulletin,* 98: 625-41.

APPENDIX 3.1

Key to Marine Benthic Habitat Classification Scheme

(modified after Greene et al., 1999)

Megahabitat – Use capital letters (based on depth and general physiographic boundaries; depth ranges approximate and can be modified according to study area).

A = Apron, continental rise, deep fan or bajada (3000-4000 m)
B = Basin floor, Borderland type (1000-2500 m)
E = Estuary (0-50 m)
F = Flank, continental slope, basin/island-atoll flank (200-3000 m)
I = Inland sea, fiord (0-200 m)
P = Plain, abyssal (4000-6000+ m)
R = Ridge, bank or seamount (crests at 200-2500 m)
S = Shelf, continental or island (0-200 m)

Seafloor induration - Use lower-case letters (based on substrate hardness).

h = hard substrate, rock outcrop, relic beach rock or sediment pavement
m = mixed (hard & soft substrate)
s = soft substrate, sediment-covered

Sediment types (for above indurations) - Use parentheses.

(b) = boulder
(c) = cobble
(g) = gravel
(h) = halimeda sediment, carbonate

(m) = mud, silt, clay
(p) = pebble
(s) = sand

Meso/Macrohabitat - Use lower-case letters (based on scale).

a = atoll
b = beach, relic
c = canyon
d = deformed, tilted and folded bedrock
e = exposure, bedrock
f = flat, floor
g = gully, channel
i = ice-formed feature or deposit, moraine, drop-stone depression
k = karst, solution pit, sink
l = landslide
m = mound, depression; includes short, linear ridges
n = enclosed waters, lagoon
o = overbank deposit (levee)
p = pinnacle, volcanic cone
r = rill
s = scarp, cliff, fault or slump
t = terrace
w = sediment waves
y = delta, fan
$z_{\#}$ = zooxanthellae hosting structure, carbonate reef
1 = barrier reef
2 = fringing reef
3 = head, bommie
4 = patch reef

Modifier - Use lower-case subscript letters or underscore (textural and lithologic relationship).

$_a$ = anthropogenic (artificial reef/breakwall/shipwreck)
$_b$ = bimodal (conglomeratic, mixed [includes gravel, cobbles and pebbles])
$_c$ = consolidated sediment (includes claystone, mudstone, siltstone, sandstone, breccia, or conglomerate)
$_d$ = differentially eroded
$_f$ = fracture, joints-faulted
$_g$ = granite
$_h$ = hummocky, irregular relief
$_i$ = interface, lithologic contact
$_k$ = kelp
$_l$ = limestone or carbonate
$_m$ = massive sedimentary bedrock
$_o$ = outwash
$_p$ = pavement
$_r$ = ripples
$_s$ = scour (current or ice, direction noted)
$_u$ = unconsolidated sediment
$_v$ = volcanic rock

Seafloor slope - Use category numbers. Typically calculated from x-y-z multibeam data. Category designations represent suggestions and can be modified by the user.

1 Flat (0-1°)
2 Sloping (1-30°)
3 Steeply Sloping (30-60°)
4 Vertical (60-90°)
5 Overhang (> 90°)

Seafloor complexity - Use category letters (in caps). Typically calculated from x-y-z multibeam slope data using neighborhood statistics and reported in standard deviation units. Category designations represent suggestions and can be modified by the user.

A Very Low Complexity (-1 to 0)
B Low Complexity (0 to 1)
C Moderate Complexity (1 to 2)
D High Complexity (2 to 3)
E Very High Complexity (3+)

Geologic Unit – When possible, the associated geologic unit is identified for each habitat type and follows the habitat designation in parentheses.

Examples:

Shp_d1D(Q/R) - Continental shelf megahabitat; flat, highly complex (differentially eroded) hard seafloor with pinnacles. Geologic unit = Quaternary/ Recent.

Fhd_d2C (Tmm) - Continental slope megahabitat; sloping hard seafloor of deformed (tilted, faulted, folded), differentially eroded bedrock. Geologic unit = Tertiary Miocene Monterey Formation.

APPENDIX 3.2

Explanation for Marine Benthic Habitat Classification Scheme

(modified after Greene et al., 1999)

HABITAT CLASSIFICATION CODE

A habitat classification code, based on the deepwater habitat characterization scheme developed by Greene et al. (1999), was created to easily distinguish marine benthic habitats and to facilitate ease of use and queries within GIS (e.g., ArcView®, TNT Mips®, and ArcGIS®) and database (e.g., Microsoft Access® or Excel®) programs. The code is derived from several categories and can be subdivided based on the spatial scale of the data. The following categories apply directly to habitat interpretations determined from remote sensing imagery collected at the scale of tens of kilometers to one meter: Megahabitat, Seafloor Induration, Meso/Macrohabitat, Modifier, Seafloor Slope, Seafloor Complexity, and Geologic Unit. Additional categories of Macro/Microhabitat, Seafloor Slope, and Seafloor Complexity apply to areas at the scale of 10 meters to centimeters and are determined from video, still photos, or direct observations. These two components can be used in conjunction to define a habitat across spatial scales or separately for comparisons between large and small-scale habitat types. Categories are explained in detail below. Not all categories may be required or possible given the study objectives, data availability, or data quality. In these cases the categories used may be selected to best accommodate the needs of the user. If an attribute characterization is probable but questionable, it is followed by a question mark to infer a lower level of interpretive confidence.

EXPLANATION OF ATTRIBUTE CATEGORIES AND THEIR USE

Determined from Remote Sensing Imagery (for creation of large-scale habitat maps)

(1) Megahabitat – This category is based on depth and general physiographic boundaries and is used to distinguish regions and features on a scale of tens of kilometers to kilometers. Depth ranges listed for category attributes in the key are given as generalized examples. This category is listed first in the code and denoted with a capital letter.

(2) Seafloor Induration – Seafloor induration refers to substrate hardness and is depicted by the second letter (a lower-case letter) in the code. Designations of hard, mixed, and soft substrate may be further subdivided into distinct sediment types, which are then listed immediately afterwards in parentheses either in alphabetical order or in order of relative abundance.

(3) Meso/Macrohabitat – This distinction is related to the scale of the habitat and consists of seafloor features ranging from one kilometer to one meter in size. Meso/Macrohabitats are noted as the third letter (a lower-case letter) in the code. If necessary, several Meso/Macrohabitats can be included either alphabetically or in order of relative abundance and separated by a backslash.

(4) Modifier – The fourth letter in the code, a modifier, is noted with a lower-case subscript letter or separated by an underline in some GIS programs (e.g., ArcView®). Modifiers describe the texture or lithology of the seafloor. If necessary, several modifiers can be included alphabetically or in order of relative abundance and separated by a backslash.

(5) Seafloor Slope – The fifth category, represented by a number following the modifier subscript, denotes slope. Slope is typically calculated for a survey area from x-y-z multibeam data and category values can be modified based on characteristics of the study region.

(6) Seafloor Complexity – Complexity is denoted by the sixth letter and listed in caps. Complexity is typically calculated from slope data using neighborhood statistics and reported in standard deviation units. As with slope, category values can be modified based on characteristics of the study region.

(7) Geologic Unit – When possible, the geologic unit is determined and listed subsequent to the habitat classification code in parentheses.

CHAPTER 4

Modelling Inshore Rockfish Habitat in British Columbia: A Pilot Study

Jeff A. Ardron and Scott Wallace

Abstract

In the absence of reliable survey data, habitat modelling can direct conservation and fishery management efforts. We have constructed a model designed to predict high-value inshore rockfish habitat, based on the variables of topographical complexity and kelp density. When applied in our pilot study area, this model showed remarkable accuracy at predicting areas previously identified by commercial fishers. Ninety-four percent of high-value fishing areas were captured by our habitat model, and 79% contained our upper three scores, which we believe to be indicative of "core" habitat areas. These upper three classes account for 28% of the study area. When used to assess some recently designated rockfish conservation areas, our model and fishers' knowledge both indicate that three out of the seven conservation areas in our study area may actually represent poor choices for rockfish conservation and restoration.

Introduction

Numerous rockfish populations along the west coast of North America are depleted (Parker et al., 2000). Of particular concern in British Columbia (BC) are six species managed collectively as "Inshore Rockfish." These species exhibit resident behaviour and are highly linked to complex rocky reef habitat (Love et al., 2002), making them suitable candidates for spatial protection. Increasing evidence over the last decade has shown that a network of well-placed spatial reserves can be an effective tool for any long-term rebuilding and management strategy (Roberts et al., 2003).

In 2002, Fisheries and Oceans Canada (DFO)[1] embarked on a strategy to rebuild inshore rockfish populations; namely, black (*Sebastes*

Jeff A. Ardron, Living Oceans Society, P.O. Box 320, Sointula, British Columbia, Canada V0N 3E0, phone 250-973-6580, fax 250-973-6581; Corresponding author: jardron@livingoceans.org

Scott Wallace, Blue Planet Research and Education, 9580 Gleadle Road, Black Creek, British Columbia, Canada, V9J 1G1 Phone: 250-337-8521, Email: scottw@island.net *melanops*), china (*S. nebulosus*), copper (*S. caurinus*), quillback (*S. maliger*), tiger (S. *nigrocinctus*), and yelloweye (*S. ruberrimus*). As part of this strategy, DFO has set the objective of setting aside 20-50% of inshore rockfish habitat in the form of rockfish conservation areas (RCAs). To date, the selection of candidate sites has been piecemeal, driven primarily by public meetings, and consultation with fishing associations. This piecemeal approach is not unusual. In North America, protected areas in general have been created in an ad hoc manner, in large part stimulated by the requirements of species-based legislation, or other singularly focussed planning (Noss et al., 1997). However, there is a growing body of literature to suggest that this approach is far from ideal, and in some cases can lead to decisions that would later be regretted (Allison et al., 1998; Margules and Pressey, 2000; Stewart et al., 2003). A non-systematic approach gives little assurance that selected areas represent optimal habitat required for an effective network of protected areas. Relying heavily on consultations with interested stakeholders can also lead to the possibility that certain important areas may be overlooked, or that unimportant habitat may be put forward (Wallace and Ardron, 2003).

Modelling as a tool for designing marine reserves has been underutilized (Ward et al. 1999; Leslie et al., 2003). Modelling rockfish habitat offers a more systematic approach to reserve area design, although it has yet to be widely applied. In the past decade, DFO has used a crude model consisting of bathymetric data (depth) coupled with a smoothing algorithm to identify inshore rockfish habitat for the purpose of stock assessment (Yamanaka and Kronlund, 1996).

In 1997, the province of BC released their Marine Ecological Classification, based on five physical variables, the intent of which was the creation of a universal marine habitat classification "... for preservation, planning and resource management purposes" (Howes et al., 1997). Since then, it has been somewhat revised, containing seven variables for benthic habitat (Axys, 2001). Both versions contain a measure of relief. However, independent dive surveys in the study area failed to correlate the classification system with reefs or other biological indicators (Haggarty, 2000). It has been suggested that the approach behind the classification is too generalized (Ardron, 2001). Furthermore, measures of relief can be unduly influenced by single large changes in depth, while not detecting smaller but clustered changes that would actually better indicate rocky reef habitat.

In 2000, we developed a unique GIS analysis to evaluate seafloor "complexity" using bathymetric data (Ardron, 2002). For BC's passages and the Strait of Georgia, complexity invariably translates as rocky reefs. Initially, to verify that complexity was relevant to known rockfish distributions, we interviewed local fishers on the southern Central Coast of BC (Fig. 4.1; see page 126) and compared their knowledge with our analysis of complexity alone. We also considered DFO fishery officers' data, and we presented the analysis for verification by independent experts. Initial results were very promising. Next, we added kelp beds into the analysis, which improved the model's predictive ability. Finally, we evaluated the utility of the recently implemented rockfish conservation areas against both our habitat model and local knowledge.

Methods: Creation of GIS Layers and Model Indices

GIS played a significant role in our analyses. We used ArcGIS (ArcInfo & ArcView) 8.2 with the Spatial Analyst and Geostatistical Analyst extensions, and ArcView 3.2 with the Spatial Analyst extension. All calculations were performed in the BC Albers Equal Area projection, which largely preserves area (though not shape or direction). Because our calculations used equal area grids of 0.2 ha per cell (44.72 m x 44.72 m), an unequal area projection (such as geographic long-lat) could have skewed the results.

Layer 1: Benthic Topographical Complexity

Complexity is the key variable in our model, identifying potential rockfish reefs. Complexity is not the same as relief, which looks at the maximum change in depth. Topographical complexity considers how convoluted the bottom is, not how steep or how rough, though these both play a role. Complexity is similar but not the same as "rugosity" as is sometimes used in underwater transect surveys, whereby a chain is laid down over the terrain and its length is divided by the straight-line distance. Rugosity, however, can be strongly influenced by a single large change in depth, whereas complexity is less so. Complexity is indicated by how often the slope of the sea bottom changes in a given area. This is the density of the slope of slope (second derivative) of the depth. We used ESRI's Spatial Analyst extension to calculate slopes, and its density "kernel" option to calculate densities using a search radius of 1 km—though this distance can vary, depending on the scale of the analysis. For more specifics on the GIS methodology behind calculating complexity, please refer to Ardron, 2002.

Line bathymetric data were purchased from Nautical Data International (Natural Resource Map series, nominal 1:250,000). These required extensive cleaning, including edge matching, and removal/correction of unattributed or erroneous segments. Due to the poor resolution of the nearshore line work (<50 m), these were digitized from nautical charts and merged with the NDI data. Ultimately, we wanted to interpolate these isobaths into a depth grid; however, the varying densities of line nodes can bias direct interpolation from lines to a grid. Thus, in order to ameliorate this issue, we first transformed the lines into evenly spaced (50 m) points, using Dr. Bill Huber's free script, Poly to Points available on the ESRI ArcScripts page (http://arcscripts.esri.com/details.asp?dbid=11407). We interpolated these points using a variety of algorithms, but found that the simplest, a TIN (triangulated irregular network—straight linear interpolation), is what worked best. The hard "creases" associated with TINs, which can be visually intrusive, actually have certain advantages when undergoing a complexity analysis, as they clearly demarcate a change in slope. The resulting depth grid was fed into the calculations of benthic topographical complexity.

The complexity algorithm used in this analysis and elsewhere invariably identifies areas of convoluted substrate (Fig. 4.2; see page 127). At this scale, most of these areas are rocky reefs or to a lesser extent, sills and ledges. The steep fjords common along BC's coastline, although high in relief, are not captured by complexity because they do not have many changes in slope over a given area; that is, they are either steep-sided, or relatively flat-bottomed. While their walls can offer some habitat to rockfish, it is generally only a narrow strip along their base. The purpose of the complexity measure was *not* to capture all rockfish habitat; rather, it was to capture exceptionally good areas that may warrant protection.

The complexity index was taken from an earlier analysis of the Central Coast (Ardron et al., 2002). The integer score ranged from 0 to 3 per grid cell, where 1 represented a varying buffer band, 2 represented moderate to high complexity, and 3 represented high to very high complexity. The buffer was used as a way to account for the varying scales of the water bodies in the Central Coast, whereby narrow inlets had no buffer, passages had a buffer of 500 m, and open sea had a buffer of 1500 m. Although all three buffers can be found in the pilot study area, generally it is 500 m.

Layer 2: Kelp Coverage

Initially, we included kelp (*Nereocystis luetkeana* and *Macrocystis intergrifolia*) in the model due to its known importance as juvenile inshore rockfish habitat (Fig. 4.3; see page 127). We felt that complex areas near kelp would exhibit higher recruitment than complex areas without kelp, and as such would represent more desirable rockfish habitat. Later, it became apparent that areas of thick kelp directly overlapped with known fishing areas. The inclusion of kelp improved the model's predictability vis-à-vis known fishing areas.

Kelp data were merged from many sources, mainly the provincial government aquaculture surveys, aerial surveys, and federal Canadian Hydrographic Service charts. Polygons were given a score of 1 (sparse) or 2 (continuous), based on notes in the surveys. If this was unknown, 1 was applied as a default. In areas of overlapping datasets, the higher values prevailed.

In our model we wanted to incorporate areas that had kelp in the vicinity of complexity. Although kelp is known juvenile habitat, there has been little published literature on the movements of juvenile rockfish, and thus no numbers were available upon which to base our "search radius." We decided to proceed with arbitrary distances in the range of 500 m to 2 km to calibrate the model to known fishing areas.

In GIS analyses, generally the use of a buffer is employed to capture areas within a given distance. However, in the case of kelp beds, which vary greatly in size, a fixed buffer zone around all kelp beds would have over-emphasized small isolated patches—of which there are many. Thus, instead we turned to a density measure, which would in its ranking reflect the size of kelp patches and neighboring patches. Even though the search radius is constant, areas of more numerous and larger kelp beds will be given a higher score, unlike a buffer.

We removed the lowest of 11 classes (Jenks natural breaks) and reclassed the remaining 10 equally into two categories (score = 1 or 2). Removing the lowest of 11 classes eliminated areas of low density, and had the effect of lowering the effective buffering distance from 1500 m to approximately 600-700 m.

Layer 3: Fishing According to Fishery Officers and Managers

There is a concern that releasing commercial fishing logbook data may breach fishers' right to confidentiality. As a result, outside researchers and other government agencies do not have accurate spatial knowledge of fishing activities in BC. In 1995, the provincial government hired a consultant to interview DFO fishery officers and managers to get a sense of where fishing activities were taking place within the Central Coast, which includes our study area. It should be noted that these DFO officers did not have access to the logbook data either, but were basing their opinions on their job experience. The same consultant returned in 2002 to update the information. For our study area, the 1995 information came from mainly one source (plus one polygon by another source) and exhibited a fairly high degree of localized precision. By 2002, however, there had been a change in staff, and the results were much more vague, with only one polygon in the study area.

Layer 4: Fishing According to Fishers

Remarks from local fishers suggested that the fishery officers' data were incomplete. In the summer of 2000, using the same interview techniques and database structure as had been used for the fishery officers, a contractor hired by Living Oceans Society interviewed 29 commercial fishers on the Central Coast. Five of the fishers fished rockfish in the pilot study area. Fishers were asked to draw their preferred fishing areas on nautical charts, which were later digitized and linked to an attributes database based on information given during the interviews.

It was emphasized at the time of the interviews that we wanted to know the fishers' preferred fishing locations so that we could take their use into account when proposing closures or protected areas. After the interview, the interviewer filled out a standard form judging the precision of the polygons, and noting how cooperative the subject appeared to be (an indication of possible accuracy). Only one fisher (of 30) refused to give information after meeting with the interviewer. Of the others that did provide information, all were judged to be honest and forthcoming. In her final report, the interviewer felt that rockfish had been one of the strongest datasets for the study area (Groff, 2002).

Figure 4.4 (see page 128) illustrates the information from the three data collections: the two 1995 fishery officers; two 2002 fishery officers (different people); and five commercial fishers. Because the intent of this pilot study was to compare good fishing grounds (as a proxy for good habitat) with our model's predictions, it was decided that if one fisher identified an area as "important," then that should be sufficient to include it as "moderate" (score = 1) potential rockfish habitat, and if another fisher(s) noted the area (either as moderate or high), it was "good" potential habitat (score = 2).

We did not want to simply add up the layers because the sample size was too low to draw out emergent trends. Fishers generally did not fish the same areas and had to some extent spread themselves out over the study area.

Layer 5: Rockfish Conservation Areas (RCAs)

GIS shapefiles of the proposed (and rescinded) RCAs were initially received from the DFO groundfish management data unit in October 2003, and later revised in March 2004. The selection of candidate sites for RCAs were the result of consultations with commercial and sports fishers in BC. Shown in Figure 4.5 (see page 128) are other closures that came into effect in 1998. These could have influenced the fishers' perceptions of importance in 2002.

Identification of Complex Areas Near Kelp

Our model's combined complexity-kelp index is weighted towards complexity. As input layers, kelp has a range of 0-2, while complexity has a range of 0-3. The two were added together to give a range of 0-5. To smooth the results and take into account neighbouring areas, we then created a density surface populated with this summed score, and ranked it based on standard deviations. We discarded the first class and scored the remaining four 1-4. Discarding the lowest class removed both the lower value kelp areas not near any complexity and lower value complexity areas not near kelp. Thus, the final score of 1-4 represents areas that have either high-rated complexity or kelp, or combinations of both kelp and complexity. The sizes of the modelled habitat areas and fishing areas are shown in Table 4.1

Text continues on page 61

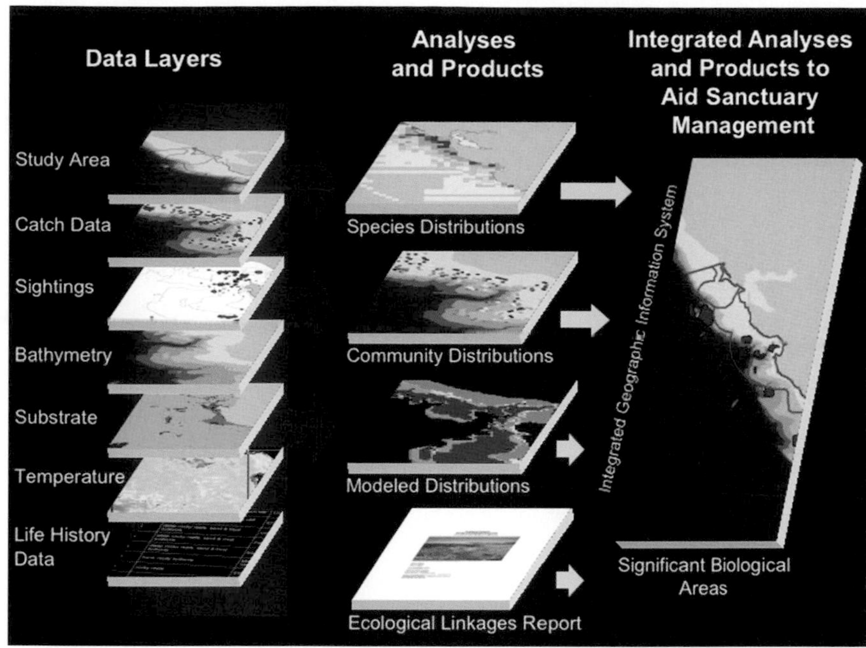

Figure 1.1. Generalized biogeographic approach to study NOAA national marine sanctuaries.

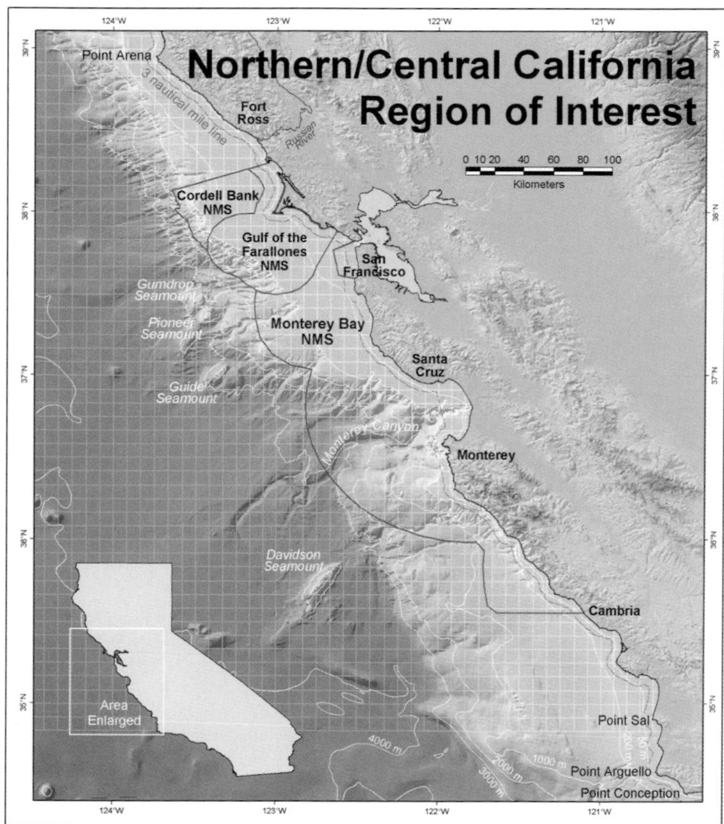

Figure 1.2. Locator map of study area from Point Arena to Point Sal. National marine sanctuary boundaries shown in red.

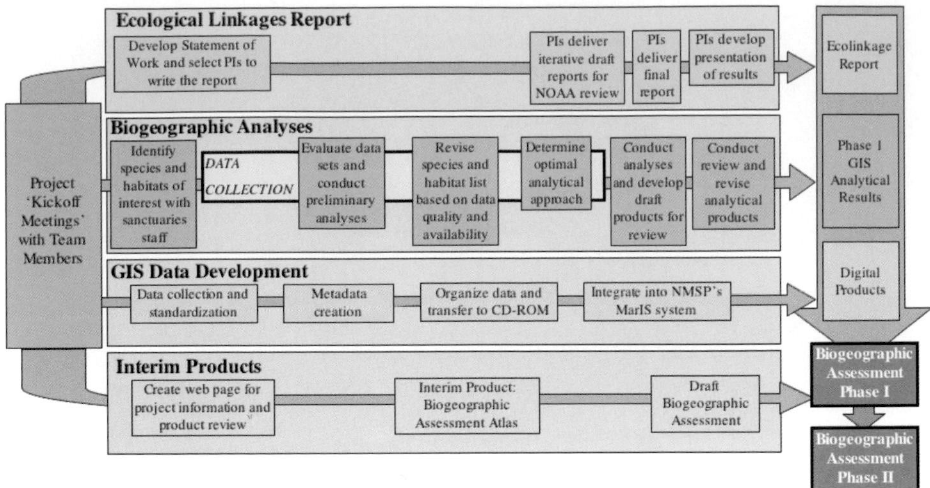

Figure 1.3. Biogeographic process for assessment of marine resources off North/Central California: supporting updates to management plans for Cordell Bank, Gulf of the Farallones, and Monterey Bay National Marine Sanctuaries

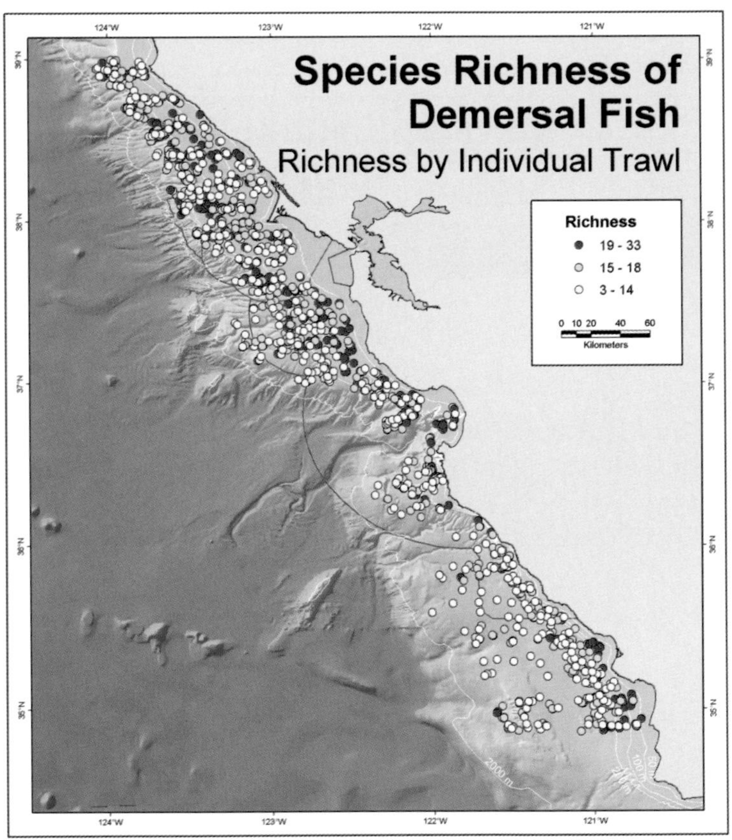

Figure 1.4. Species richness of rockfish from individual NMFS shelf and slope trawls.

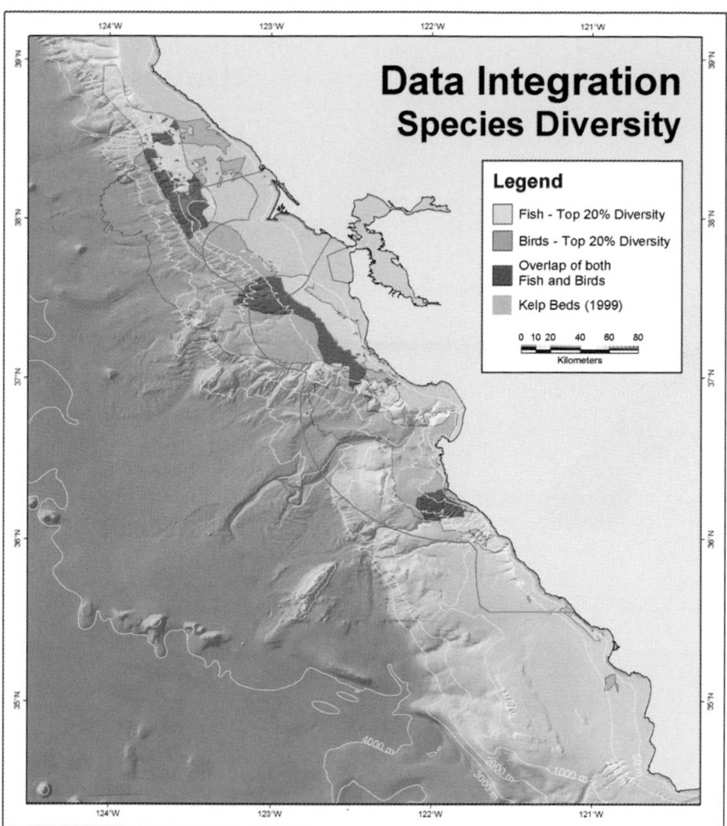

Figure 1.5. Potential distribution of habitat suitability for adult and juvenile Dover sole. Map inset contains validation statistics, and Suitability Index values for bathymetry and substrate are displayed below the maps.

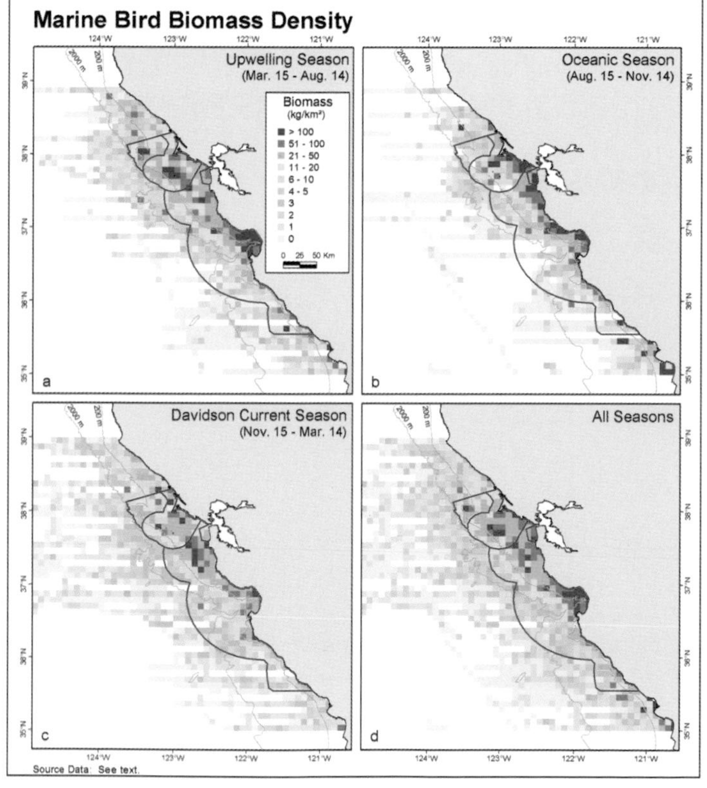

Figure 1.6. Marine bird biomass, by season and for all seasons in study area.

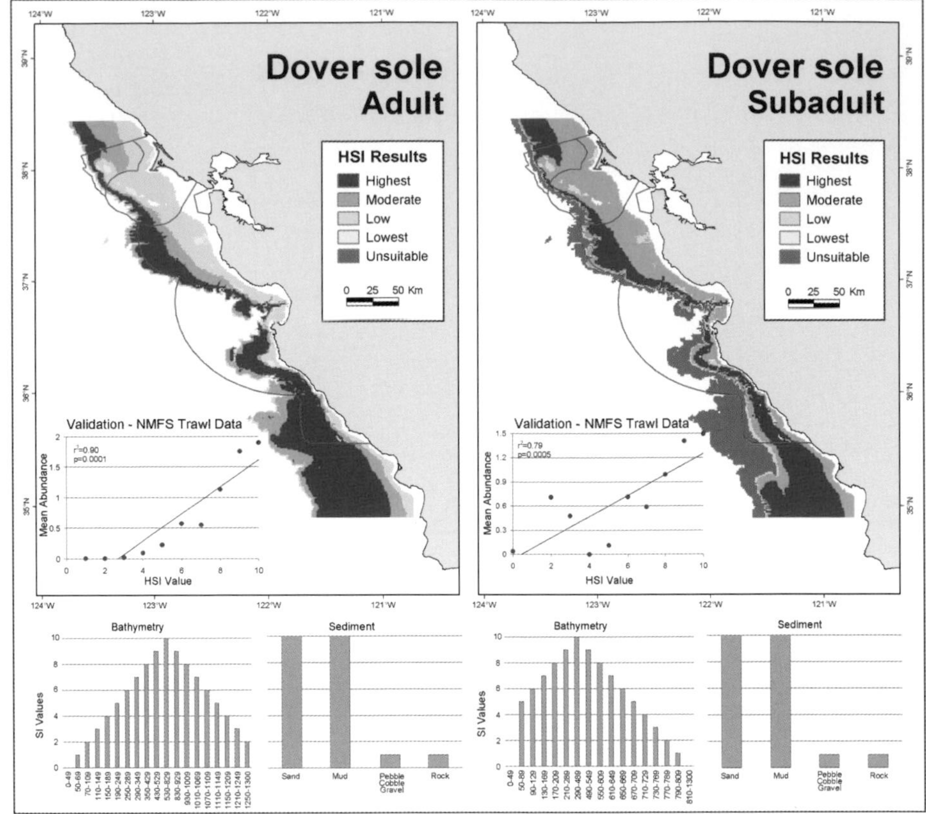

Figure 1.7. Data Integration: diversity hot spots (top 20%) for fish and marine birds. Coastal kelp bed areas are also shown.

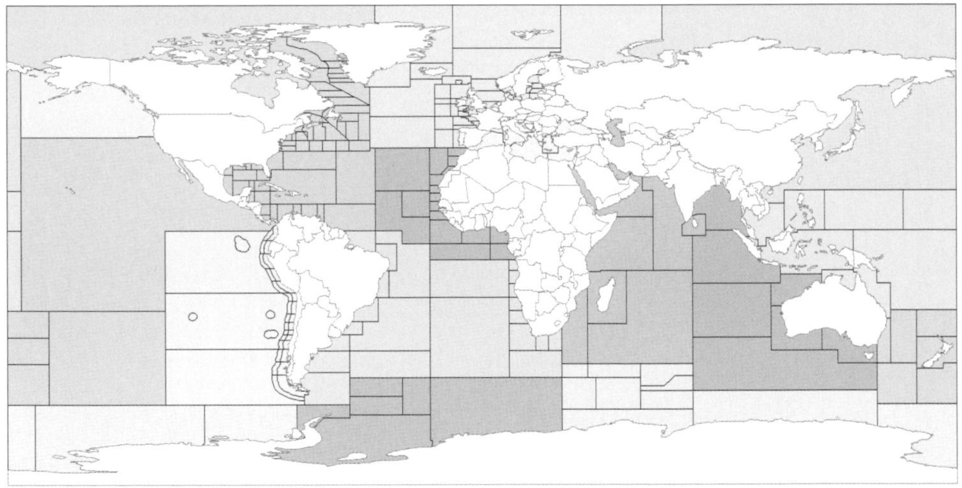

Figure 2.1. Reporting areas used for fisheries statistics by Food and Agriculture Organization (FAO) and its regional bodies, the International Council for the Exploration of the Sea (ICES), and the Northwest Atlantic from the Northwest Atlantic Fisheries Organization (NAFO), currently used

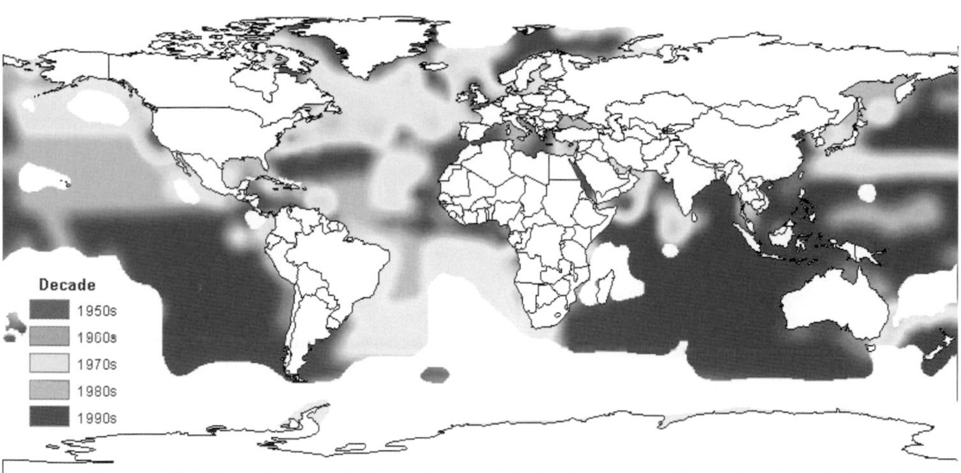

Figure 2.2. Decade of maximum commercial landings.

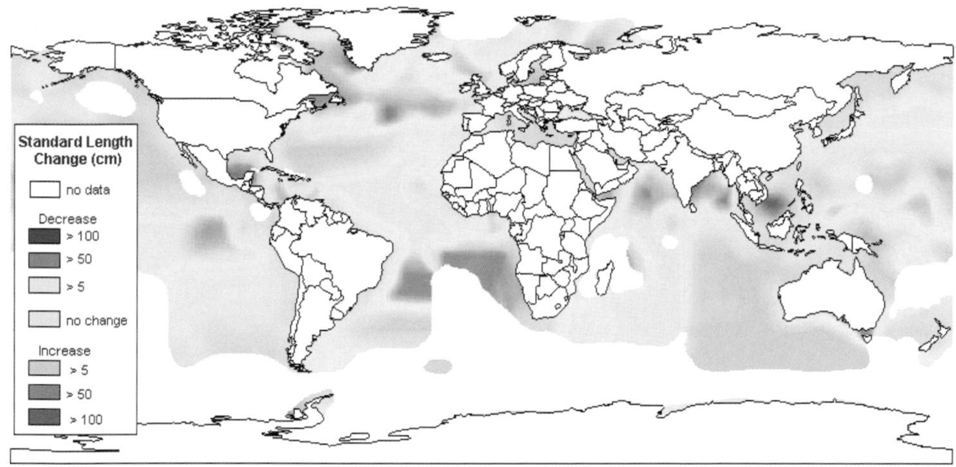

Figure 2.3. Global change in the mean length of commercial landings.

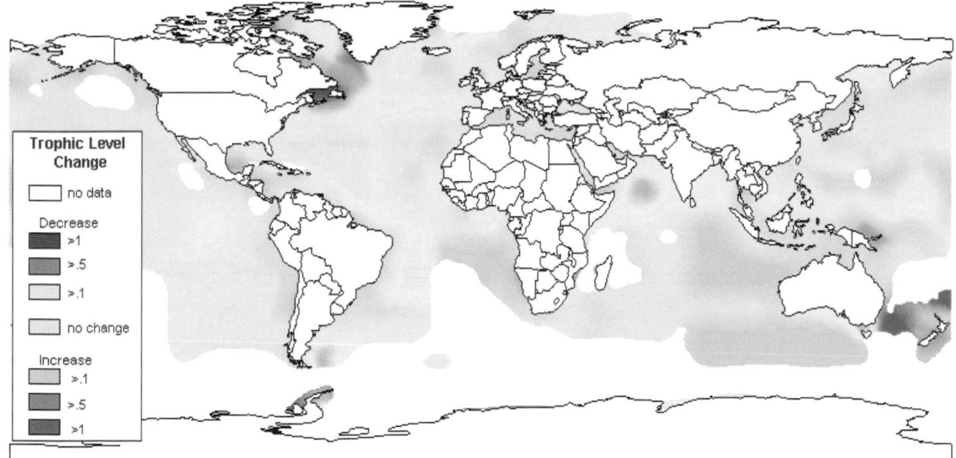

Figure 2.4. Global change in the mean trophic level of commercial landings.

Figure 3.3. Map of potential marine benthic habitats constructed from Simrad EM 300 (30 kHz) bathymetric and backscatter imagery. See Figure 3.1a for location and Appendices 3.1 and 3.2 for explanation of habitat code. All habitat types are located on the upper continental slope, or flank (F).

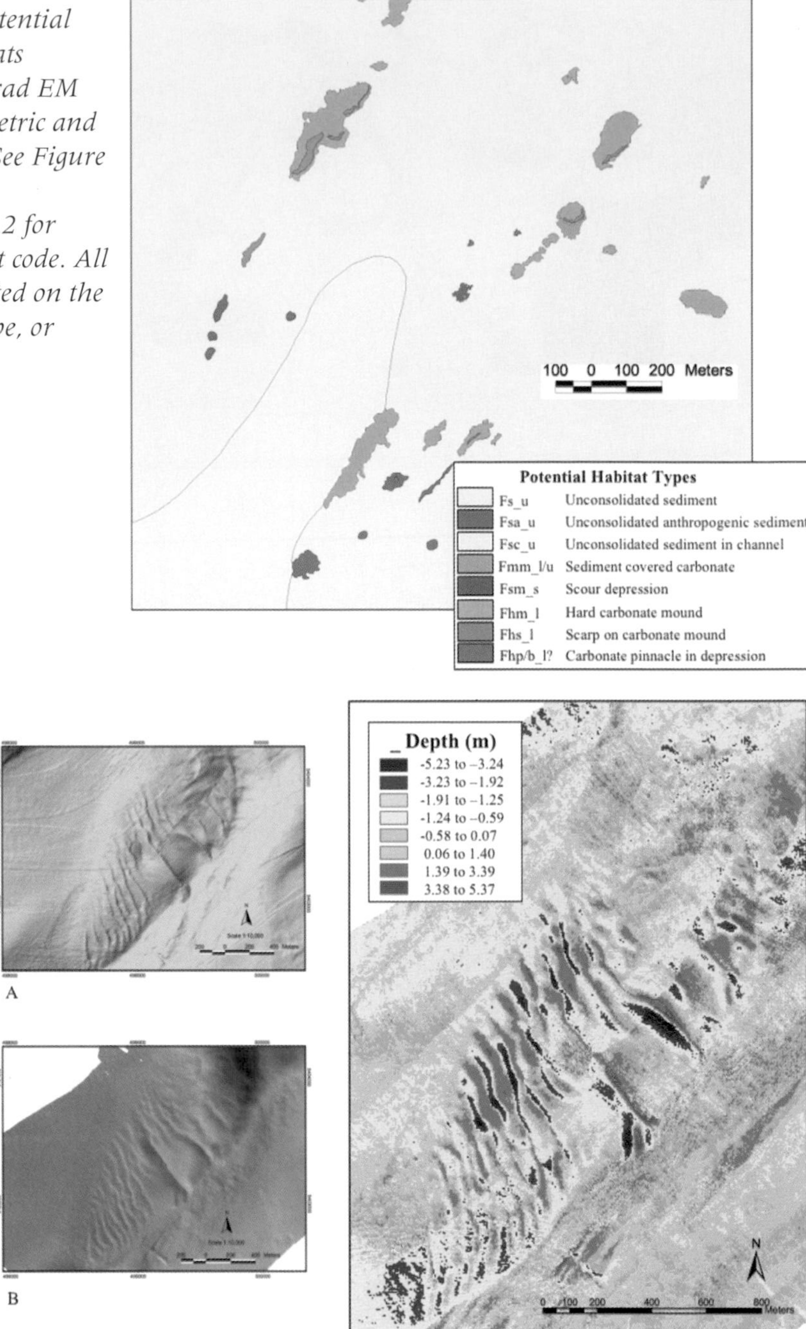

Figure 3.4. Dynamic sand waves in the Boundary Pass region, Canada, collected with a Simrad EM 1002 95 kHz system by the Geological Survey of Canada and the Canadian Hydrographic Service. Image 3.4a displays multibeam bathymetric data collected in 2001 and Image 3.4b displays data collected in 2003 in the same area. Both images were created with the GIS program, ArcMap® and are shown at a scale of 1:50,000. Using ArcMap®'s Raster Calculator, which calculates depth differences at each pixel location, two 5 m grids were subtracted (2001 grid - 2003 grid). Results are displayed in Image 3.4c. The red and green colors represent probable migration of the sand waves, where red = accumulation of sand and green = loss of sand.

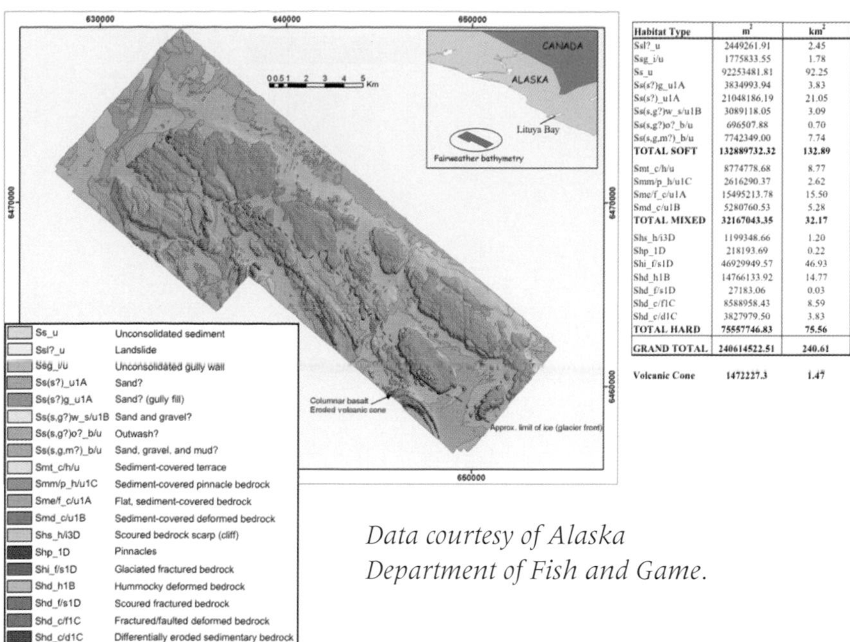

Habitat Type	m^2	km^2
Ssl?_u	2449261.91	2.45
Ssg_i/u	1775833.55	1.78
Ss_u	92253481.81	92.25
Ss(s?)g_u1A	3834993.94	3.83
Ss(s?)_u1A	21048186.19	21.05
Ss(s,g?)w_s/u1B	3089118.05	3.09
Ss(s,g?)o?_b/u	696507.88	0.70
Ss(s,g,m?)_b/u	7742349.00	7.74
TOTAL SOFT	**132889732.32**	**132.89**
Smt_c/h/u	8774778.68	8.77
Smm/p_h/u1C	2616290.37	2.62
Sme/f_c/u1A	15495213.78	15.50
Smd_c/u1B	5280760.53	5.28
TOTAL MIXED	**32167043.35**	**32.17**
Shs_h/i3D	1199348.66	1.20
Shp_1D	218193.69	0.22
Shi_f/s1D	46929949.57	46.93
Shd_h1B	14766133.92	14.77
Shd_f/s1D	27183.06	0.03
Shd_c/f1C	8588958.43	8.59
Shd_c/d1C	3827979.50	3.83
TOTAL HARD	**75557746.83**	**75.56**
GRAND TOTAL	**240614522.51**	**240.61**
Volcanic Cone	**1472227.3**	**1.47**

Data courtesy of Alaska Department of Fish and Game.

Figure 3.5. Map of potential marine benthic habitats of Fairweather Ground, a heavily fished area in SE Alaska. Refer to Appendices 3.1 and 3.2 for explanation of habitat code. Associated table displays the area of habitat and induration (hardness) types calculated in ArcView® using the Feature Geometry Extension.

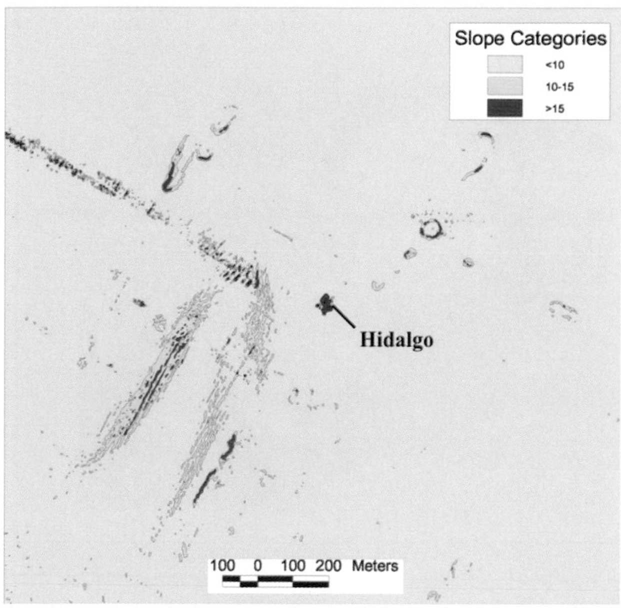

Figure 3.6. Slope inclination map generated in ArcView® using the Spatial Analyst Extension and Simrad EM 300 (30 kHz) bathymetry. Slope category values are listed in degrees.

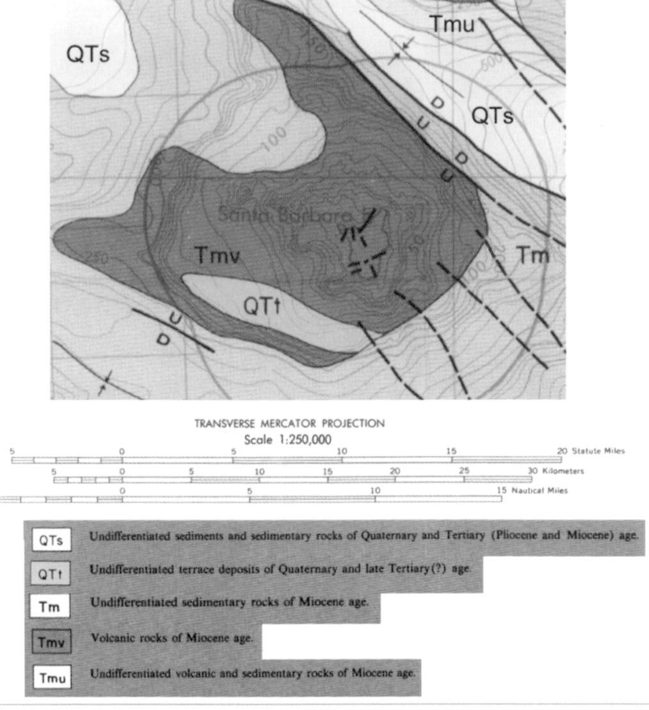

Figure 3.7. Geologic map and legend of Santa Barbara Island offshore region illustrating geologic data that can be used in mapping of potential marine benthic habitats. After Vedder et al. 1986.

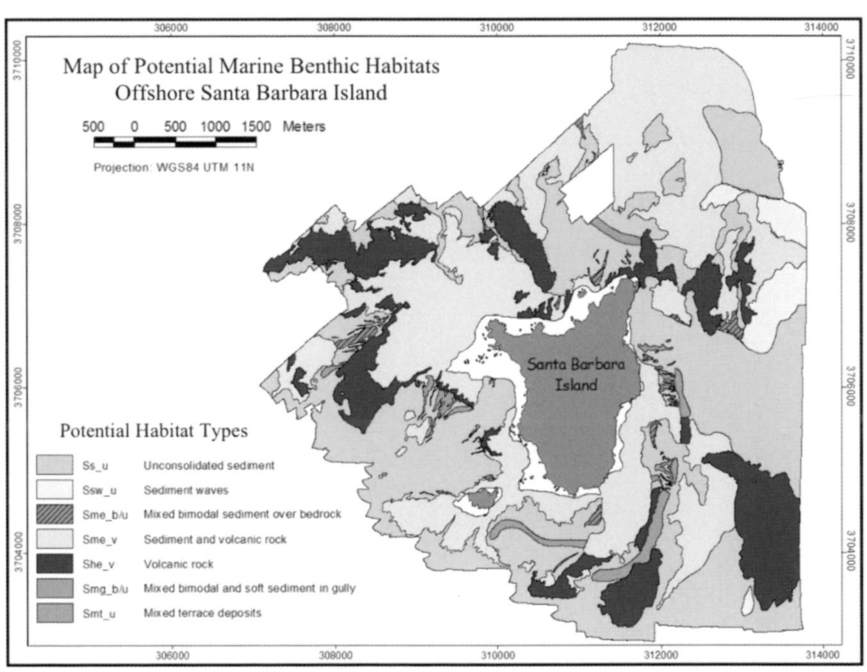

Figure 3.8. Map of potential habitats constructed in GIS from Reson 8101 (240 kHz) bathymetry collected off Santa Barbara Island. Refer to Appendices 3.1 and 3.2 for explanation of habitat code.

Results

Evaluating the Model's Predictions

The habitat model displayed very good overlap with areas identified by local fishers. The four classes of the model overlapped 94.0% of the high-value fishing areas, while accounting for 46.5% of the study area. The top three classes of the model (scores 2-4) overlapped 78.9% of the high-value fishing areas, while accounting for 28.1% of the study area. Thus, the top three habitat classes were three times more common in high-value fishing areas than would be expected due to random chance (Table 4.1). In some places, the predicted habitat and fishing areas overlap with very high precision (see inset map, Fig. 4.6; see page 129). There was less overlap with the medium-value fishing areas, though still much higher than would be expected by chance. This may indicate that medium-value fishing areas represent less valuable habitat.

Looking at the summed scores, there are clear and strong trends correlating fishers' preferred areas with our model's predictions, and vice versa; that is, each can predict the other, though there is greater predictive power going from habitat to fishing area, which may be a function of the relatively sparse fishing data. The summed scores of the fishing areas applied to each of the three highest habitat classes are well above the mean value for the study area, with the highest class being about three times higher. Within the highest value habitat areas, the fishing score is 72%. That is, the combined scores (1s and 2s) of the fishing areas added up to being 72% of what they would have been had the habitat area overlapped entirely with high-value fishing areas (2s). Conversely, of the areas represented by the lowest value habitat class, the cumulative fishing score is just 5.6%. The unidentified (value = 0) fishing areas contain a cumulative habitat score of 11%, while the high-value fishing areas have a cumulative habitat score five times greater (56%; Fig. 4.7).

Again looking at the cumulative fishing scores we find that the top three habitat classes are much above the study area mean, while the fourth class is at about the mean. This would indicate that the fourth

Table 1: Sizes of fishing areas, RCAs, and modelled habitat.

	Total Area (Km²)	*Proportion*
Fishing 2 (high)	427.16	17.7%
Fishing 1 (med.)	360.93	15.0%
RCA	156.33	6.5%
Rescinded	267.51	11.1%
Habitat 4 (v. high)	81.08	3.4%
Habitat 3 (high)	203.02	8.4%
Habitat 2 (med-high)	392.98	16.3%
Habitat 1 (med.)	444.49	18.4%
Study Area, less Inlets	2410.37	100.0%

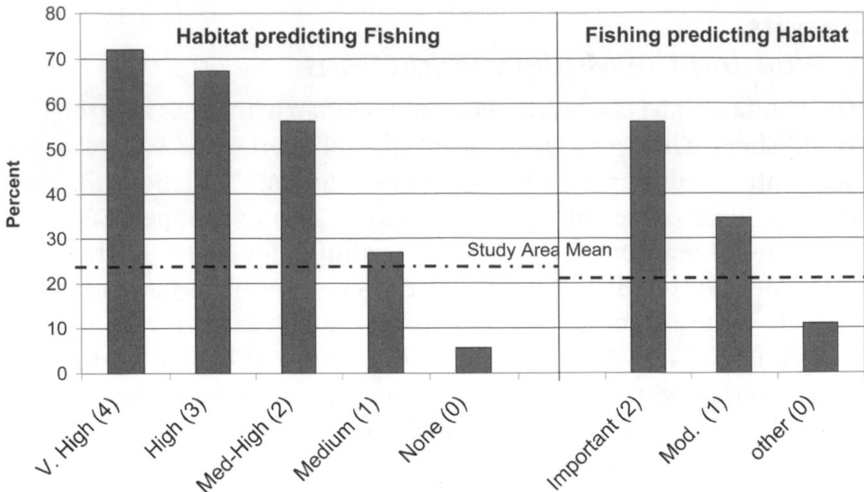

Figure 4.7. The left panel plots how well the habitat index predicts the valued fishing areas. The right panel looks at how well the fishing areas can predict the habitat identified in the model. In both cases the trends are clear. Habitat, however, is somewhat better at predicting fishing than vice versa. Note that the "none" class is actually quite good at predicting where fishing will not occur.

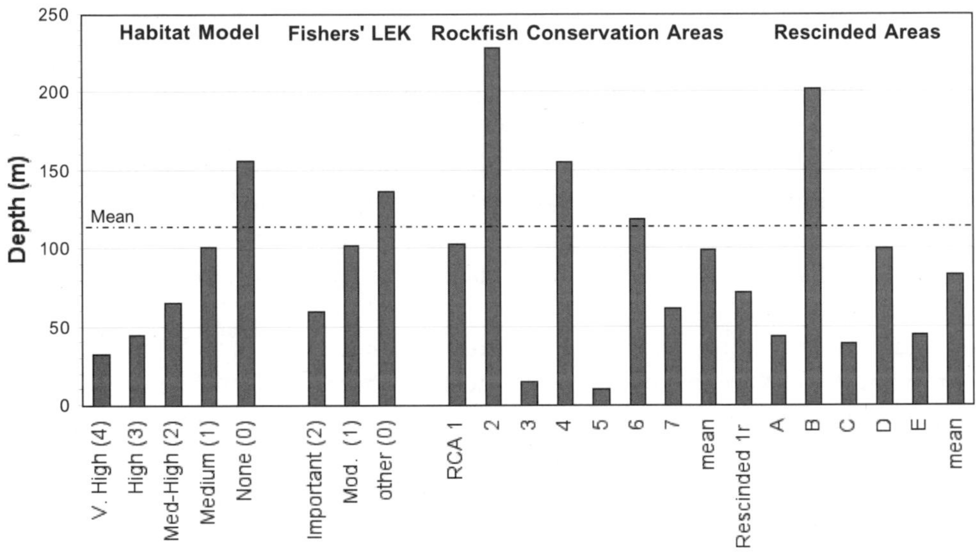

Figure 4.8. The high-value classes of the habitat model predictably contained a greater portion of shallower habitats than did the lower value classes (left panel). Likewise, higher value fishing areas ("LEK" = local ecological knowledge; also referred to as "TEK" or traditional ecological knowledge) also contained a greater portion of shallower habitats. On the other hand, the RCAs only sometimes follow this trend. Overall, the rescinded RCAs are somewhat shallower than the selected RCAs (left panels).

class does not add any predictive strength to the model (Fig. 4.7). Thus, we suggest that the top three habitat classes represent the "core" potential rockfish habitat in the study area, and should be used when making predictions. The fourth class, however, can remain included for scoring the utility of proposed areas. While the fishery officer data were too sparse to feed into any formal habitat evaluation, visually they generally overlapped with the areas identified by the model (compare Fig. 4.4 with Fig. 4.6)

Depth versus Habitat Index, Fishing Areas, and RCAs

The inshore rockfish are found in depths of less than 200 m, and commonly in depths of less than 100 m (Love et al., 2002). As discussed above, the two variables, complexity and kelp were found to capture the fishing polygons quite well. However, reefs and kelp are both associated with shallower depths, and one might expect there to be a cross-correlation. Indeed, looking at the habitat classes, there is a clear trend from shallower, high-value areas to deeper, low-value areas (left section, Fig. 4.8). Similarly, there is a trend in the fishers' ranking of areas.

With the RCAs, we see a great deal of variability in mean depth, with the final RCAs being more variable than those that were rescinded. There is some correspondence of shallower areas scoring better than deeper ones, especially in extreme examples (e.g. #2 vs. #5), but there are also notable exceptions (e.g., #3 vs. B). Thus, while it is true that shallower areas appear to capture more features of known rockfish habitat, we would not say that depth alone is sufficient to measure this; that is, our model is not simply a depth model in disguise.

Evaluating the Rockfish Conservation Areas

Rockfish conservation areas were announced in March 2004 by Fisheries and Oceans Canada (DFO). These represent a subset of initial areas originally put forward in 2003. We mapped the 2004 RCAs, as well as the ones put forward in 2003 and rescinded in 2004, for the purpose of examining how well they capture known important fishing areas and our modelled habitat. Furthermore, we were curious to see how the 2004 RCAs compared with those that were rescinded.

Visually, it is clear that some RCAs overlap better with high-value (potential) habitat than others; however it is difficult to discern any particular trends (Fig. 4.9; see page 129). Unlike Fig. 4.6, which shows habitat and fishing areas, any correlation between the RCAs and either habitat or fishing is not immediately obvious. In Tables 4.1 and 4.2, we find that the RCAs overlap less potential rockfish habitat than the medium-value fishing areas, and much less than the high-value fishing areas. The rescinded RCAs, however, overlap more predicted habitat than the medium-value fishing areas, and more than the final selection of RCAs.

Table 4.2. Predicted habitat vs. fishing areas and RCAs. While high-value fishing areas contain 79% of "core" potential habitat, the Rockfish Conservation Areas contain about half that, 41%. The rescinded conservation areas, however, contain more core habitat, and at 65% fall in between the two.

	Total Habitat 1-4	*Total Habitat 2-4 "Core"*	*Habitat 4 (v. high)*	*Habitat 3 (high)*	*Habitat 2 (med-high)*	*Habitat 1 (med.)*
Fishing 2 (high)	94.0%	78.9%	11.8%	27.2%	39.9%	15.2%
Fishing 1 (med.)	72.7%	43.8%	4.5%	11.3%	28.0%	28.9%
RCA	61.1%	41.3%	7.4%	9.6%	24.3%	19.8%
Rescinded	86.8%	65.1%	10.1%	21.3%	33.7%	21.7%
Study Area, less Inlets	46.5%	28.1%	3.4%	8.4%	16.3%	18.4%

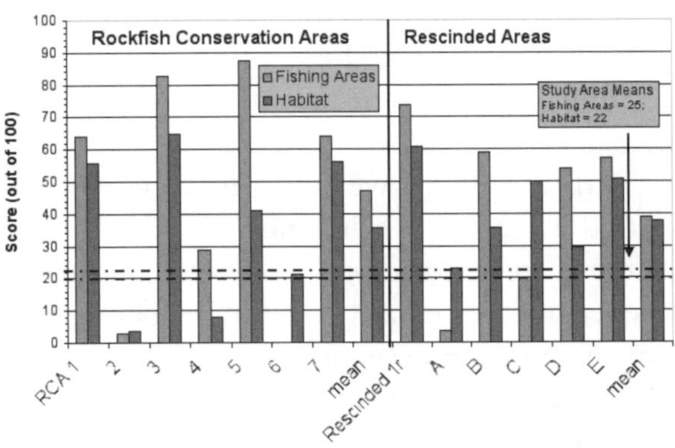

Figure 4.10. The RCAs are extremely variable in how well they capture known fishing areas (light gray) or modelled habitat (dark gray). The variability is greater in the actual RCAs than those that were rescinded. Their mean scores are very similar, with the final RCAs capturing a little more of known fishing areas, but a little less of predicted habitat.

Tabulating the cumulative scores ("zonal statistics"), we find the RCAs are extremely variable in how well they do at capturing known fishing areas or our modelled habitat (Fig. 4.10). Fishing area scores (out of 100) were created by calculating the mean fishing values (0-2) for each RCA, and standardizing to a score out of 1-100, where 100 would represent perfect overlap with high-value (2s) fishing areas. These RCA scores range from zero (no fishing area in RCA #6) to 87 (RCA #5). Cumulative habitat scores were likewise calculated and range from 4 (RCA #2) to 65 (RCA #3). The variability is greater in the recently announced RCAs than those that were rescinded, and have both the highest and lowest scores in the study area. Their mean scores are very similar, however.

One RCA (#1) in the study area was initially proposed to be much larger (#1r). In this case, the earlier option had higher scores than the final version; i.e., the section that was rescinded appears to have been more valuable than the section that was left behind (inset map, Figs. 4.9 and 4.10).

Discussion

Effectiveness of Complexity Modelling

We believe that our model has shown itself to be a powerful predictor of high-value fishing areas. Because of the relatively sedentary nature of inshore rockfish adults, we suggest that high-value fishing areas should often overlap with high-value habitat, and the habitat-based model would appear to support this.

Initially, we were curious to see how well complexity alone could predict rockfish habitat. A comparison of Fig. 4.2 with Fig. 4.4 does indicate that there is indeed a great deal of predictive power in that variable alone; but there are certain gaps (e.g., the strip along the shore including RCA #5). Many of these could be filled when taking kelp into account. While complexity and kelp are no doubt correlated, including these two indicators in the model allow for one dataset to fill in for weaknesses in the other, as well as indirectly taking into account other complementary ecological attributes such as primary production. Although we did not use depth as a variable in our model, it could be examined afterwards, to sort out habitats for particular species with known depth preferences.

The habitat model is somewhat better at predicting fishing areas than vice versa; that is, the fishing areas tended to be larger than the associated modelled habitat. This can be explained in three non-exclusive ways: (1) the fishers drew their polygons a little more generally than where they actually fished; (2) the model's search radii in the various density analyses were a bit too short ("tight"); and (3) the fish wander somewhat from core habitat areas and are still caught in good numbers. The first and third explanations would suggest that the model could already be doing what it should (predicting habitat), while the second explanation would suggest broadening the constraints of the analysis somewhat. Surveying actual fish distributions would answer this and other questions. In the meantime, we feel it is prudent to err on the side of caution, whereby the areas selected represent a conservative estimation of important rockfish habitat. That way, we can feel more confident that their protection would be beneficial to the species.

Rockfish Conservation Areas

While high-value habitat areas consistently overlapped with high-value fishing areas, the rockfish conservation areas were much less consistent. We feel this high variability of (modelled) habitat quality within the RCAs could be a reflection of the stakeholder-driven RCA selection process.

Three of the seven RCAs appear to have been poor choices: #2 (Goletas Channel), #4 (Numas Island), and #6 (Salmon Channel). Five of the six of the rescinded RCAs (1r, B, C, D, E) scored better than any of these three RCAs. In light of these findings, we would suggest that the five higher-scoring but rescinded RCAs could be re-considered as replacements for the three low-scoring RCAs currently under protection. However, this would be our second choice.

Our preferred methodology, as stated in the introduction to this paper, would be to adopt a systematic approach in identifying all possible rockfish habitat. Providing numerous options from the onset offers the opportunity to include other essential elements into reserve design (e.g., source-sink relationships, connectivity, enforcement). Once all habitats are identified, and other necessary design criteria are considered, only then should socioeconomic criteria be applied for removing contentious areas. However, we would like to emphasize that the opinions and concerns of fishers should be taken very seriously. In collecting our data from fishers, we stressed that we wanted to find solutions that optimized both conservation and fishing.

Future Research

The analysis presented in this paper is a first step towards a better understanding of the utility of complexity-based habitat modelling. The measure of complexity is heavily dependent on the quality of available bathymetry data. Unfortunately, affordable good quality bathymetry data are still difficult to acquire in Canada, and remains an issue that needs to be addressed. Nonetheless, bathymetry is often the best dataset when considered next to other options. Similar data to those presented in this paper exist for other regions of BC and would allow for further testing of the model.

We are presently engaged in the collection of additional local fishers' knowledge, both commercial and recreational, for the Central Coast, with plans to have surveyed most of coastal BC by 2005. Also, we are examining divers' data to see if these can be incorporated, though there are some issues of scale.

We believe this same model may prove useful in the management of other species (e.g., lingcod), which utilize complex habitat, and we are presently looking for data with which to calibrate and verify such models. If successful, modelling would be used to direct conservation and survey efforts at a fraction of the cost presently required through broader, less directed surveys.

Finally, the results of the rockfish conservation area analysis clearly show the need for a systematic habitat-based approach for reserve selection. Our analysis covered only a small section of the BC coast containing seven RCAs. At present, 90 RCAs have been designated coastwide. Our results suggest that a significant proportion (perhaps 40%) of the RCAs are not actually protecting valuable rockfish habitat. We feel this warrants a larger coastal analysis and survey verification. Results from our complexity-based habitat modelling demonstrate an effective approach to narrowing down possible rockfish conservation sites, and could assist with the selection of high quality sites, while avoiding the selection of poor choices in the future.

Acknowledgements

The authors would like to thank the 30 anonymous commercial fishers who took hours to explain their fishing practices and to draw their areas on charts. Big thanks are also owed to Maja Groff who conducted the interviews and entered the results into the database. We would also like to acknowledge the David and Lucile Packard Foundation for supporting the initial phase of this project, and the Homeland Foundation for supporting the second phase of this project.

Notes

1. Fisheries and Oceans Canada is the current name of the department, but they elected to retain the acronym from their original name, Department of Fisheries and Oceans.

References

Allison, G. W., Lubchenco, J., and Carr, M. H., 1998. Marine reserves are necessary but not sufficient for marine conservation, *Ecological Applications*, 8(1) Supplement: S79-S92.

Ardron, J. A., 2001. Review of: British Columbia Classification Update, Final Report. Part 1: Issues of Scale, Resolution, & Accuracy, unpublished report submitted to the BC provincial government design team, Sointula, British Columbia, Canada, Living Oceans Society.

Ardron, J. A., 2002. A recipe for determining benthic complexity: An indicator of species richness, in Breman, J. (ed.), *Marine Geography: GIS for the Oceans and Seas*, Redlands, CA, ESRI Press, 169-75.

Ardron, J. A, Lash, J., and Haggarty, D., 2002. Modelling a Network of Marine Protected Areas for the Central Coast of British Columbia. Version 3.1, July 2002, British Columbia, Canada, Living Oceans Society, www.livingoceans.org/library.htm, 125 pp.

AXYS Environmental Consulting Ltd, 2001. British Columbia Marine Ecological Classification Update, final report submitted to Land Use Coordination Office, in association with John Roff, Ellen Hines.

Groff, M., 2002. Final report on the Local Ecological Knowledge Collection Project, Part 1: Data and Database Comments, Sointula, British Columbia, Canada, Living Oceans Society.

Haggarty, D., 2000. *The Biological Significance of the BC Marine Classification System Parameters*; and, *A Preliminary Test of a Species-Habitat Model for the Central Coast of British Columbia*. Reports to Living Oceans Society. (Data available upon request.)

Howes, D. E., Zacharias, M. A., and Harper, J. R., 1997. British Columbia Marine Ecological Classification: Marine Ecosections and Ecounits. *Report for the Resources Inventory Committee Coastal Task Force*. Last accessed June 2004: http://srmwww.gov.bc.ca/risc/pubs/coastal/marine/index.htm.

Leslie, H., Ruckelsaus, M., Ball, I. R., Andelman, S., and Possingham, H. P., 2003. Using siting algorithms in the design of marine reserve networks. *Ecological Applications* 13(1): S185-S198.

Love, M. S., Yoklavich, M., and Thorsteinson, L., 2002. *The Rockfishes of the Northeast Pacific, Berkeley*, CA, University of California Press, 405 pp.

Margules, C. R., and Pressey, R. L. 2000. Systematic conservation planning, *Nature*, 405: 243-53.

Noss, R. R., O'Connell, M. A. and Murphy, D. D., 1997. *The Science of Conservation Planning: Habitat Conservation Under the Endangered Species Act*, Washington, DC and Covelo, CA, Island Press.

Parker, S. J., Berkeley, S. A., Golden, J. T., Gunderson, D. R., Heifetz, J., Hixon, M. A. Larson, R., Leaman, B. M., Love, M. S., Musick, J. A., O'Connell, V. M, Ralston, S., Weeks, H. J., and Yoklavich, M. M., 2000. Management of Pacific rockfish, *Fisheries*, 25:22-29.

Roberts, C. M., Andelman, S., Branch, G., Bustamente, R., Castilla, J. C., Dugan, J., Halpern, B., Lafferty, K., Leslie, H., Lubchenco, J., McArdle, D., Possingham, H., Ruckleshaus, M., and Warner, R., 2003. Ecological criteria for evaluating candidate sites for marine reserves, *Ecological Applications*, 13(1): S199-S214.

Stewart, R. R., Noyce, T., Possingham, H. P., 2003. Opportunity cost of ad hoc marine reserve design decisions: An example from South Australia, *Marine Ecology Progress Series*, 253: 25-38.

Wallace, S., and Ardron, J., 2003. Submission on the Proposed Rockfish Conservation Strategy, Sierra Club of Canada, British Columbia Chapter, Living Oceans Society, and Canadian Parks and Wilderness Society.

Ward, T. J., Vanderklift, M. A., Nicholls, A. O., and Kenchington, R. A., 1999. Selecting marine reserves using habitats and species assemblages as surrogates for biological diversity, *Ecological Applications*, 9:691-98.

Yamanka, K. L., and Kronlund, A. R.,1996. Inshore rockfish stock assessment for 1996 and recommended yield options for 1997, Canadian Technical Report of Fisheries and Aquatic Sciences, 2175, 1-80.

PART II

Science in Action: Working Examples of Marine GIS

Chapter 5

The OCEAN Framework—Modeling the Linkages between Marine Ecology, Fishing Economy, and Coastal Communities

Astrid Scholz, Mike Mertens, Charles Steinback

Abstract

Many of the most vexing problems and challenges of marine resource management and conservation turn on the interaction of human activities and communities with the marine environment. In the case of fisheries, recent management measures, such as habitat protection, fishing restrictions, and alternative area usages, are precipitated by fisheries declines and ecosystem-based management mandates, and require an integrated understanding of the resource and its users. Analytically, this challenge reduces to the need to link—through data and analysis—ocean ecosystems and human communities. In addition to the physical, geological and biological data collected and compiled by scientists to understand the ocean floors and water column, any analysis seeking to link the ocean environment to fishing activities and coastal communities must also include information on use patterns, economic statistics, and human behavior. With fisheries, essentially, the question becomes "where in the ocean are the resources, the fleets that harvest them, and the communities that depend on them?"

Ocean Communities "3E" Analysis (OCEAN) refers to a suite of geographic information systems, databases, and analyses designed to answer this question. Using the case of the West Coast groundfish fishery, we illustrate how multiple, heterogeneous datasets can be linked and interpreted for marine management applications, notably area-based management. OCEAN operates at an intermediate, regional scale, with explicit consideration of the socioeconomic impacts of management measures on coastal communities. The system can be queried from within any one data layer, for example, to find particular vessels or gear groups fishing in a habitat of

Astrid Scholz, Mike Mertens, and **Charles Steinback**, Ecotrust, 721 NW 9th Avenue, Suite 200, Portland, OR 97209

Corresponding author: ajscholz@ecotrust.org, phone 503-467-0758 interest, which we illustrate with the trawl fishery in the coral and sponge habitat of the Monterey Bay National Marine Sanctuary. Information can be manipulated both in database formats and map-based user interfaces, and results are plotted on maps.

Introduction

The oceans are the "final frontier" of science, a vast, largely uncharted expanse of the Earth's surface that poses innumerable challenges to mapping and exploration. No less formidable are the challenges to science and policy arising from human uses of the oceans for activities such as oil and gas extraction, transportation and shipping, commercial and recreational fishing, and recreational uses. Increasingly, these uses impact ecosystem health and functions, often with deleterious effects on marine environments and organisms from climate change, pollution, habitat degradation, and extinction.

In this chapter, we focus on fisheries and the potential for geographic information systems (GIS), spatial analysis, and new software tools to support the management and conservation of marine resources. Specifically, we address the potential of GIS for integrating socioeconomic information into models and tools used for managing marine resources. The Ocean Communities "3E" Analysis (OCEAN) framework is so called because it facilitates the consideration of ecological, economic and equity considerations in marine resource management. In particular, OCEAN is a set of geographic information systems and databases that link the economic behavior of fishing fleets with habitat and other oceanographic data, and relate them to coastal communities.

Globally, it is estimated that 47% of the world's fish stocks are fully exploited, and another 28% are either overexploited or so significantly depleted that they require drastic and long-term reductions in fishing pressure (FAO, 2002). The United States, with 11 million km^2 (4.5 million mi^2), has the largest Exclusive Economic Zone (EEZ) of any nation (NRC, 1998), and is the fourth-largest producer of capture fisheries, after China, Peru, and Japan (FAO, 2002). In 2000, U.S. fisheries produced 4.7 million tons of fish (FAO, 2002), the vast majority of which comes from the abundant waters of the Northeast Pacific.

In the Pacific Northwest of the United States, many fisheries are quite healthy, with Alaska groundfish and salmon accounting for the majority of Alaska landings, which averaged 2.5 million tons in recent years (NMFS, 1999). Farther south, however, off the coasts of Washington, Oregon, and California, several salmon stocks have been listed as endangered, and the groundfish fishery—a complex of 89 different species of flatfish, roundfish and rockfish, nine of which are considered overfished and the status of most of the others is unknown—was declared a federal disaster (PFMC, 2002).

The management of these and other U.S. fisheries is regulated by the Magnuson-Stevens Fishery Conservation and Management Act (MSFCMA). The Act requires the eight regional councils charged with managing fisheries in the U.S. EEZ to not only to develop and execute fishery management plans that generate optimal yields while preventing overfishing (National Standard 1, MSFCMA, 1996), but to do so in ways that protect the marine environment (particularly habitats that are considered essential for fish at various life stages) and consider the socioeconomic impacts of management decisions on fishing communities (National Standard 8, MSFCMA, 1996). On a political level, this dual mandate of environmental conservation and community economic viability is the backdrop for many contentious policy issues, including marine protected areas, reduced harvest quotas, and disagreement over the siting of aquaculture facilities (for overviews of these issues see, respectively: NRC, 2001; Weber, 2002; and Naylor et al., 2003). Each of these and other marine resource management issues entail policy decisions based on scientific assessments that link marine ecology, fishing economy, and coastal communities. In this chapter, we focus on the analytical challenges of making these linkages, and novel GIS tools developed for this purpose.

Data Sources, Data Limitations and Ways to Overcome Them

Many of the data used to build OCEAN are relatively readily available. Typically, physical, geological, and biological data describing the ocean floors and water column have the best spatial resolution. We obtained bathymetry and other data on oceanographic characteristics from the National Oceanic and Atmospheric Administration (NOAA), the United States Geological Survey (USGS), as well as from state agencies such as the California Department of Fish and Game. The continental shelf in our study area has been the subject of considerable habitat mapping efforts, such as the USGS habitat GIS for the Monterey Bay National Marine Sanctuary (Wong and Eittreim, 2001; Greene et al., this volume). Using known habitat associations for various fish species, as well as the depth constraints on particular types of fishing gear, habitat data can also be used to relate fishing effort to particular ocean areas.

Serious data limitations emerge when considering the human aspects of the marine environment. One of the main limitations of fishery management today is that many routine data collection efforts are outpaced by evolving analytical needs. Emerging issues such as ecosystem-based management, Essential Fish Habitat (EFH) or spatiotemporal zoning reduce, analytically, to the need to link areas of the ocean where commercial and recreational fishing activities take place with the human communities that depend on the resulting landings, tourism economy, or other marine-related businesses. With fisheries, essentially, the question becomes "where in the ocean are the resources, the fleets that harvest them, and the communities that depend on them?"

Fish, Fleets and Fishing Communities

Although a plethora of data are routinely collected that document these three dimensions—fishery resources, fishing fleets, and coastal communities—they are not usually considered in an integrated, spatially explicit format. Rather than devising an entirely new data collection effort, OCEAN illustrates that existing data sources can be mined and interpreted in spatially explicit ways. Consider, in turn, the three dimensions of fishery management, resources, fleets and communities.

In terms of the fishery resource, little is known about the distribution of fish in the sea. Historically, fishery science has focused on estimating the biomass of stocks targeted in fisheries, developing several different schools of thought and models for inferring the volume of fish in the sea from samples and surveys (Smith, 1994). While these are generally considered to be the best available science for estimating the current status of a stock, the rate of removal due to fishing, or the abundance needed to sustain the stock in the future, a recent National Academy Study also cautioned that current stock assessment models may not be adequate for precautionary management as required by the Magnuson-Stevens Act (NRC, 1998).

This disconnect between current stock assessments and the information collected for them is borne out, for example, by the MSFCMA mandate to protect Essential Fish Habitat for species at their different life stages. To do this, fishery managers need to know where, both geographically and biologically, species spend their various life stages. The same dataset used for stock assessments provides an important starting point for this inquiry. Scientists employed by NOAA Fisheries conduct periodic trawl surveys, focusing on species that are commercially harvested. These surveys can be interpreted spatially to generate associations of species and biogeographic regions (see Monaco et al., this volume). Since the trawl surveys adhere to a strict sampling protocol, they essentially suggest where in the ocean scientific vessels are likely to encounter the various species and various life stages. Furthermore, the trawl surveys have been found to overlap well with fishing locations as reported in trawl logbooks (Fox and Starr, 1996), so it seems reasonable to use them to generate probability surfaces of where fishing vessels are likely to encounter the species they record in their landing tickets. In this chapter, we present another use of this same dataset. Adapting techniques developed by Monaco et al. (this volume), we use the West Coast fisheries survey data to infer spatial distributions of various marine organisms, both to constrain a model of fishing behavior and to infer distributions of non-targeted, but habitat-forming invertebrates.

The fact that we are discussing a *model* of fishing behavior is indicative of information challenges in the second dimension of marine resource management, the fleets. Although the logbooks and landing receipts generated by both commercial and recreational fishing vessels are integral to data collection efforts for fisheries management on the West Coast, they provide remarkably little information as to where fish are caught. To date, only a handful of U.S. fisheries employ electronic vessel monitoring systems (VMS) or on-board fisheries observers (NRC, 2000), and the Pacific Northwest is no exception. There is also little comprehensive observer coverage that would provide another fishery-independent source of location data. Where vessels fish, however, is increasingly important for management issues designed to protect Essential Fish Habitat, conserve marine biodiversity and ecosystems, or to facilitate the rebuilding of overfished stocks.

Again, existing datasets provide a solution to this quandary, albeit imperfect. For more than 20 years, operators of fishing vessels have filled out logbooks and landing receipts, with varying degrees of accuracy as to the geographic provenance of the catch. The quality of location-specific data on fishing activities varies significantly across fisheries. If vessels are required to record fishing locations at all, these tend to be reported in large statistical areas. In some fisheries and states on the West Coast, logbook and landings data are spatially coded in blocks that range from 5 to 30 nautical miles at the sides.[1] Even at the finest available resolution, however, the recorded data are too coarse to allow meaningful inferences about, for example, the interaction of fishing gear and sensitive habitat.

Airamé (this volume) details a process that entailed using fishers' knowledge to generate maps of fishing effort at a much smaller scale (1 n.m.) for a marine reserves planning process in Southern California. While this scale cannot realistically be attained along the entire coast, the logbook and landings data can be spatially interpreted to make them considerably more useful for marine management applications by relating them with other data sources in a GIS.

For most marine management issues, an intermediary scale between point locations and regional generalization is indicated. For example, in the context of the MSFCMA mandate to protect Essential Fish Habitat, resource managers are interested in the extent and intensity of fishing activity on various habitat types, as well as the interaction of fishing gear with marine habitats. In order to assess the potential damage to benthic habitats from gear interactions, managers have to know where and how intensely fishing effort is distributed. We discuss this in more detail below, but the basic rationale is to constrain the landing receipts and logbooks with other data, notably on species-habitat associations and inferences on the likely distance a particular fishing vessel will travel for a reported trip.

Finally, the human dimension of fisheries, notably fishing communities, is central to marine resource management, yet relatively sparsely documented. In many parts of the West Coast, fisheries are an important part of the economic base and cultural fabric of communities, and any management measure that affects the amount or extent of fishing in an area will tend to have local impacts on employment and income. In addition to National Standard 8 of the MSFCMA, various federal and state laws and regulations also require that managers take these effects into consideration, yet available data are often problematic. For example, census data are recorded at the county level, which can make it difficult to distinguish coastal communities from inland areas. Similarly, most employment statistics tend to lump fishing activities with forestry and agricultural employment, thus making it hard to differentiate effects on the fishery sector. Since these statistics also tend to be based on unemployment insurance data, they may significantly under-report small fishing businesses that are exempt from paying unemployment insurance.

In OCEAN, we used an existing regional economic model to approximate the effects of alternative management measures, the Fisheries Economic Assessment Model (FEAM) (Jensen, 1996). The FEAM belongs to a class of regional input-output models that treat the economic activity in a region as a set of interconnected sectors (Hewings, 1985). Each dollar generated in one sector has a "multiplier effect" because it generates economic activity in other sectors. For example, fish are landed and the vessel is paid a price per pound for its catch. Out of this ex-vesselrevenue, crew shares, maintenance and moorage costs and other expenses are paid, which in turn generate personal income, and revenues for the port district and other marine-related businesses.

The FEAM estimates these effects for the two primary sectors affected by fishing activity, i.e., harvesters (fishers and their families) and processors. We summarized these model outputs in a set of Excel spreadsheets, which we integrated into OCEAN. This allowed us to consider the income impacts of changes in landings in a port resulting from particular management scenarios. A key limitation of the FEAM analysis is that it is static in nature and only provides an incomplete snapshot in time. It is based on the landings and revenues generated by the fishing fleet, but remains silent on alternative sources of revenues in coastal communities such as tourism. Unlike other regional input-output models, FEAM is not designed to assess employment effects.

Furthermore, there are a host of considerations over and beyond economic impacts that are of importance to coastal communities and managers, but are not yet routinely assessed. For example, the lifestyle aspects of fishing communities are important (The H. John Heinz III Center for Science Economics and the Environment, 2000), as are concerns about the social and cultural resilience of ports and towns in response to the structural changes in the fishery (Langton-Pollock, 2004). Researchers have been using qualitative approaches to profile fishing communities on the West Coast (see, for example: Gilden, 1999; Pomeroy and Dalton, 2003; Package and Sepez, 2004), which are, in principle, compatible with GIS tools (Airamé, this volume; Scholz et al., 2004). By way of addressing these concerns, and to lay the groundwork for more in-depth analyses of coastal communities in future applications of OCEAN, we incorporated census statistics as well as qualitative information derived from port visits and interviews.

The Ocean Communities "3E" Analysis (OCEAN) Approach

Conceptually, OCEAN is a multi-layered information system comprising geographic and other data in a set of linked, "smart" maps. It is rooted in the growing literature of marine GIS that are being developed to address a host of oceanographic, coastal, and fisheries issues and problems (Kruse et al., 2001; Breman, 2002; Valavanis, 2002; Green and King, 2003). OCEAN is essentially a meta-analytical tool for combining a range of data, using a relational database architecture and spatial analysis as the common currency.

Analytically, the OCEAN approach is centered on the spatial association of multiple, heterogeneous datasets. This kind of analysis has been used in other marine applications of GIS, for example, to assess the location of fishery efforts close to shore (Caddy and Carocci, 1999), or to detect trends in global fishery statistics (Watson and Pauly, 2001; Watson et al., this volume). The OCEAN approach operates at an intermediate, regional scale, with explicit consideration of the socioeconomic impacts on coastal communities. The system can be queried from within any one data layer, for example, to find particular vessels or gear groups fishing in a habitat of interest, or to generate the ex-vessel revenues associated with a particular species or gear type. Information can be manipulated both in database formats and map-based user interfaces, and results are plotted on maps.

The centerpiece of this approach is the modeling of data that are already available in spatially explicit formats, and combining them with other, newly spatially interpreted information. The challenge is to organize and standardize data from diverse sources, recorded in diverse formats, and of varying quality, and to integrate them into a single framework. We began work on OCEAN by reviewing existing sources of data pertaining to fish populations, fishing activities and coastal communities, and compiling them into a single relational database. Where necessary, we built new models to spatially interpret data (further detailed in this chapter), especially those pertaining to the distribution of fishing effort. Combining bathymetry and habitat information with fishing effort and species distributions then formed the basis for analyzing where vessels fish, stratified by gear type and target species. To this we added a regional economic model for assessing the relative socioeconomic impacts of different management scenarios on coastal communities. The result is a set of smart maps that are linked through a relational database, and that allow the user to investigate jointly ecological, economic, and social questions; Figure 5.1 shows a schematic of OCEAN.

Modeling Fishing Effort, or Where in the Ocean Is the Fleet?

By way of illustrating a central piece of spatial modeling involved in OCEAN, we discuss here the methods developed for interpreting existing landing records spatially in order to arrive at distributions of fishing effort. Effort maps are central for estimating the amount and degree of interaction between fishing gears and marine habitats, or for estimating the relative economic cost to fishing businesses of spatiotemporal closures of the fishing grounds, e.g., for temporary or permanent marine reserves.

We use the West Coast groundfish fishery off the coasts of Washington, Oregon, and California as an example, since it illustrates several linked management issues that center on the relationship between fishing and habitat. Managers are involved in processes to identify EFH areas and reduce by-catch of overfished species, both of which may lead to area restrictions on trawl, pot and line gear. Current management measures include large rockfish conservation areas on the continental shelf within which targeting of groundfish is prohibited to facilitate the rebuilding of several overfished stocks of rockfish. Five national marine sanctuaries off the coast of California and Washington

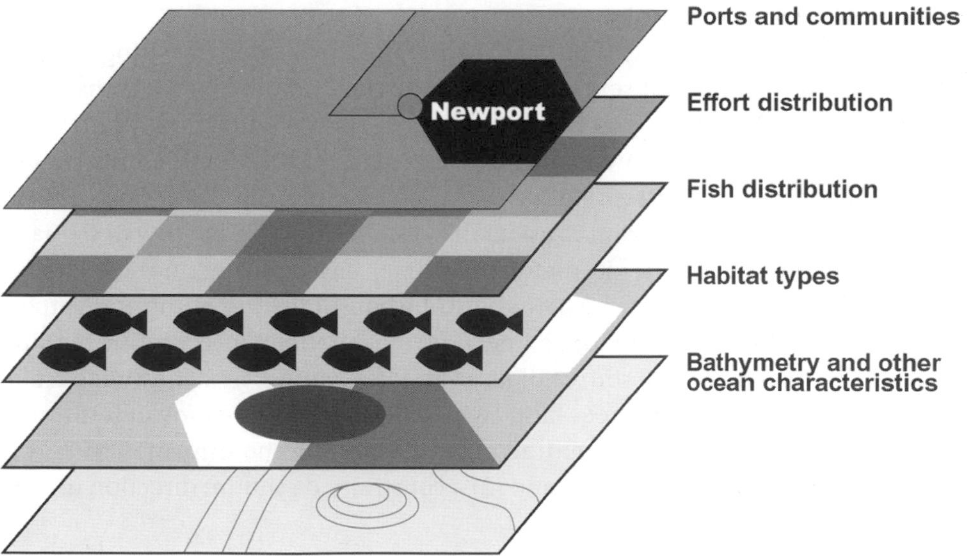

Figure 5.1. Linking smart maps in OCEAN.

are undergoing periodic management plan review processes that may result in changing regulations in some sanctuary waters.

There are several important aspects to the West Coast groundfish fishery. The fleet consists of around two thousand vessels that target groundfish using both fixed and mobile gear. Fixed gear includes hook and line equipment such as benthic longlines, pots, and traps. Mobile gear refers to gear that is towed through the water, typically on or just off the bottom. Although only a small (and, since a recent vessel buyback, shrinking) fraction of the fleet, the now roughly one hundred trawlers on the coast (2004) account for nearly 90% of all groundfish landed. All sectors of the fishery are required to report their landings, in the form of receipts filled out at the processor or buying station in the landing port. These landing receipts contain little geographical information, since the only spatial reference is to one of several large, statistical areas off the coast, requiring considerable creativity when making inferences about the spatial distribution of the fixed gear fleet.

For the trawl sector, however, there is a second data source, on which we focus in this chapter. Vessels submit federally mandated logbooks to fish and wildlife management agencies in the three coastal states, which in turn process and forward them to a regional database (Sampson and Crone, 1997), the Pacific Fisheries Information Network (PacFIN). The logbooks include vessel identifiers;[1] landing port and home ports; the date, location (set points and block number), number of tows and duration of each tow; species caught; gear used; and the amount landed (both hailed and adjusted by landing receipts). From the latitude and longitude recorded for the set points it is relatively straightforward to map the distribution of trawl effort. We summarized fishing effort in terms of tow duration, tow intensity (number of tows), and landings by the 10-min (approximately 20-km) statistical blocks used in the logbooks.

More accurately, what can be mapped from the trawl logbooks is the distribution of set points, since the haul points are not transcribed into PacFIN. This, and the fact that only the tow duration, but not the direction of each tow is recorded, introduces considerable uncertainty into the spatial interpretation of the logbook records. Since trawl vessels are capable of covering considerable distances, with each tow potentially covering dozens of kilometers, any maps based purely on the logbooks likely misrepresent the actual distribution of fishing effort and catch.

We therefore developed one model for spatially interpreting landing receipts—the only source of location information, at a coastwide scale, for the fixed gear fleet—and two additional methods for determining trawl activity: (1) one extracting and mapping the information as it is recorded in the logbooks, and (2) a constrained random direction model.

Interpreting Landing Receipts

The OCEAN landing receipts (also known as "fish tickets") model essentially consists of a sequence of steps, programmed in ArcInfo, which successively constrain each landing record and subsequently apportion catch and revenue to equal area units (9-km by 9-km blocks) based on the probability of fishing activity occurring in an area. In contrast to multivariate analysis used in terrestrial applications, which generally predicts what happens in a particular location (e.g., Hargrove

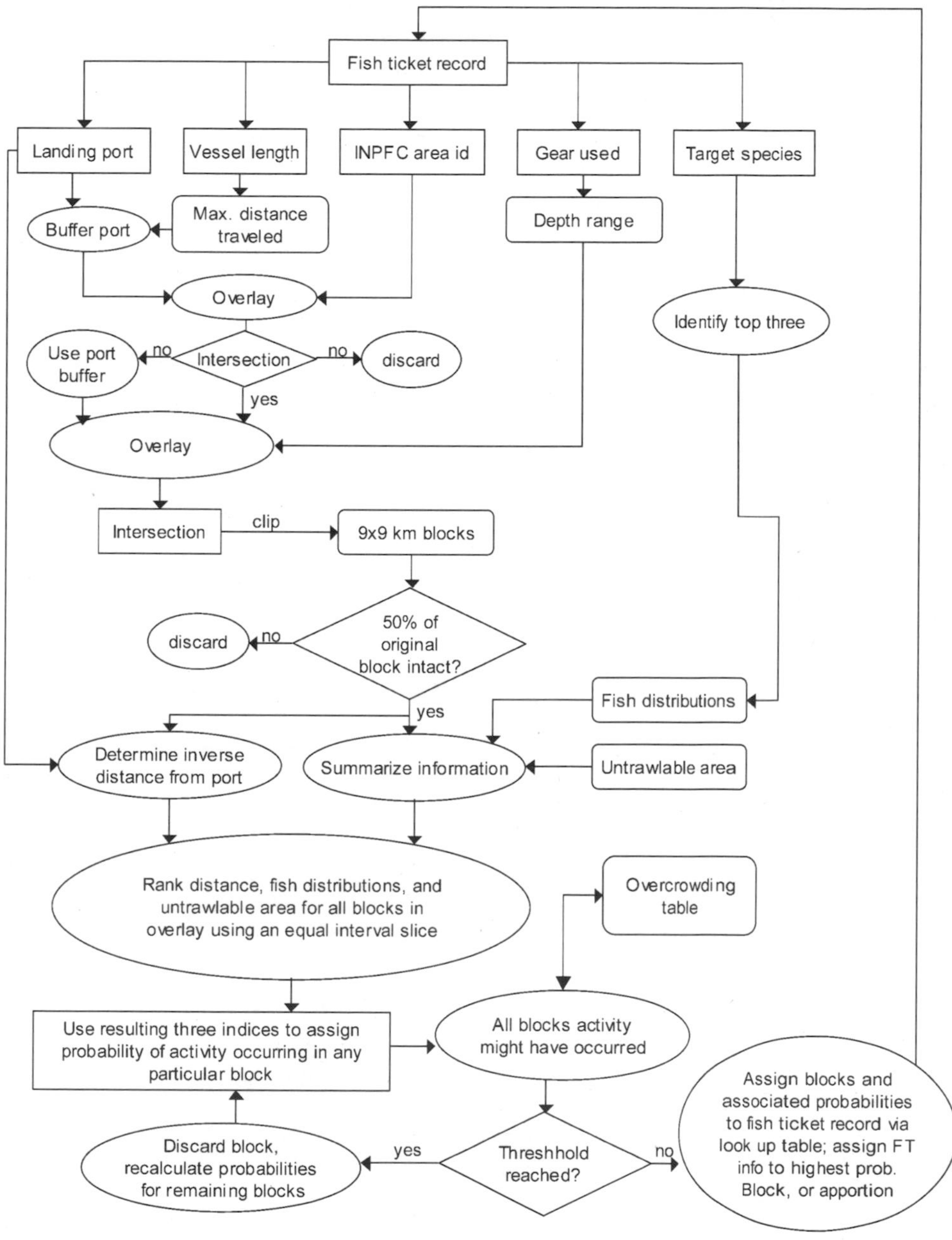

Figure 5.2. Flow chart for fish ticket model (Ecotrust)

and Hoffman, 2000), we try to predict the location for known entities. The following steps characterize this process; Figure 5.2 shows a flow chart of the model:

1) Each PacFIN record contains information on the gear used, species caught, landing port, vessel information, and one of 12 statistical management areas where the catch originated;

2) Impose a maximum range from the landing port that a vessel is likely to have fished, given its length and gear type used—this is currently derived from expert witness testimonies, pending more formal studies of fishing behavior on the West Coast;

3) Impose depth restrictions on fishing gear used and target species—there are limits to the depth from which West Coast trawlers can haul their nets, or in what depth various fixed gear types are used; similarly, different species of fish have known ranges of bathymetric associations;

4) Compare this to the species distribution densities derived from the fishery-independent surveys—some areas are associated with higher frequencies of the target species in question, making it more likely that a fishing vessel would have gone there for its catch;

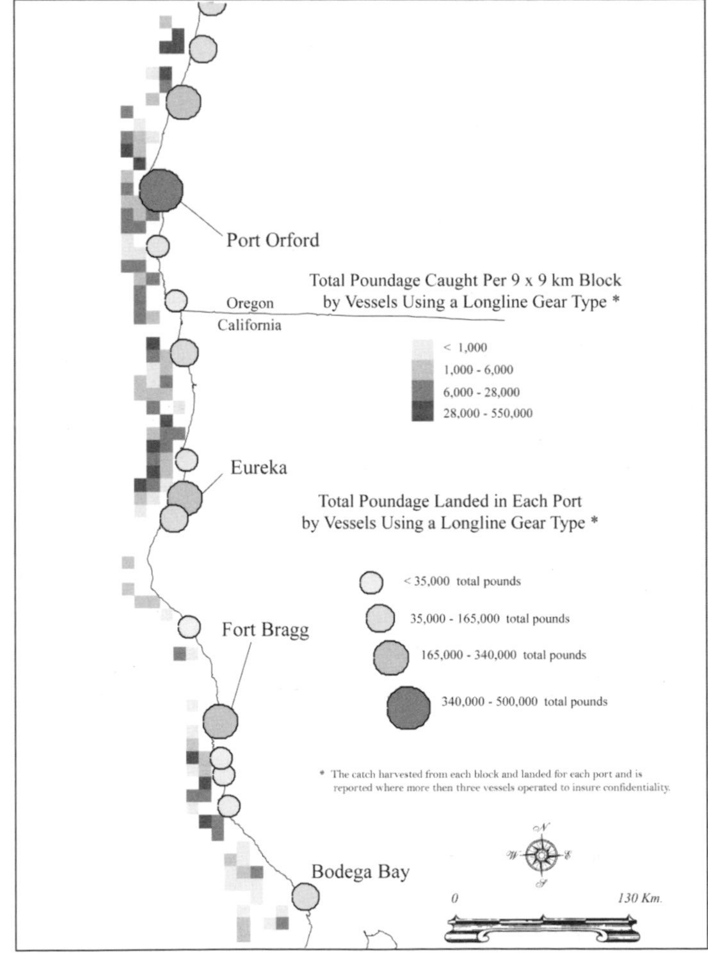

Figure 5.3. Probability distribution of fishing effort. Estimated pounds caught and landed summarized for all longline gear, southern Oregon and northern California, 2000.

5) Within that maximum range, weight the species density clusters inversely by distance from port—this is a "friction of distance" idea: because travel is costly, vessels tend to fish closer to port even if they are slightly less likely to encounter the target species;

6) Impose habitat restrictions on fishing gear used—trawlers do not operate in high relief areas, while these same areas tend to be frequented differentially by vessels using hook and line gear;

7) Apportion pounds caught and associated revenue from fish tickets. This can be done either deterministically, associating the entire catch and revenues with the block that has the highest likelihood of fishing having occurred there; or probabilistically, apportioning catch and revenues to fishing blocks within the maximum range based on probabilities derived from distance from port, targeted species densities, habitat restrictions and previous activity.

8) Repeat for all records and map the resulting distribution of fishing activity; in principle, this can be normalized by number of records associated with an area, or—in the case of trawlers—number and duration of tows made there, to provide a measure of effort.

The maps resulting from this algorithm are probability surfaces of the distribution of fishing effort and the associated catches and revenues (see Fig. 5.3). The results shown here are derived from an earlier, deterministic version of the model. We discuss the probabilistic model and its sensitivity to various assumptions in another publication (Scholz et al., 2004). In general, however, the model is most sensitive to assumptions about the maximum range of vessels from port and about the associations of gear types with particular habitats, as well as to the weight given to the overcrowding parameter. We are in the process of validating the model with fishermen along the coast, and eliciting additional information for further refining this approach.

Mapping the Trawl Logbooks

Fish tickets are available for all fisheries and all fleets on the West Coast. In addition, the trawl fleet, like several other fisheries, is also required to fill out and submit logbooks that record fishing locations. Typically, these logs record the latitude and longitude of the set points of individual tows, as well as the duration of each tow. It is relatively straightforward to map the resulting distribution of set points, and associate catches and revenues with them.

We determined the duration of each tow using database procedures. Essentially, we assigned the tow duration recorded in the logbooks to the block into which the recorded latitude and longitude of the set point fell. This generated a new set of records based on information about the duration of each tow (in hours); x, y location of set point; block id; and gear used—as recorded in the logbook data. Species information is omitted as multiple species are recorded for each tow. This newly created record set is then used to determine the cumulative

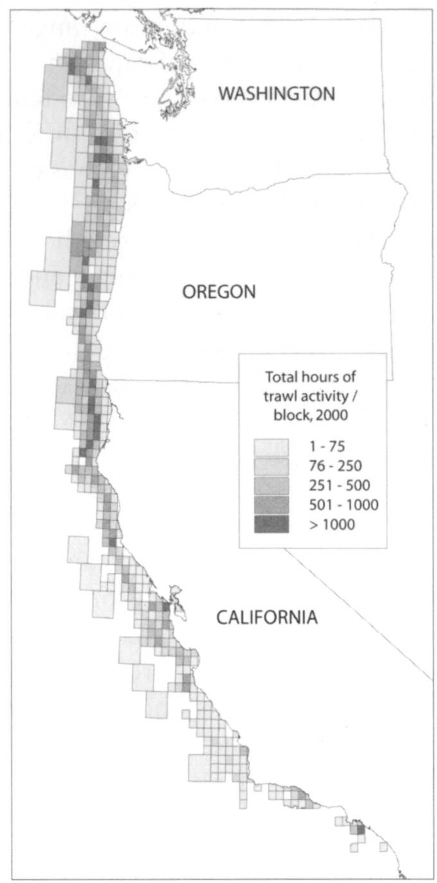

Figure 5.4. Trawl activity off the West Coast.

number of tows and cumulative tow duration for each block, generating a measure of trawl effort per area (Fig. 5.4).

Figure 5.4 illustrates the extent of trawl activity off the West Coast in terms of hours per unit area. It is somewhat problematic to associate trawl duration with the block in which the coordinates of the set points fall, since trawl vessels routinely cover distances larger than the 10 n.m. (roughly 20 km, the length of each block), or a set point might have fallen on the corner of a block. These confounders not withstanding, it should be apparent that some areas of the EEZ are more heavily trawled than others, since tow duration is directly related to bottom contact. Given the limitations of the logbooks as recorded, however, it would not suffice to simply overlay a nautical chart and expect to be able to identify the areas where tow activity occurs. The random direction model discussed in the next section is an attempt to remedy the uncertainty involved in interpreting logbooks.

A Constrained Random Direction Model for Trawl Vessels

Using the set points and additional information extraneous to the logbooks, we developed a model that generates a set of possible directions a vessel might have traveled. Based on the unique tows identified from the logbooks, we extract from each record the x and y-coordinates of each set point for each vessel, tow-date, and trip combination. We then derive a tow distance by multiplying the (recorded) tow duration by a constant speed. In our model, we assumed an average speed of 3 knots.[1] This was derived from interviews with expert witnesses, all participants in the West Coast groundfish fishery. It is likely an upper limit for actual tow speeds, which are influenced by local variables such as weather, depth at which the tow occurs, and the experience of the skipper.

The tow distance is then added to the set point's "y" location to get a secondary y-coordinate. The start point and secondary "y" coordinate (associated with the "x" coordinate) are used to create a vertical line representing the distance the vessel would have covered while trawling at the constant speed of 3 kt. from the recorded set point. Since the logbook record contains no information about the direction traveled, this line is then copied and rotated 360° in 11.5° increments (for a total of 32 such increments, or possible directions) around the start point of the tow. Each rotated line is put into a comprehensive data layer. The resulting data layer represents 32 lines radiating from the tow start point each 11.5° apart (Fig. 5.5). This data layer is then converted to a raster model.

Given what we know about trawling, in particular that vessels tend to avoid rocks and other obstructions to avoid gear entanglements, it is possible to reduce the number of possible towpaths. Specifically, we excluded areas that we identified as "untrawlable," adapting a method for interpreting the NMFS trawl survey data (Zimmermann, 2002). Trawl and non-trawl fisheries are, by and large, mutually exclusive in that trawl gear is not used on high-relief substrates and many of the non-trawl gear types target species that live in rocky habitats. Interviews with expert witnesses confirmed this; our technique effectively designates around 80-90% of the fishing grounds as trawlable.

The data derived in the previous steps are overlaid with untrawlable areas and bathymetry. If any given tow line intersects untrawlable areas,

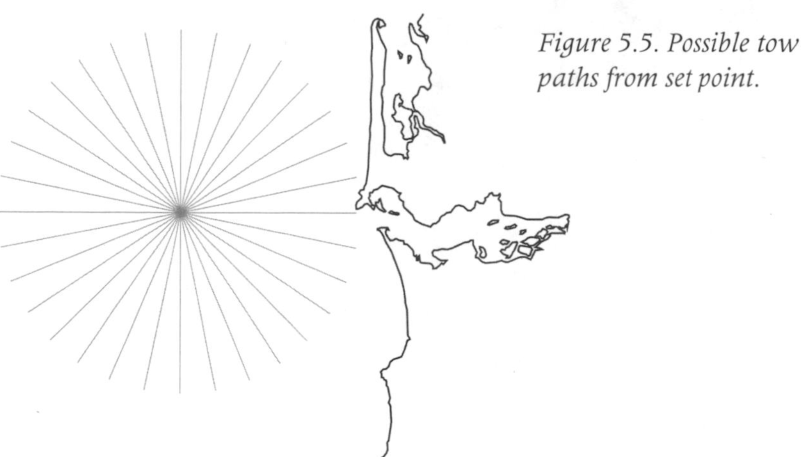

Figure 5.5. Possible tow paths from set point.

which—for the purposes of constraining direction—includes regulated areas and non-marine areas (islands, mainland), that line is removed from the analysis (Fig. 5.6; see page 130). If all lines fall within untrawlable areas, the record is removed from the analysis. The total number of records removed from the analysis are tracked.

We further constrained these tow paths by factoring in the slope of the terrain, using bathymetry information. Considering the slope (rise over run) of each tow line remaining, any line with a slope greater than 1% is removed from the analysis. This is again based on expert testimonials, and may be a conservative constraint. If all lines have a slope greater than 1%, then the line with the lowest slope is selected as the most likely tow line. Otherwise a random function is applied to determine the tow line. All other lines are removed from the data layer and the resulting line is copied into a master tow-lines dataset. This further reduces the number of potential tow paths originating from each set point (Fig. 5.7; see page 130).

Of the remaining potential tow paths, we then picked one at random for each set point and summarized the information from the logbook record by area. Once all tow lines have been delineated, these are overlaid with a grid, and total distance towed is then summarized for each block (Fig. 5.8). These are properly interpreted as probabilistic estimates of trawl activity—essentially a density map of possible tow tracks—rather than a literal map of where trawling takes place.

Repeating this procedure for each year of data generates a time series of effort distribution maps—a kind of movie—that illustrates the changes in fishing effort over time in terms of pounds-landed associated with each fishing block. The trawl effort distribution also details the species

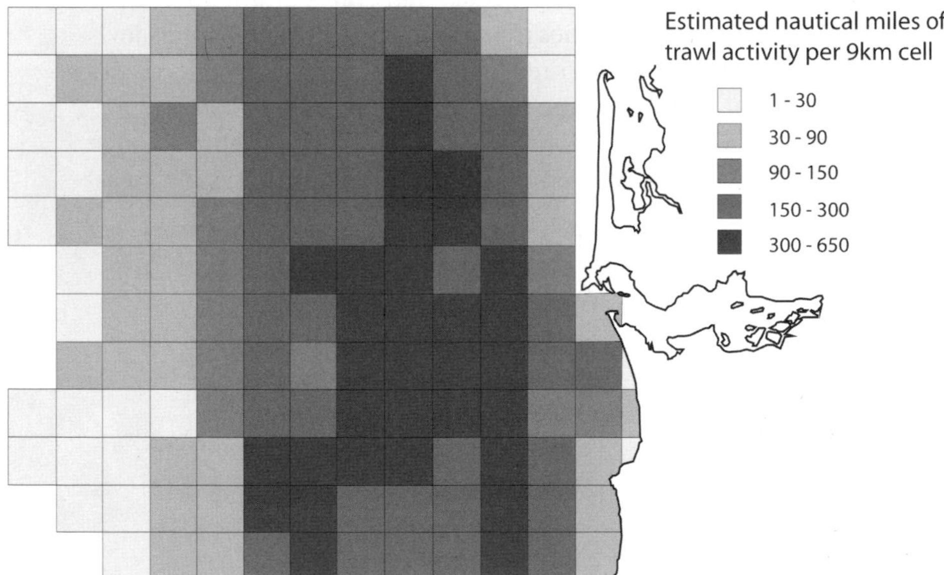

Figure 5.8. Estimated miles of trawl activity.

caught per block, and pounds landed per port. Applying average prices for each species (another data set available from PacFIN), it is possible to derive the revenues generated per block. We normalized the catch per area by tow duration and number of tows, thus deriving a measure of catch per effort and area.

The technique for interpreting trawl set points outlined in the above steps could be further improved. Specifically, the treatment of tow speed and terrain constraints (trawlable incline) could be improved by using a range of speeds and slopes, as well as conducting further expert consultations. Also, the constraints on tow directions could be improved with more detailed habitat information than the trawlable/untrawlable distinction used to date. Finally, there is emerging research using a more detailed set of logbooks to suggest that the trawl fleets on different parts of the coast exhibits patterns of predominant directions, which would help further constrain the model.

OCEAN Applications to Marine Resource Management

Once fishing activities are modeled geospatially, it is possible to tell complex stories and to investigate issues that span marine ecosystems, economic activities, and coastal communities. Consider, for example, some of the issues faced by the managers of the three central California national marine sanctuaries (NMS)—Cordell Bank, Gulf of the Farallones, and Monterey Bay—in the context of the Joint Management Plan Review (JMPR) process they are undergoing. Together, they cover 7,000 mi^2 (approximately 18,200 km^2) of ocean and occupy almost one third of the state's coastline (Fig. 5.9). Like the other ten NMS in U.S. waters, the three central California sanctuaries were designated to protect natural and cultural resources, while encouraging multiple uses of sanctuary waters, including fishing and shipping. Historically, the regulatory authority over fishing activities in federal waters resides with NOAA Fisheries (formerly the National Marine Fisheries Service), while the sanctuaries are administered by NOAA Oceans.

With emerging concerns over ecosystem-based management, habitat impacts of fishing, and Essential Fish Habitat, the line between fishery management and ecosystem conservation is beginning to blur. Given their size, any management measure that the sanctuaries consider to protect the marine resources in their charge are likely to have socioeconomic impacts on the fishing fleet operating in, and coastal communities adjacent to, the sanctuaries. Similarly, existing fishery regulations are already affecting commercial and recreational fishermen operating in sanctuary waters and cannot be ignored in the sanctuaries' JMPR process.

In partnership with the sanctuaries, we are using the OCEAN framework to build an integrated database and GIS for profiling fishing activities in and communities adjacent to sanctuary waters. This provides

Figure 5.9. Map of three central California sanctuaries and the Exclusive Economic Zone.

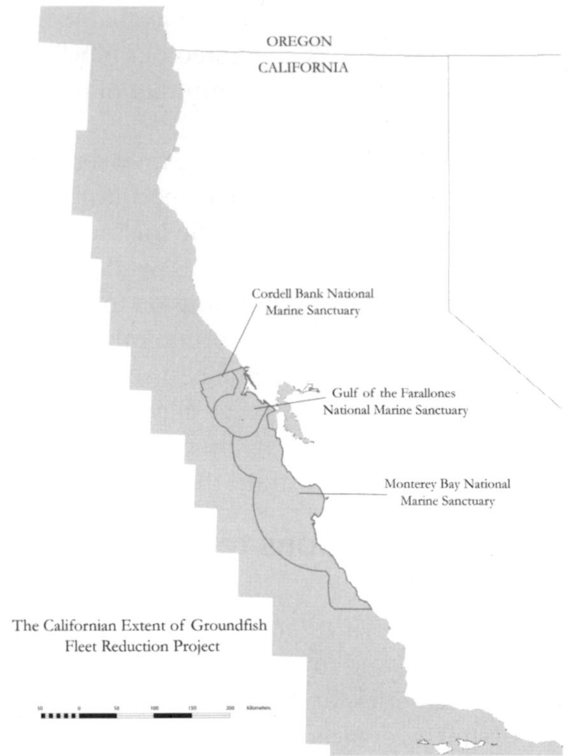

socioeconomic baseline information for a variety of management measures that the sanctuaries and the various stakeholder groups participating in the process are considering. Since this is still a work in progress, and without wanting to prejudice these deliberations, we present here a hypothetical example based on concerns about interactions between fishing gears and sensitive marine habitats in the three central California sanctuaries.

Within the sanctuaries lie areas with some of the highest coral and sponge concentrations on the West Coast, notably along the Monterey Canyon break (Morgan et al., 2005). Corals create important habitat for marine fishes, providing shelter, nursery and feeding areas, including for some of the rockfish species that are currently considered overfished, or in danger of being overfished, on the West Coast. While it is well established that various fishing gears touch or otherwise interact with benthic substrates and any structures on them (NRC, 2002; Morgan and Chuenpagdee, 2003), the specific interactions between fishing gear and coral aggregations have not yet received wide-spread management attention on the West Coast.

A first step in assessing the extent of the potential interactions, and the effects on coral distributions, is to plot the distribution of fishing effort and that of corals, and to overlay the resulting maps. Figures 5.10a-d (see pages130-32) illustrate the results from this analysis for the area of the three central California sanctuaries.

As should be apparent from Figure 5.10a, some of the highest volumes of trawl-caught groundfish comes from areas of high coral aggregations, e.g., the block due west of Moss Landing, or an area approximately 40 mi. (75 km) off Bodega Bay. Not surprisingly, these areas also coincide with the largest numbers of tows per unit area (Fig. 5.10b) and thus presumably the most bottom contact. That is not to say, however, that there is necessarily habitat damage from this trawl activity. There are several complicating factors. One is that the trawl logbooks from which this information is derived only contain the set points of individual tows, and so the actual footprint of the fishing activity may not overlap as directly as suggested by these maps. Further investigation, ideally using fishermen's local knowledge of the fishing grounds and independent observer data (that also record the amount of invertebrates caught incidentally) is needed for a comprehensive assessment.

Figures 5.10c and 5.10d illustrate another important aspect of the sanctuaries. Both in terms of miles towed and tow duration—two important measures of fishing intensity—the sanctuaries appear to be very important for the local fishing fleets. The possibility that the set points overstate this effect, and that actual activity could be taking place just outside sanctuary boundaries, notwithstanding, these maps have important implications for the JMPR process and fishery management in the area more generally. Given the relative importance of the sanctuary waters as fishing grounds, any management measures restricting fishing in sanctuary waters is likely to have significant impacts on the economic viability of fishing vessels and the ports adjacent to the sanctuaries where they land their catch.

This issue is further complicated by recent rockfish conservation areas implemented on large sections of the continental shelf by NOAA Fisheries starting in 2002. Designed to aid the rebuilding of several overfished rockfish species, the closure areas prohibit the targeting of groundfish and have forced the fleet to relocate farther inshore and offshore, likely altering the spatial pattern of the resulting fishing footprint in sanctuary waters. With some of the rebuilding plans measured in decades rather than seasons, these closures may well become a permanent feature on the West Coast, and require careful consideration in the sanctuaries management process. As part of the socioeconomic analysis for the sanctuaries, we are assessing the actual and potential future effects of this shift in spatial behavior, and anticipate that the nexus of issues around gear impacts and area-based management measures will take center stage in the JMPR process and subsequent marine management processes.

Using the OCEAN framework, these and other questions can be explored, mapped and analyzed, and inform the decision-making process for marine management. The example of the sanctuaries, as well as other potential applications on the West Coast, demonstrate the utility of marine GIS for integrating socioeconomic concerns and issues into the policy process. Whether at the regional scale or at smaller, local scales as elaborated in the other chapters in Part II of this book, such integrated systems greatly enhance, and—to the extent that fishermen and other stakeholders participate in the creation of the spatial information—even help smooth out otherwise contentious marine management processes.

Acknowledgements

The OCEAN framework emerged out of our 2001-2002 Groundfish Fleet Restructuring Project, and we are indebted to our project partners at the Pacific Marine Conservation Council, research assistant Marlene Bellman, and over 100 fishermen, scientists, managers and industry observers who generously provided background information, design suggestions, and innumerable reviews and suggestions that helped improve our modeling and thinking. We thank the Cordell Bank, Gulf of the Farallones, and Monterey Bay National Marine Sanctuaries for the opportunity to refine OCEAN in the context of their Joint Management Plan Review Process, and participants of that process for providing invaluable input. We would also like to thank the California Department of Fish and Game and the Pacific States Marine Fisheries Commission for providing data. Funding for OCEAN was provided by the David and Lucile Packard Foundation, NOAA Fisheries' Northwest Region, NOAA Oceans' National Marine Protected Area Science Center, the Marisla Foundation, and Oregon Sea Grant.

Notes

1 Technically, distance on water is measured in nautical miles (n.m.). One n.m. is the angular distance of 1 min of arc on the Earth's surface. One min of latitude equals 1 n.m., and degrees of latitude are 60 n.m. apart. At the equator, 1 n.m. equals 1,852 m. The distance between degrees of longitude is not constant, since they converge at the poles.

2 The logbooks contain actual vessel identifiers, which in our case were fictionalized to maintain confidentiality. There are also several other constraints on which data are made available to researchers, and how they can be displayed.

3 A "knot" (kt.) is a nautical measure of speed, and equals 1 n.m. per hr. Modern fishing vessels travel at up to 10 kt. (roughly 20 km/h), and trawl tow speeds vary, depending on bathymetry, from 1.5 to 4 kt. (approximately 3 to 8 km/h).

References

Airamé, S., this volume. Channel Islands National Marine Sanctuary: Advancing the science and policy of marine protected areas, in Wright, D. J., and Scholz, A. J. (eds.), *Place Matters–Geospatial Tools for Marine Science, Conservation and Management in the Pacific Northwest,* Corvallis, OR, Oregon State University Press.

Breman, J. (ed.), 2002. *Marine Geography: GIS for the Oceans and Seas,* Redlands, CA, ESRI Press. 224 pp.

Caddy, J. F., and Carocci, F., 1999. The spatial allocation of fishing intensity by port-based inshore fleets: a GIS application, *ICES Journal of Marine Science*, 56: 388-403.

Food and Agriculture Organization (FAO), 2002. *The State of World Fisheries and Aquaculture 2002*, Rome, United Nations.

Fox, D. S., and Starr, R. M., 1996. Comparison of commercial fishery and research catch data, *Canadian Journal of Fisheries and Aquatic Science*, 53: 2681-94.

Gilden, J. (ed.), 1999. *Oregon's Changing Coastal Fishing Communities*. Corvallis, OR, Oregon Sea Grant. 73 pp.

Green, D. R., and King, S. D. (eds.), 2003. *Coastal and Marine Geo-Information Systems: Applying the Technology to the Environment*, Dordrecht, Kluwer Academic Publishers. 616 pp.

Greene, G. H., Bizzarro, J. J., Tilden, J. E., Lopez, H. L., and Erdey, M. D., this volume. The benefits and pitfalls of GIS in marine benthic habitat mapping, in Wright, D. J., and Scholz, A. J. (eds.), *Place Matters–Geospatial Tools for Marine Science, Conservation and Management in the Pacific Northwest*, Corvallis, OR, Oregon State University Press.

Hargrove, W. W., and Hoffman, F. M., 2000. *An analytical assessment tool for predicting changes in a species distribution map following changes in environmental conditions*, Banff, Canada, Cooperative Institute for Research in Environmental Sciences. 22 pp.

The H. John Heinz III Center for Science Economics and the Environment, 2000. *Fishing Grounds: Defining a New Era for American Fisheries Management*, Washington, DC, Island Press. 241 pp.

Hewings, G., 1985. *Regional Input-Output Analysis*, Beverly Hills, CA, Sage Publications. 96 pp.

Jensen, W., 1996. *Pacific Fishery Management Council West Coast Fisheries Economic Assessment Model*, Vancouver, WA, William Jensen Consulting, 68 pp.

Kruse, G. H., Bez, N., Booth, A., Dorn, M. W., Hills, S., Lipcius, R. N., Pelletier, D., Roy, C., Smith, S. J., and Witherell, D. (eds.), 2001. *Spatial Processes and Management of Marine Populations*, Fairbanks, AK, University of Alaska Sea Grant. 720 pp.

Langdon-Pollock, J., 2004. *West Coast Marine Fisheries Community Descriptions*, Seattle, WA, Pacific States Marine Fisheries Commission. 153 pp.

Magnuson-Stevens Act, 1996. Magnuson-Stevens Fishery Conservation and Management Act. Public Law 94-265.

Monaco, M. E., Kendall, M. S., Higgins, J. L., Alexander, C. E., and Tartt, M. S., this volume. Biogeographic assessments: The integration of ecology and GIS to aid in marine management boundary delineation and assessment, in Wright, D. J., and Scholz, A. J. (eds.), *Place Matters–Geospatial Tools for Marine Science, Conservation and Management in the Pacific Northwest*, Corvallis, OR, Oregon State University Press.

Morgan, L. E., Etnoyer, P., Scholz, A. J., Mertens, M., and Powell, M., 2005. Conservation and Management Implications of Deep-Sea Coral Distributions and Fishing Effort in the Northeast Pacific Ocean in Freiwald A., and Roberts J. M. (eds.), Cold-water Corals and Ecosystems. Springer-Verlag, Berlin and Heidelberg.

Morgan, L. E., and Chuenpagdee, R., 2003. *Shifting Gears: Addressing the Collateral Impacts of Fishing Methods in U.S. Waters*, Washington, DC, Island Press. 42 pp.

Naylor, R., Eagle, J. and Smith, W., 2003. Salmon aquaculture in the Pacific Northwest: A global industry, *Environment* 45(8): 16-39.

National Marine Fisheries Service (NMFS), 1999. Our Living Oceans, report on the status of U.S. living marine resources, Silver Spring,

MD, Department of Commerce National, Oceanic and Atmospheric Administration, National Marine Fisheries Service.

National Research Council, 1998. *Improving Fish Stock Assessments*, Washington, DC, National Academy Press. 188 pp.

National Research Council, 2000. *Improving the Collection, Management, and Use of Marine Fisheries Data*, Washington, DC, National Academy Press. 236 pp.

National Research Council, 2001. *Marine Protected Areas: Tools for Sustaining Ocean Ecosystems*. Washington, DC, National Academy Press.

National Research Council, 2002. *Effects of Trawling and Dredging on Seafloor Habitat*, Washington, DC, National Academy Press. 272 pp.

Pacific Fishery Management Council (PFMC), 2002. *Status of the Pacific Coast Groundfish Fishery through 2001 and Acceptable Biological Catches for 2002: Stock Assessment and Fishery Evaluation*, Portland, OR, Pacific Fishery Management Council. 250pp.

Package, C., and Sepez, J., 2004. "Fishing Communities of the North Pacific: Social Science Research at the Alaska Fisheries Science Center." AFSC Quarterly Report, April-June 2004, 11 pp.

Pomeroy, C., and Dalton, M., 2003. *Socioeconomics of the Moss Landing Commercial Fishing Industry*. Report to the Monterey County Office of Economic Development. Monterey, Monterey County Office of Economic Development, 134 pp.

Sampson, D. B., and Crone, P. R., 1997. Commercial Fisheries Data Collection Procedures for U.S. Pacific Coast Groundfish. NOAA Technical Memorandum NMRS-NWFSC-31, Silver Spring, MD, NOAA, 189 pp.

Scholz, A., Mertens, M., Sohm, D., Steinback, C., and Bellman, M., 2004. Place matters: Spatial tools for assessing the socioeconomic implications of marine resource management measures on the Pacific Coast of the United States, in Barnes, P. W., and Thomas, J. (eds.), *Benthic Habitats and the Effects of Fishing*. Proceedings of the American Fisheries Society, Bethesda, MD, American Fisheries Society.

Scholz, A., Bonzon, K., Fujita, R., Benjamin, N., Woodling, N., Black, P., and Steinback, C., 2004. "Participatory socioeconomic analysis: drawing on fishermen's knowledge for marine protected area planning in California." *Marine Policy* 28(4): 335-49.

Smith, T. D., 1994. *Scaling Fisheries: The Science of Measuring the Effects of Fishing, 1855-1955*. Cambridge, Cambridge University Press. 392 pp.

Valavanis, V. D., 2002. *Geographic Information Systems in Oceanography and Fisheries*, London, Taylor & Francis. 209 pp.

Watson, R., and Pauly, D., 2001. Systematic distortions in world fisheries catch trends. *Nature*, 414: 534-36.

Watson, R., Alder, J., Christensen, V., and Pauly, D., this volume. Mapping global fisheries patterns and their consequences, in Wright, D. J., and Scholz, A. J. (eds.), *Place Matters–Geospatial Tools for Marine Science, Conservation and Management in the Pacific Northwest*, Corvallis, OR, Oregon State University Press.

Weber, M. L., 2002. *From Abundance to Scarcity: A History of U.S. Marine Fisheries Policy*, Washington, DC, Island Press. 245 pp.

Wong, F. L., and Eittreim, S. E., 2001. *Continental shelf GIS for the Monterey Bay National Marine Sanctuary*, USGS Open File Report, 01-170, Menlo Park, CA, U.S. Geological Survey, http://geopubs.wr.usgs.gov/open-file/of01-179/.

Zimmermann, M., 2001. Calculation of untrawlable areas within the boundaries of a bottom trawl survey, Seattle, WA, NOAA Alaska Fisheries Science Center, 44 pp.

CHAPTER 6

Channel Islands National Marine Sanctuary:

Advancing the Science and Policy of Marine Protected Areas

Satie Airamé

Abstract

A network of marine protected areas was established in the Channel Islands National Marine Sanctuary in April 2003. Geospatial modeling tools were used to advance the science and policy underlying the network design. Scientists used a modeling tool, SPEXAN, to develop and evaluate design options. This modeling tool was used to process a large amount of spatially explicit data and to produce a set of solutions to the complex problem of where to establish marine protected areas. Each solution generated by SPEXAN included all conservation targets identified by planners, such as important habitats and species. The most efficient solutions, comprised of many small patches with high conservation value, generally were not practical for maximum compliance and enforcement. Thus, solutions with fewer and larger reserves were presented to planners for discussion. One of the most useful outputs of the SPEXAN analysis was a map of conservation value showing the number of times each planning unit was included in a final solution. Planners used the map of conservation value to begin the discussion about where to establish marine protected areas in the Sanctuary. Another geospatial modeling tool, the Channel Islands Spatial Support and Analysis Tool (CI-SSAT) was developed by NOAA's Coastal Services Center specifically for the Channel Islands process. CI-SSAT was used primarily as a tool for visualization and querying of ecological and socioeconomic data. Planners developed design options in CI-SSAT and then used the tool to query the data to determine their potential benefits and impacts. Planners made adjustments to the design options in order to increase ecological benefits and decrease socioeconomic impacts. In June 2001, planners

Satie Airamé, Channel Islands National Marine Sanctuary, 113 Harbor Way, Suite 150, Santa Barbara, CA 93109
airame@msi.ucsb.edu, 805-893-3387

submitted two alternative designs to the California Department of Fish and Game and the Channel Islands National Marine Sanctuary. The network of marine protected areas established in 2003 was a compromise, developed by the agencies, between the two alternative designs. Throughout the process, geospatial modeling tools advanced the science and policy, eventually leading to the carefully considered management action.

Introduction

In April 2003, the California Fish and Game Commission established the largest network of marine protected areas (MPAs) in California state waters (Title 14, California Code of Regulations, Sections 27.82, 530, and 632). The network of MPAs is located within the California state waters of the Channel Islands National Marine Sanctuary. The eight Channel Islands are located off of southern California. Five of the islands (San Miguel, Santa Rosa, Santa Cruz, Anacapa, and Santa Barbara islands) are located within the Channel Islands National Marine Sanctuary. The network of marine protected areas includes 10 no-take marine reserves and 2 marine conservation areas, which allow limited take of certain fisheries[1] (Fig. 6.1). The design of the MPA network was based on scientific and socioeconomic data assembled by experts, and policy developed by a working group of local stakeholders and agency representatives (California Department of Fish and Game, 2002).

The group of local stakeholders and agency representatives (working group) was appointed by the Channel Islands National Marine Sanctuary Advisory Council in July 1999. The formation of the working group was driven by growing pressure from local communities to improve local ocean management[2]. The working group included representatives from fisheries, kelp harvesting, recreational industries,

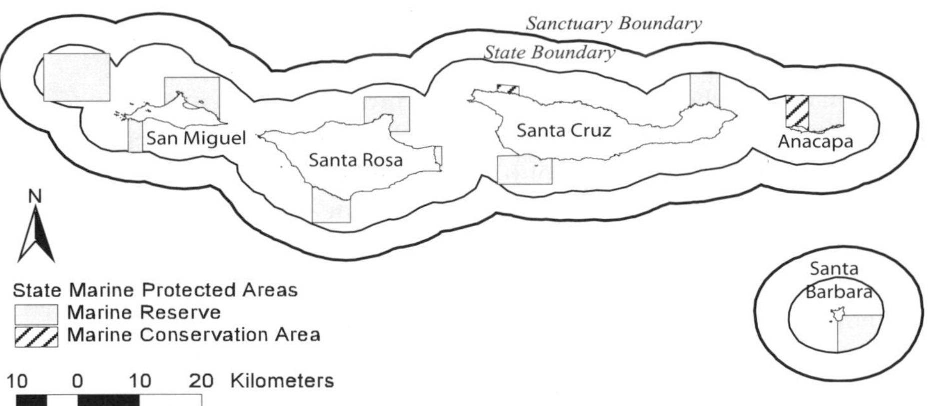

Figure 6.1. Network of state marine protected areas in the Channel Islands National Marine Sanctuary, established April 9, 2003. Solid polygons represent no-take marine reserves. Hatched polygons represent limited-take marine conservation areas.

Table 6.1. Goals for marine reserves developed by the Marine Reserves Working Group (Jostes and Eng 2001).

Biodiversity conservation	To protect representative and unique marine habitats, ecological processes, and populations of interest.
Sustainable fisheries	To achieve sustainable fisheries by integrating marine reserves into fisheries management.
Socioeconomic viability	To maintain long-term socioeconomic viability while minimizing short-term socioeconomic losses to all users and dependent parties.
Natural and cultural heritage	To maintain areas for visitor, spiritual, and recreational opportunities which include cultural and ecological features and their associated values.
Education	To foster stewardship of the marine environment by providing educational opportunities to increase awareness and encourage responsible use of resources.

and conservation organizations as well as state and federal agencies. The working group was given the task of considering the use of MPAs for management within the Channel Islands National Marine Sanctuary. The working group determined by consensus that no-take reserves should be used to conserve biodiversity, help sustain fisheries, contribute to opportunities for education and research, and preserve natural and cultural heritage (Table 6.1; Jostes and Eng, 2001). The working group agreed to design a network of no-take reserves to achieve these goals, with the minimum social and economic impacts to the community of users. The primary objectives of the design process were to conserve marine biodiversity and habitats in a network of reserves at a minimum cost in terms of area, boundary length, and economic impacts.

Two advisory panels were appointed by the Sanctuary Advisory Council to assist the working group with acquisition and evaluation of information. The science advisory panel, a body of seventeen marine scientists, was given the task of developing ecological criteria and options for reserve design based on the goals for biodiversity conservation and sustainable fisheries. The science advisory panel also evaluated potential ecological impacts of alternatives developed by the working group. The socioeconomic advisory panel included two economists from the National Oceanic and Atmospheric Administration (NOAA) and four contractors with expertise in economic and social sciences. The socioeconomic advisory panel was given the task of gathering data and evaluating potential social and economic impacts of marine reserves.

Table 6.2. Conservation targets for marine reserves, developed by the science advisory panel to the Marine Reserves Working Group.

Coastline characteristics
- Sandy beach
- Rocky coast (low exposure)
- Rocky coast (high exposure)

Substrate type and depth
- Soft sediment (0-30 m)
- Hard sediment (0-30 m)
- Soft sediment (30-100 m)
- Hard sediment (30-100 m)
- Soft sediment (100-200 m)
- Hard sediment (100-200 m)
- Soft sediment (>200 m)
- Hard sediment (>200 m)

Additional features
- Emergent rocks
- Submerged rocky features (pinnacles, ridges, seamounts)
- Submarine canyons

Dominant plant communities
- Giant kelp
- Surfgrass
- Eelgrass

From Airamé et al. (2003).

The ecological criteria, which guided the design of reserves, were based on the goals established by the working group and the scientific literature. To achieve objectives for biodiversity conservation, marine reserves must include representative habitats in each biogeographic region of the study area (Roberts et al., 2003a). Reserve design also should consider species (or populations) of particular interest, such as endangered or threatened species and species of economic importance (Roberts et al., 2003a). Additional criteria are necessary to address the question of sustainable biodiversity and fisheries. To be sustainable, reserves must be large enough to protect viable habitats and populations of interest. To contribute to sustainable fisheries, reserves must include a portion of the critical habitats and vulnerable life-history stages of targeted species. Connectivity among reserves must be considered in the design of reserves if they are expected to contribute to fisheries through spillover and export. In addition, the design of reserves must consider the distribution of human and natural threats, which could prevent reserves from achieving their objectives (Allison et al., 2003). Using these criteria, a set of conservation targets was identified (Table 6.2) to guide reserve design.

Historically, protected areas have been established using ad hoc approaches on a case by case basis. Geospatial modeling has made it possible to develop more systematic approaches to designing protected areas. As tools for design, geospatial models help decision-makers find solutions that include all targets through a process that is transparent and defensible (Pressey et al., 1996; Margules and Pressey, 2000).

Methods

Two geospatial tools were used to evaluate the available data and advance the design process. The science advisory panel used a streamlined derivative of SPEXAN (Version 3.1; Ball and Possingham, 2000), to evaluate spatial data and develop options for MPA design. SPEXAN (SPatially EXplicit ANnealing) utilizes several algorithms for selecting reserves, including simulated annealing, which was used in the Channel Islands process. The program runs within the framework of ArcView 3.2 (Environmental Systems Research Institute, Redlands, California), allowing easy visualization of data and solutions. Versions of this tool[1] have been applied to locate terrestrial reserves for The Nature Conservancy and marine reserves in Australia (Lewis et al., 2003), Canada (Ardron et al., 2002), Mexico (Sala et al., 2002), the northern Gulf of Mexico (Beck and Odaya, 2001), and Florida (Leslie et al., 2003).

There are a large number of solutions to the complex problem of reserve design. SPEXAN applies a process known as simulated annealing to identify sites within the study areas that contribute to management goals (Possingham et al., 2000). To begin the process, the algorithm adds sites until a set of conservation goals are met. Sites are added randomly, but sites that do not contribute to the goals can be rejected. With each change, the algorithm selects the solution that meets the greatest number of goals established by the user (Possingham et al., 2000). During the initial runs of the model, the algorithm explores a broad range of possibilities, including suboptimal solutions. As the analysis proceeds, the algorithm becomes more selective, leading to a final solution that meets the conservation targets within the minimum area. In the Channel Islands process, the program repeated 1,000,000 annealing iterations per run.

SPEXAN requires the division of a study area into a set of planning units, each with a unique identification number. Planning units can be added together to produce a reserve and subsets of planning units may be aggregated into a network of reserves. Planning units may be regular, as in a grid of squares (Leslie et al., 2003) or hexagons (Ardron et al., 2002). Alternately, planning units may be irregular polygons that reflect natural barriers, such as watersheds or habitat types. Lewis et al. (2003) utilized a combination of regular and irregular polygons of different sizes to describe the Great Barrier Reef Marine Park. Large hexagons (30 km^2) were used offshore where homogeneous habitats were prevalent. Smaller hexagons (10 km^2) were used closer to shore in non-reef areas. In reef areas, the actual boundaries of the reefs were used as planning units.

Leslie et al. (2003) varied the size of planning units in an analysis of the Florida Keys National Marine Sanctuary. They divided the study area into planning units of 1 km^2 and 100 km^2. Leslie et al. (2003) concluded that the small planning units were on the scale of the habitat patches themselves and that solutions based on analysis with large planning units contained considerably more area (and were less efficient) than solutions based on small planning units.

The size of the planning unit should be at the same scale as the management effort. In the state of California, waters are divided into 10 x 10 square n.m. "fish blocks." Data on fishery landings are collected by the Department of Fish and Game at the scale of fish blocks. Because of their large size, fish blocks were not useful planning units for marine zoning in the Channel Islands National Marine Sanctuary. Discussions about where to establish marine reserves, at a scale of 10 x 10 n.m., would likely break down in conflict over different resources within a single fish block. Smaller planning units that better approximate the scale of resource heterogeneity may facilitate more efficient solutions because they allow a more detailed representation of the biophysical environment. In the Channel Islands planning process, fish blocks were divided into 100 1 x 1 square n.m. planning units (Fig. 6.2). Within the Sanctuary boundary, 1,535 planning units were defined. All of the ecological and economic data were scaled to 1 x 1 square n.m. planning units or, if new data were collected, they were gathered at this scale.

During the process of simulated annealing, the algorithm seeks the minimum cost to achieve all conservation targets. The cost of marine reserves can be identified in various ways, including area and boundary length, or the opportunity or management cost incurred by establishing reserves (Leslie et al., 2003). The demands of conservation goals and minimum cost can be resolved by minimizing the following objective function (McDonnell et al., 2002).

$$\text{Objective Function (t)} = \text{BLM x Boundary(t)} + \sum(Penalty[i]) + \text{Cost(t)}$$

(Eqn. 1) where t is time as the algorithm proceeds, BLM is the Boundary Length Modifier (discussed below), Boundary(t) is the length of the outer boundary of the selected sites at time t, and Penalty(i) is the penalty for not meeting conservation goal i. Penalty(i) is zero when the conservation goal for target i is included in the reserve network. Cost(t) is the cost value, in terms of area or boundary length and/or missed opportunity, of all sites included in the network at time t.

The boundary length modifier, BLM, determines the relative importance placed on minimizing the boundary length relative to minimizing area. A network of reserves of minimum area may be highly fragmented because the algorithm selects only the planning units that contribute to the conservation goals. A fragmented network of reserves may be undesirable from the perspectives of management, enforcement and monitoring. The boundary length modifier is introduced in order to cluster the reserves. The boundary length modifier forces the algorithm to consider the relationship between the perimeter and area of the reserves.

As the boundary length modifier increases, the importance of minimizing the perimeter of the reserve system also increases. If the boundary length modifier is 0, then the algorithm selects the subset of planning units to meet conservation targets at a minimum total cost. If the boundary length modifier is greater than 0, the algorithm selects the subset of planning units to both meet conservation targets and reduce the ratio of boundary length around the network of reserves to total area in reserves. The larger the boundary length modifier, the more aggregated the planning units within individual reserves. In the Channel Islands process, the boundary length modifier was set at a range of values (0, 0.2, and 1) to explore the behavior to the model. The science advisory panel selected a boundary length modifier of 1, in which the perimeter and area of the reserves in the network were jointly minimized, to generate a set of alternatives for consideration by the working group.

SPEXAN provides an opportunity for users to input the "cost" of each planning unit. The cost may be equal to the total value of commercial and recreational activities within the planning unit. Alternatively, the cost may be the sum of relative contributions of each unit to each commercial and recreational activity, thus normalizing the scale so that activities that generate low revenue are valued equally with activities that generate high revenue. If no cost is included in the

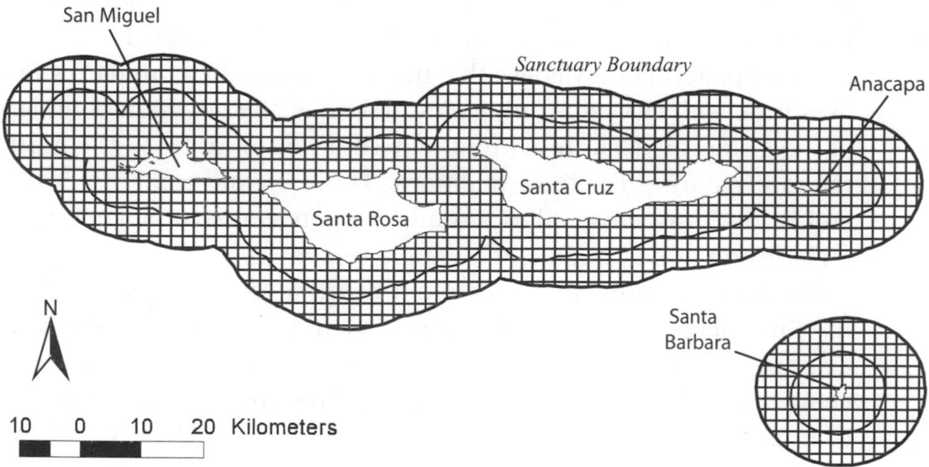

Figure 6.2. Planning units (1 x 1 square nautical mile) within the Channel Islands National Marine Sanctuary.

dataset, then the cost function defaults to the area of each planning unit. In the Channel Islands case, the default cost function was used in the SPEXAN analysis and the cost of each planning unit was equal to its area.

Conservation Targets for Reserve Design

To generate a suite of reserve designs, SPEXAN requires continuous data, a list of explicit conservation targets, and goals for representation of each target. In the Channel Islands case, scientists organized ecological data according to biogeographic patterns. Three primary regions were identified based on sea surface temperature (The Institute for Computational Earth System Science (ICESS) at the University of California, Santa Barbara, unpublished data) and composition of marine communities. The northwestern Channel Islands, including San Miguel and Santa Rosa islands, are bathed in the cool waters of the California Current. The species in this region are similar to those found along the west coasts of California, Oregon, and Washington. The eastern Channel Islands, including Anacapa and Santa Cruz islands, are influenced by the California Countercurrent, which carries warmer waters north along the coast. The species in the region influenced by the countercurrent are similar to those along the coast of Baja California. The two currents collide in the Channel Islands region, mixing in a transition zone around Santa Barbara Island and southern Santa Rosa and Santa Cruz islands. The species composition of the transition zone is a unique blend of the communities from the northern and southern biogeographic provinces.

Because of the uniqueness of each of the biogeographic regions, the SPEXAN model was used independently within each of the three regions to identify potential reserves. The science advisory panel used information on average sea surface temperature and bathymetry to draw working boundaries for the biogeographic regions. In the areas of sharpest transition in temperature, the boundaries were drawn along the deepest bathymetric contours (Airamé et al., 2003). The science advisory panel acknowledged that the locations of the boundaries vary over time with climate. During El Niño cycles, the northern boundary of the California Countercurrent may shift tens of miles to the north, retreating during La Niña conditions. The dynamic zone of advance and retreat was classified as a transition between two biogeographic regions and was evaluated as a unique region.

The biogeographic regions are not distributed equally throughout the Channel Islands National Marine Sanctuary. Most of the northern Channel Islands are bathed in the cool, nutrient-rich waters that characterize the California Current. Only Anacapa and east Santa Cruz islands are situated within the warmer waters of the California Countercurrent. Because the biogeographic regions differ in size, the SPEXAN analysis could not be applied in the same way to each region. Instead, the number of runs of the model was proportional to the size of each bioregion and varied from 300 to 800. By adjusting the number of runs for the total area within each region, the opportunity for selecting a planning unit for a reserve was relatively equivalent across biogeographic regions. The results across all three biogeographic regions were pooled for the purposes of visualization and discussion.

Scientists identified conservation targets of different habitats and species in each biogeographic region (Table 6.2). Conservation targets included coastline characteristics, substrate type and depth, unique physical features, dominant plant communities, seabird colonies, and pinniped haul-out sites (Airamé et al. 2003). In the Channel Islands region, approximately 70% of the coastline is rocky whereas 30% is sandy. About 40% of the benthic substrate in shallow subtidal zone (0-30 m) is rocky. Giant kelp (*Macrocystis pyrifera*) is the dominant alga in shallow subtidal rocky habitats. Surfgrass (*Phyllospadix* spp.) is common in shallow rocky subtidal habitats. In a few sheltered locations, shallow sandy substrate supports populations of eelgrass (*Zostera* spp.). In deeper waters on the continental shelf, the sediment is primarily sand, silt or unconsolidated rock with only a few rocky features scattered throughout the area. A deep submarine canyon divides the southern edges of Santa Rosa and Santa Cruz islands. Initially, all submerged rocky features were identified as unique. However, this classification of submerged rocky features constrained the reserve options identified by SPEXAN because each unique habitat was selected as a potential reserve area in every run of the model. Some flexibility was introduced by generalizing the habitat classification to include broader groups, such as emergent rocks, submerged pinnacles, and submarine canyons, rather than identifying each individual feature.

The choice of a particular habitat classification scheme can significantly influence the outcome of the model (Leslie et al., 2003). A simple habitat classification was developed for the Channel Islands process, based on available data and ecological differences between habitats. Ecological communities were characterized by sediment type and bathymetry. Benthic sediments were divided into two groups: soft (mud, silt, sand, cobble, and unconsolidated rock) and hard (boulders, rocky reefs, and bedrock). Bathymetry was divided into four groups: (1) the euphotic zone (0-30 m depth), (2) the shallow continental shelf (30-100 m depth), (3) the deep continental shelf (100-200 m depth), and (4) the continental slope (>200 m depth). The outer boundary of the Sanctuary falls near the continental shelf break, and therefore, most of the Sanctuary is in depths shallower than 200 m. For case studies in much deeper waters, additional depth zones should be specified. Dominant algae and plant communities, including giant kelp, surfgrass, and eelgrass, also were considered important conservation targets because these species provide shelter and food for distinct marine communities. The conservation targets are summarized in Table 6.2.

Table 6.3. Selected species of interest in the Channel Islands National Marine Sanctuary (from Airamé et al. 2000).

Breeding seabirds	*Scientific Name*
California Brown Pelican	*Pelecanus occidentalis californicus*
Pelagic Cormorant	*Phalacrocorax pelagicus*
Double-crested Cormorant	*Phalacrocorax auritus*
Brandt's Cormorant	*Phalacrocorax penicillatus*
Common Murre	*Uria aalge*
Pigeon Guillemot	*Cepphus columba*
Xantus's Murrelet	*Synthliboramphus hypoleucus*
Rhinoceros Auklet	*Cerorhinca monocerata*
Cassin's Auklet	*Ptychoramphus aleuticus*
Leach's Storm-petrel	*Oceanodroma leucorhoa*
Ashy Storm-petrel	*Oceanodroma homochroa*
Black Storm-petrel	*Oceanodroma melania*
Black Oystercatcher	*Haematopus bachmani*
Snowy Plover	*Charadrius alexandrinus*
Western Gull	*Larus occidentalis*
Pinnipeds	*Scientific Name*
California sea lion	*Zalophus califonianus*
Northern fur seal	*Callorbinus ursinus*
Northern elephant seal	*Mirounga Angustirostris*
Harbor seal	*Phoca vitulina*

In addition to developing the habitat classification, the science advisory panel assisted the working group with the identification of species of interest in the Channel Islands (Airamé et al. 2000). The list of species of interest includes species of economic and recreational importance; keystone or dominant species; candidate, proposed, or species listed under the Endangered Species Act; species that have exhibited long-term or rapid declines in harvest and/or size frequencies; habitat-forming species; indicator or sensitive species; and important prey species. The list excludes species that are incidental, at the edge of their ranges, or highly migratory.

Distributions of breeding seabirds and haul-out sites for pinnipeds were utilized as conservation targets for the process of locating potential reserves (Table 6.3). Fifteen species of seabirds, including the endangered California brown pelican (*Pelecanus occidentalis californicus*) and Western snowy plover (*Charadrius alexandrinus nivosus*), roost and nest along the coastline of the Channel Islands. Important locations for breeding seabirds include Anacapa and Santa Barbara islands, Prince Island (off of San Miguel Island), Arch Rock (off of northeastern Santa Cruz Island), and Sutil Island (off of Santa Barbara Island).

Table 6.4. Selected species of commercial importance in the Channel Islands National Marine Sanctuary (from Leeworthy and Wiley 2002).

	Scientific Name	*Average Value (1996-1999)*
Top invertebrate fisheries		
Market squid	*Loligo opalescens*	$11,249,837
Sea urchin	*Strongylocentrotus franciscanus*	$5,265,233
California spiny lobster	*Panulirus interruptus*	$922,098
Prawn	*Pandalus platyceros*	$703,186
Abalone (historical)	*Haliotis* spp.	$178,027
Crab	*Cancer, Loxorhynchus* spp.	$343,664
Sea cucumber	*Parastichopus* spp.	$167,700
Top commercial fisheries		
Rockfish	*Sebastes* spp.	$549,319
Anchovy and sardine	*Engraulis mordax, Sardinops sagax*	$234,367
California sheephead	*Semicossyphus pulcher*	$235,928
Flatfish	Pleuronectidae, Bothidae	$183,871
Mackerel	*Auxis, Scomber Trachurus* spp.	$67,119
Sculpin and bass	*Leptocottus, Icelinus, Paralabrax, Stereolepis, Atractoscion*	$60,327
Tuna	*Thunnus, Katsuwonus, Sarda, Euthynnus*	$205,884
Swordfish	*Xiphias gladius*	$39,090
Shark	Chondrichthyes	$34,751
Other commercial activities		
Kelp harvesting	*Macrocystis pyrifera*	$5,000,000

Four pinniped species, including the California sea lion, northern elephant seal, harbor seal, and northern fur seal, commonly haul out on beaches in the Channel Islands. San Miguel Island is the most important haul-out for pinnipeds, supporting populations of up to 80,000 California sea lions and 50,000 northern elephant seals (DeLong and Melin, 1999). Harbor seals are found throughout the Channel Islands region.

Information about the distribution and abundance of fishes and invertebrates in the Channel Islands region was available primarily from fisheries records. Fisheries data were not used as a part of the design process because of the potential conflict between conservation targets and fisheries interests. To meet conservation goals, the network of reserves must include all species of interest, including those that are fished. To minimize the impact on commercial and recreational activities, the reserves must not overlap the areas of greatest use. If fisheries data are used to define species distributions, then these goals will conflict during the process of locating potential reserves. The fisheries data were excluded from the SPEXAN analysis in order to produce a suite of alternative solutions that were clearly responsive to the conservation goals. Fisheries data (Table 6.4) were used in a separate economic impact analysis to evaluate potential costs of reserve designs (Leeworthy and Wiley, 2002).

Conservation Goals for Reserve Design

The science advisory panel used SPEXAN to explore scenarios for reserve networks that represent the conservation targets efficiently with respect to both the total area and perimeter of the network. To run SPEXAN, the user must specify conservation goals, or fractions of each conservation target that must be represented in the final set of reserve sites. In some cases users may weight conservation targets on the basis of ecological importance, rarity, or vulnerability/threatened status. For example, Sala et al. (2002) required that 100% of the coral communities, seagrass beds and spawning aggregations be included in a network of reserves in the Gulf of California. Other habitats, such as rocky or sandy substrate, were included at levels of 20% or more. Another approach is to set aside conservation targets in proportion to their abundance in the study area (Roberts et al., 2003b). This approach is used when no information is provided about the relative importance of different targets. In the Channel Islands process, the science advisory panel established conservation goals in proportion to the abundance of each target in the study area.

Different conservation goals produce solutions that differ in their spatial extent and potential locations of reserves. To explore the model's behavior, conservation goals were set at three different levels: 30, 40, and 50%, based on an evaluation of the total reserve size needed to achieve goals established by the working group (Airamé et al., 2003).

Evaluation of Reserve Designs

As standard output, SPEXAN produces an evaluation of each solution. The evaluation indicates the degree to which a particular design meets conservation goals. The best solution is the one with the lowest value of the objective function (Eqn. 1), which balances conservation goals with a weighted sum of area and boundary length. For each solution, the program provides information about the total area of the network of reserves, the area of each conservation target within reserves, and the proportion of the target met. In some cases, particularly if the conservation target is common, it may be overrepresented in the network because of constraints imposed by rare targets.

After solutions were generated and evaluated, the science advisory panel conducted an "irreplaceability analysis" (Leslie et al., 2003) to determine how many times each planning unit was included in a final solution. The program generated this information as a list of summed solutions indicating the number of times planning units were included in a final solution. The summed solutions were normalized to the percent of the total number of runs and converted to a map showing the percent use of each planning unit in the portfolio of final solutions.

From the hundreds of alternatives generated using SPEXAN, the science advisory panel identified similar designs using a clustering program, PRIMER v.4, Plymouth Routines in Mulitvariate Ecological Research (Clarke and Warwick, 2001). The 100 top ranking solutions were selected from the total runs for each biogeographic region and the Bray-Curtis similarity between solutions was calculated. Clusters of solutions were grouped together at 60% similarity, creating five clusters of solutions for each biogeographic region at each conservation goal. If the grouping algorithm produced more than five groups, the group with the lowest high score was removed from the analysis. From each cluster, scientists selected those that included the greatest number of habitats and species of interest, represented in patches of adequate size, in the most efficient configuration (or smallest boundary length and area). The set of solutions, which all met conservation goals and provided a range of spatial options, were delivered to the working group for consideration.

Application of CI-SSAT to Reserve Design

Members of the working group contributed their own expertise to modify designs or generate alternatives to the designs developed by the science advisory panel. The working group utilized a geospatial tool, known as the Channel Islands Spatial Support and Analysis Tool (CI-SSAT; Killpack et al., 2000), to advance the policy discussion. The tool was designed by NOAA's Coastal Services Center to facilitate interaction between working group members and provide a platform to conduct spatial analysis (Killpack et al., 2000). CI-SSAT provided a common framework for visualization, manipulation, and analysis of data for the purpose of designing marine reserves. CI-SSAT was tested during this process and the tool will be refined for general use.

CI-SSAT is a computer-based environment for viewing and evaluating information (Killpack et al., 2000). The tool was developed with Environmental Systems Research Institute (ESRI) ArcView 3.2 and Microsoft Visual Basic 6.0 software. The user interface resembles a Geographic Information System (GIS) with spatially explicit data relevant to the problem. Data can be selected or hidden by checking a box beside the data label. Once the data have been selected, the user can zoom in or out to obtain broader or more detailed views.

In the Channel Islands process, CI-SSAT contained both ecological and socioeconomic data. The map of conservation "hot spots," generated by irreplaceability analysis in SPEXAN, was included in the CI-SSAT. Other ecological data, including distributions of sediments, giant kelp, seagrasses, seabirds, and marine mammals, were available for viewing and basic querying. Ten options for marine reserves, generated by

SPEXAN, were available for purposes of comparison. The tool also contained maps showing revenue gained from the most important commercial industries and usage by various recreational activities. Proprietary data describing the economic value of each planning unit to each fishery was not released. However, maps of the revenue for aggregate fishing, squid fishing and kelp harvesting were included in the tool (Leeworthy and Wiley, 2002). Additionally, the tool contained maps of usage by recreational fishing and diving industries and private recreational fishers and divers. Ethnographic data showing the distributions of habitats and species, based on interviews with local citizens who lived, worked, and recreated around the Channel Islands for many years, also were included (Kronman et al., 2000). Anecdotal descriptions of declines or increases in abundance and shifts in distribution of species of interest also were captured in the dataset.

All datasets were referenced to a common base map consisting of a 1 x 1 square n.m. grid. These planning squares were used as the common units for all the data layers and the SPEXAN analysis. The data were displayed in raster format in CI-SSAT.

Selection and Weighting of Criteria

CI-SSAT offers the opportunity for users to weigh various criteria for reserve design (Fig. 6.3). In the Channel Islands process, the criteria were based on goals established by the working group. The tool offered two primary criteria: ecological and socioeconomic. In CI-SSAT, these conceptual criteria were tied to a dataset for evaluation in the form of evaluation layers. Each evaluation layer is a synthesis of many individual datasets representing the probability that a particular goal will be met in a particular area. Each planning unit in the evaluation layer uses a common ranking unit that describes how well that unit satisfies a

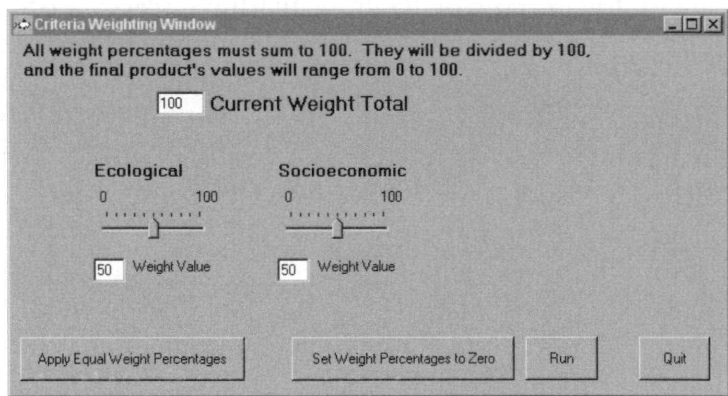

Figure 6.3. Dialog box for the criteria weighting tool for CI-SSAT. Two criteria were provided for the Channel Islands process: ecological and socioeconomic. The weights for all criteria must sum to 100%. The user may adjust the weight of a particular criterion by moving the slider bar to the desired percent.

particular goal. The probability values range between 0 and 100, where 100 is the highest rank for a particular criterion.

The science advisory panel developed the ecological criterion layer using SPEXAN (see earlier discussion). The ecological criterion layer was the map of conservation value resulting from the irreplaceability analysis (Fig. 6.4). The ecological evaluation layer input to CI-SSAT therefore utilized all of the ecological data and conservation targets considered in SPEXAN analysis.

The socioeconomic criterion layer was based on revenue and usage (Fig. 6.5). For some commercial industries, including squid fishing and kelp harvesting, the total revenue per species fished was available at a scale of 1 x 1 square n.m. For other commercial industries and all recreational activities, the total usage was estimated in person-days per year for each planning unit. Because industry values were represented in different units, it was not possible to calculate a straight summation of the total revenue per planning unit. Instead, each individual socioeconomic dataset was normalized from 0 to 100 and the values for different commercial industries and recreational activities were summed. Normalization of the summed data created a relative ranking of planning units based on revenue or usage. Areas of high revenue or usage were identified as undesirable locations for marine reserves. Locations of low revenue or usage were ranked highly in the section of marine reserves. Thus, a value of 100 for the socioeconomic

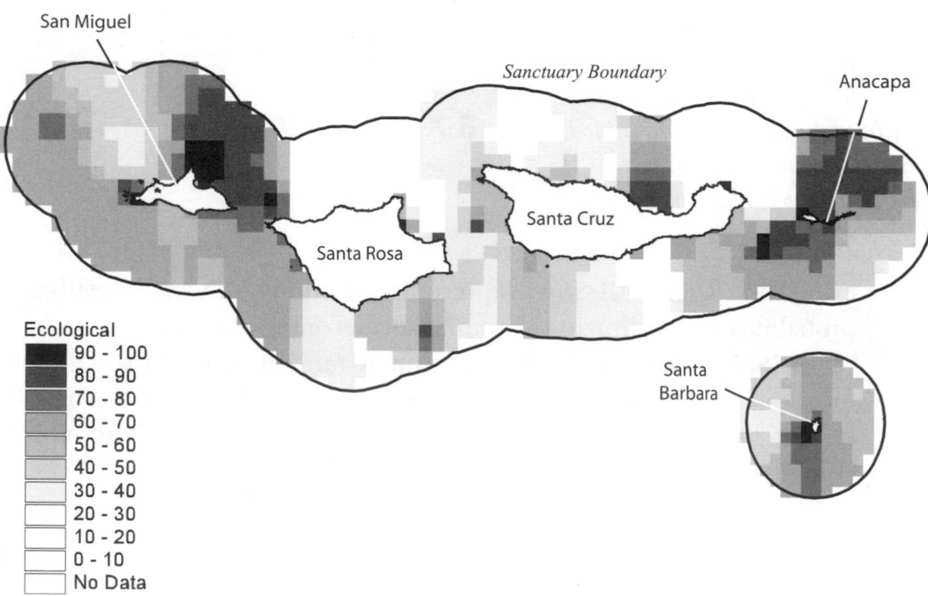

Figure 6.4. Ecological criterion layer for CI-SSAT. The data were derived from the irreplaceability analysis in SPEXAN. The grayscale represents the conservation value, or the number of times each planning unit was included in a final solution. Areas shaded in black or dark gray were included more often than areas shaded in light gray or white.

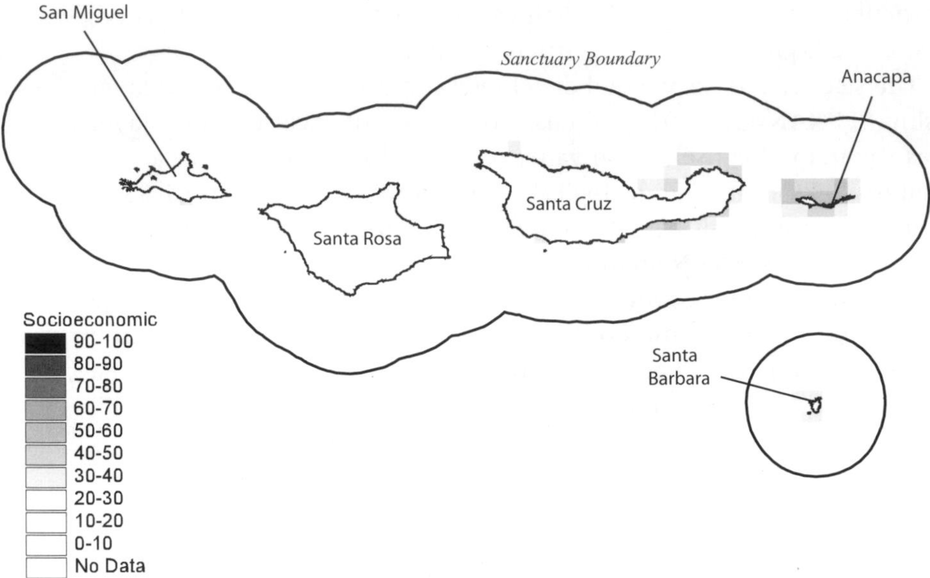

Figure 6.5. Socioeconomic criterion layer for CI-SSAT. The data were derived from the economic impact analysis conducted by Leeworthy and Wiley (2002). The grayscale represents the relative economic value estimated as a percentage of each of the commercial and recreational activities conducted in the Channel Islands National Marine Sanctuary. Areas shaded in black or dark gray are more valuable, in terms of commercial and recreational activity, than areas shaded in light gray or white.

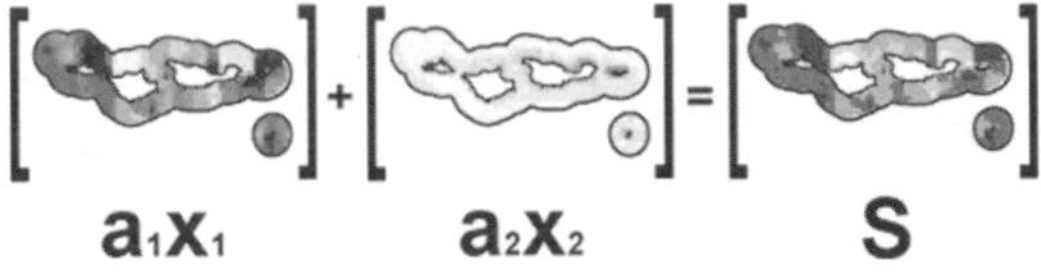

Figure 6.6. Criteria weighting analysis in CI-SSAT. Each criterion data layer (x_n) is weighted (a_n) according to user values and added to other weighted criterion data layers to produce a base map (S).

evaluation layer in CI-SSAT represents a planning unit with the highest probability of minimizing potential socioeconomic impacts.

CI-SSAT performs a simple algorithm to rank the suitability of various locations for marine reserves (Fig. 6.6). The algorithm, which is usually known as simple additive weighting or weighted linear combination, combines the criteria layers according to Eqn 2,

$$a_1x_1 + a_2x_2 + a_3x_3 + \dots a_nx_n = S$$

where a_n is the weight value ($\Sigma\, a_n = 1$), x_n is the criterion values for the planning unit bearing a value between 0 and 100, and S is the resultant outcome data cell value ranking between 0 and 100. A high value represents a suitable area for a marine reserve based on the chosen weights and criteria (Killpack et al., 2000).

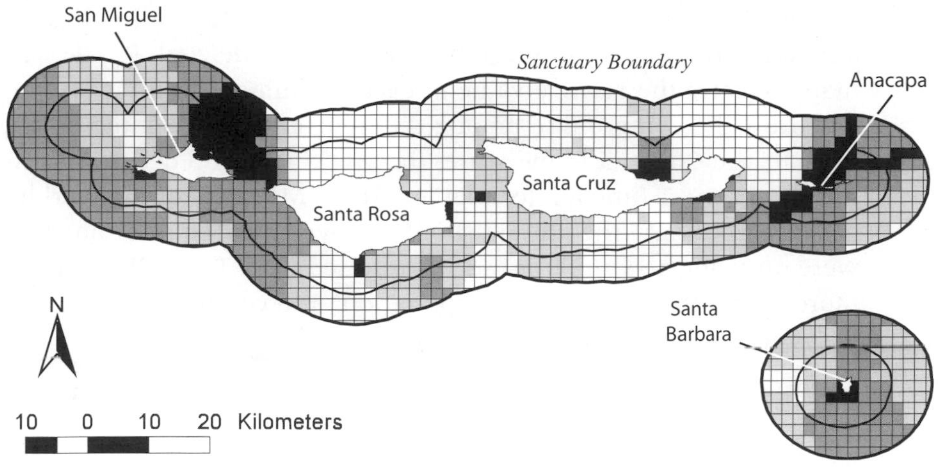

Figure 6.7. Results of the criteria weighting analysis in CI-SSAT. Each criterion data layer (x_n) is weighted (a_n) according to user values and added to other weighted criterion data layers to produce this base map (S). In this example, the ecological and socioeconomic criteria were weighted, each at 50%, to produce the resulting base map. High values (shaded in dark gray) represent suitable areas for reserves based on the chosen weights and criteria.

Each weighting process is unique and based on values held by the user. If the user desires to produce a zoning plan based entirely on ecological criteria, the analysis will reflect only ecological data and the conservation "hot spots" will be identified based on habitat heterogeneity, species diversity, rare habitats and species, or other criteria identified in the ecological evaluation. If the user desires to minimize economic impact of a zoning plan, then CI-SSAT selects areas that have low overlap with existing commercial and recreational consumptive activities. If the user desires to balance ecological with economic criteria, the areas of conservation value will be selected in the sites that minimize economic impacts.

The user selects a number between 0 and 100 to represent the weighting for each criterion, with the sum of all weights no greater than 100. For purposes of analysis, these values are divided by 100 to standardize to values between 0 and 1. Then, the weight value and criteria data grid are multiplied and all weighted criteria grids are summed together using raster addition (Fig. 6.6). The resulting values for each planning unit, ranging between 0 and 100, indicate the potential of each planning unit to achieve the desired outcome based on user specifications. The results of the weighting process are displayed as a raster or grid map with high numerical values representing areas that meet the criteria and the weighting scheme of the user.

Once the analysis is completed, the user can use the output layer as a base map (Fig. 6.7) to develop a plan for marine zoning. The base map indicates the relative value of each planning unit for a reserve, based on the compromise among different criteria weighted by the user. In the Channel Islands case, the working group decided not to use the CI-SSAT function to weight criteria. Members of the working group agreed that the criteria should be weighted equally, but they were unwilling to work from a compromised map. Thus, CI-SSAT was more useful for visualization, exploration, and comparison of zoning plans developed by working group members.

Exploration and Comparison of Options in CI-SSAT

The user also has the ability to query some of the data in CI-SSAT. The tool contains two predefined query functions. Ecological queries focus on information provided by the science advisory panel and on habitat types. Data layers such as kelp, surfgrass, eelgrass, coastline geomorphology, benthic sediments, and bird and mammal densities were available for query. The socioeconomic queries focus on the consumptive industries operating in the waters near the Channel Islands. Revenues from aggregate commercial fishing, squid fishing and kelp harvesting were available for query. Usage of recreational fishing and diving industries, and private recreational fishing and diving also were available for query.

Queries begin by identifying a particular location in the study area. Simple drawing features in CI-SSAT allow users to create rectangles, circles, or irregular polygons to represent potential reserves. Additionally, the user may indicate an area that should be excluded from further analysis. Examples include areas that already are protected in reserves, areas of particularly high commercial or recreational value, or areas that are particularly vulnerable to human threats or natural disturbances (Allison et al., 2003). If the exclusion areas are located in a GIS shapefile, they can be integrated easily with the data in CI-SSAT. Once exclusion areas have been identified, the data within those areas will be excluded from further analysis.

Once the user has located a potential reserve, a quick evaluation provides the user with (1) information about the amount of each habitat or portion of species' range captured within the reserve boundaries; and (2) the potential impact of the reserve on major commercial industries and recreational activities. The query window returns a table of values for the analyzed area and compares these values with the total Sanctuary area.

By allowing the user to iteratively adjust the boundaries to include more of a particular habitat or species, or to reduce the impact to a particular industry or activity, CI-SSAT facilitates development of a marine zoning plan to meet the user's criteria. The tool supports rapid modification and real-time evaluation of alternatives. The analysis facilitates negotiation among users by providing quantitative information to supplement personal knowledge.

In the Channel Islands process, the working group utilized CI-SSAT to visualize and query data. The working group used the tool at public meetings on several occasions to develop and evaluate alternative designs. Initially, the working group divided into four subgroups to consider alternatives to satisfy the collective goals of the group. Each subgroup was charged with the task of creating three alternatives. A technical facilitator was assigned to each subgroup to assist with computer operations so that the subgroups would not be distracted by the complexity of the support tool. Each group considered three standard weighting schemes to begin their analysis: (1) 50% ecological and 50% socioeconomic; (2) 75% ecological and 25% socioeconomic; and (3) 25% ecological and 75% socioeconomic (Killpack et al., 2000). However, the weighting process was discarded due to concerns that results from the criteria weighting analysis were compromised. Working group members expressed strong conviction that all goals should be considered equally and that compromises should be avoided, if possible. The working group proceeded with development of designs by using the data viewing and querying functions in CI-SSAT. Some working group members drew alternatives on paper maps. If paper maps were used, a technical facilitator recreated the design concepts in CI-SSAT for further evaluation. All maps were evaluated and results returned to working group members for discussion (Killpack et al., 2000).

Results from SPEXAN

The science advisory panel utilized SPEXAN to evaluate ecological data for reserve design. Different scenarios were run to explore the behavior of the model. Changes in the boundary modifier (at conservation goals of 30% of each target) affected the spatial configuration of the network of reserves (Fig. 6.8). Application of SPEXAN with no boundary modifier resulted in highly fragmented network (Fig. 6.8a). Without a boundary length modifier, the final solutions generally are so highly fragmented that they would be impossible to implement and enforce. Increases in the boundary modifier to 0.2 and 1 contributed to clustering of potential reserve sites (Fig. 6.8b and 6.8c). The science advisory panel selected a boundary length modifier of 1 to produce solutions that were more likely to meet standards for management, enforcement, and monitoring as described in Roberts et al. (2003a).

The most efficient reserve system (BLM=0) at a conservation goal of 30% of each target included 404 planning units. A change in the boundary length modifier from zero to 0.2 resulted in an increase of twelve planning units and a decrease in the total perimeter of the best reserve system. Further increasing the boundary length modifier to 1 resulted in an increase of forty-one planning units to the area of the network and a further decrease in the perimeter (Table 6.5). The most

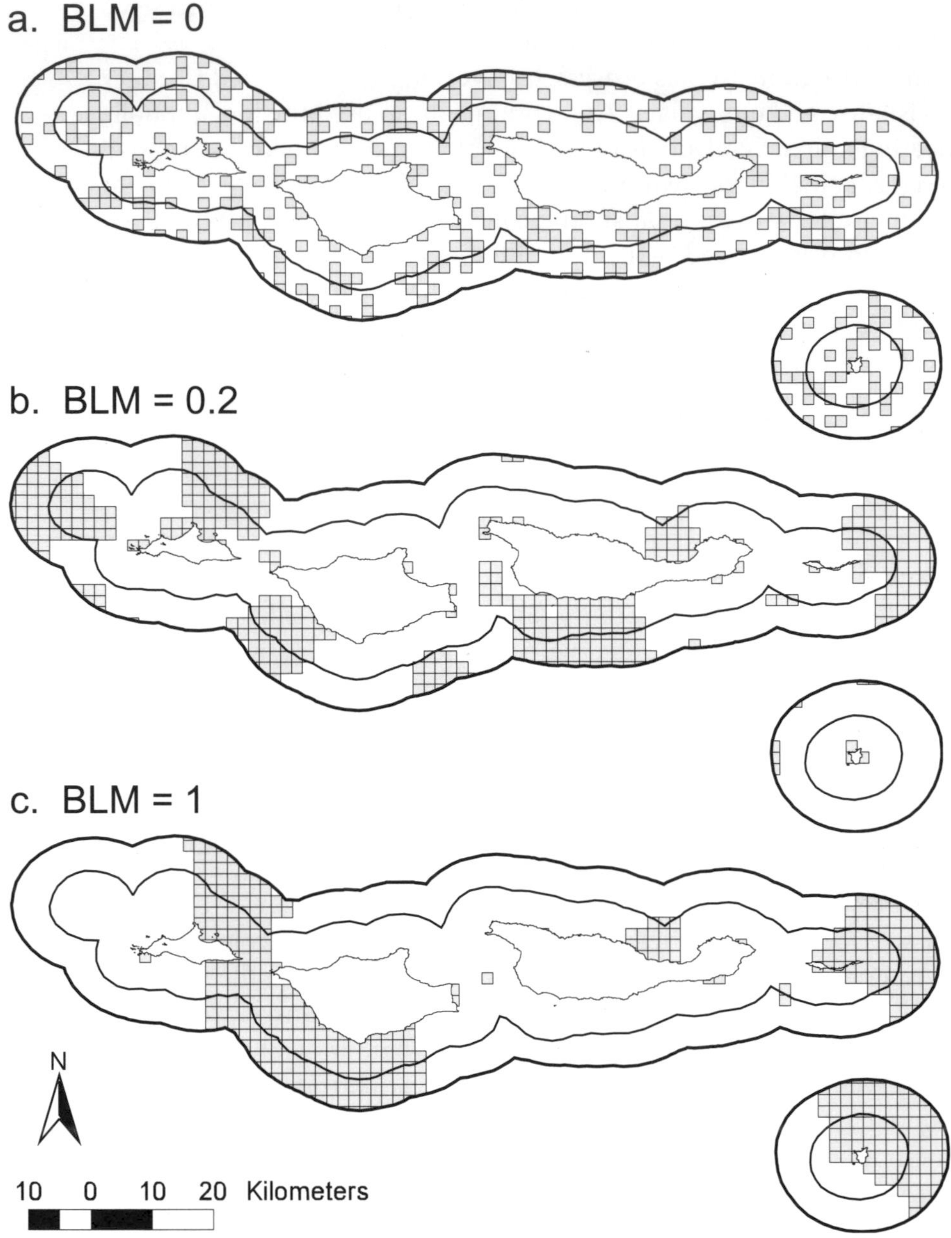

Figure 6.8. Changes in the boundary modifier allow the analyst to control the spatial configuration of the network of reserves. Application of SPEXAN with no boundary modifier results in a highly fragmented network (a). Increases in the boundary modifier contribute to clustering of potential reserve sites (b and c).

highly connected networks (BLM = 1) had just 10% of the total perimeter of solutions generated without regard to spatial clustering (BLM = 0). The increase in the boundary length modifier resulted in a more clustered set of reserves (Fig. 6.8c).

A change in the conservation goal for each target affected the total area and perimeter of the network of reserves. Figure 6.9 shows the total network size and perimeter for conservation goals of 30, 40, and 50% of each target. The most connected reserve system (BLM=1) at a conservation goal of 30% of each target included 457 planning units. A change in the conservation goal from 30 to 40% resulted in an increase of 152 planning units and a decrease in the total perimeter of the best reserve system. Further increasing the conservation goal to 50% resulted in an increase of 146 planning units to the area of the network and a further decrease in the perimeter (Table 6.6). The conservation goals were met in all scenarios because no constraints were placed on the total area of the network of marine reserves. With an increase in the conservation goal for each target, the total area and perimeter of the network increased (Fig. 6.9).

Through the process of evaluating different conservation goals, the science advisory panel determined that larger conservation goals generated solutions with more flexibility in potential locations of marine reserves than smaller conservation goals. Analyses conducted at the

Table 6.5. Reserve system solutions generated by simulated annealing at three different levels of clustering with conservation goals of 30 percent of each target.

BLM	*Best Area (PU)*	*Min Area (PU)*	*Max Area (PU)*	*Best Perimeter (n.m.)*	*Minimum Perimeter (n.m.)*	*Maximum Perimeter (n.m.)*
0	404	403	409	1089.8	990.0	1165.4
0.2	416	409	422	246.9	236.3	427.5
1	457	435	483	102.7	89.6	196.3

BLM, Boundary Length Modifier; PU, Planning Units; n.m., nautical miles

Table 6.6. Reserve system solutions generated by simulated annealing at three different conservation goals with a boundary length modifier of 1.

Con. Goal	*Best Area (PU)*	*Min Area (PU)*	*Max Area (PU)*	*Best Perimeter (n.m.)*	*Minimum Perimeter (n.m.)*	*Maximum Perimeter (n.m.)*
30%	457	446	474	150.9	144.2	306.6
40%	609	595	627	155.5	154.8	321.9
50%	755	747	777	165.1	157.3	317.2

Con. Goal, Conservation Goal; PU, Planning Units; n.m., nautical miles

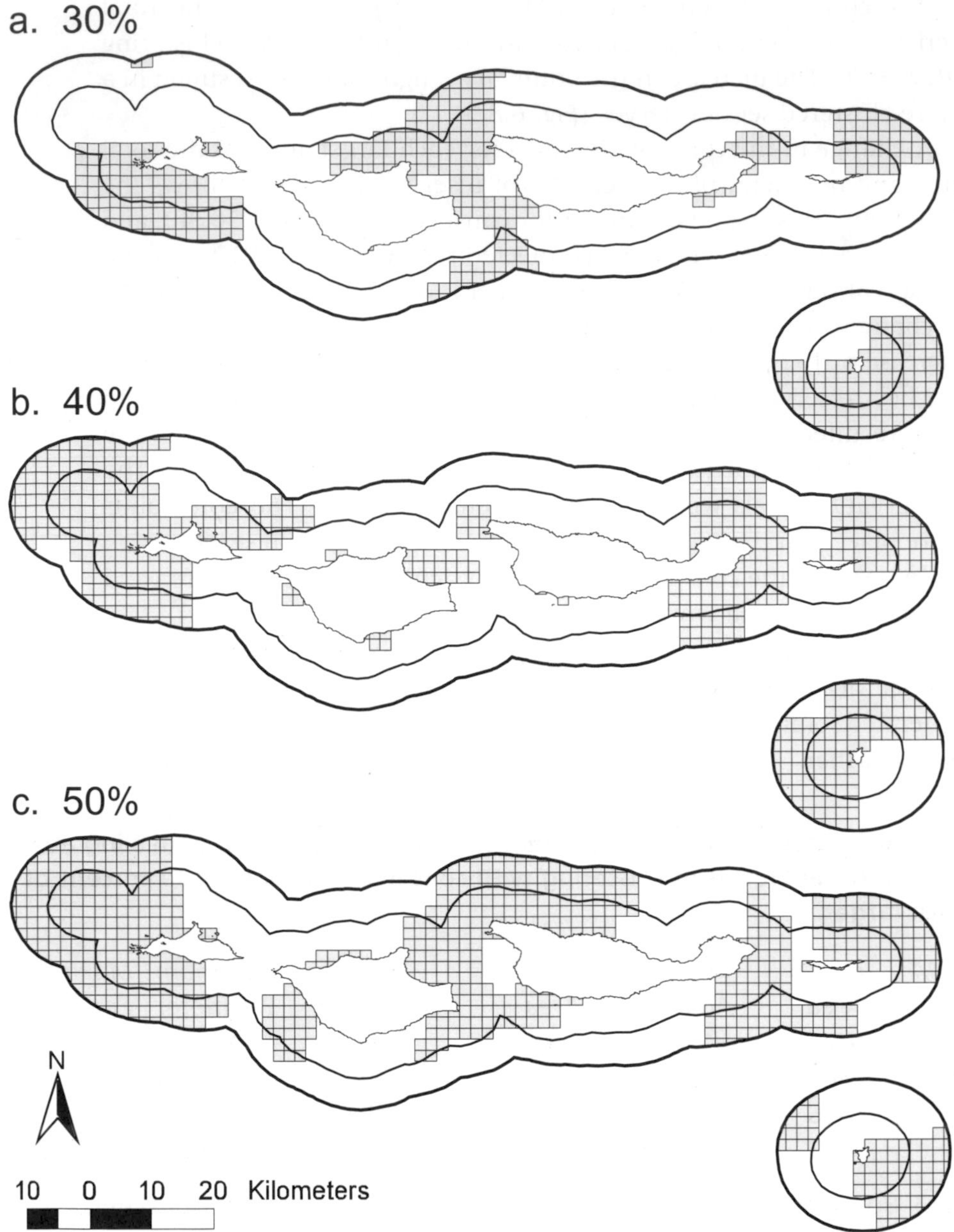

Figure 6.9. Changes in the conservation goals alter the spatial extent of reserves. Conservation goals were set at (a) 30%, (b) 40%, and (c) 50%. Increasing the conservation goals altered the spatial extent and configuration of reserves.

lowest conservation goals produced the least amount of flexibility about potential sites. At low conservation goals, the algorithm always selected the rarest conservation targets for core locations for marine reserves. The analyses conducted at the highest conservation goal produced the greatest number of alternative locations for reserves. In the process of developing a bigger network of marine reserves, the algorithm compared hundreds of similar planning units that contained common habitats or species. The map of conservation value not only highlighted the areas where rare habitats or species were found, but also indicated alternatives for protecting more common habitats and species in a variety of locations (e.g., not just in the area adjacent to rare habitats).

Rarity of conservation targets constrains the potential locations of reserves, particularly at low conservation goals. If a conservation target is rare, but required in the solution, the algorithm has no choice but to select a portion of the rare target. If the conservation goal is small and the total reserve size is minimized, common targets (e.g., soft benthic substrate) generally are added to reserves adjacent to rare targets.

In the Channel Islands process, eelgrass was one of the rarest habitats, found only in sheltered coves on sandy substrate (Engle, J., unpublished data). All of the final solutions from the model included a substantial portion of the eelgrass beds. Other—more common—conservation targets, such as soft benthic substrate, were added to reserves in the vicinity of the eelgrass beds in order to minimize the overall area and perimeter of the network of reserves.

With larger conservation goals, the selection process is more flexible. Rare habitats are included, but some common habitats may be included in areas that are not adjacent to the rare habitats. The flexibility of evaluating the problem with high conservation goals is useful in discussions about policy because users can select from a range of options rather than being constrained to the regions around the rarest habitats.

Using SPEXAN, the science advisory panel generated hundreds of solutions to the complex problem of reserve design. The solutions were sorted, using a cluster analysis in PRIMER v. 4 (Clarke and Warwick, 2001) and solutions were grouped at 60% similarity. By sorting the solutions, the science advisory panel was able to identify the range of geospatial alternatives that meet the same set of conservation goals. The most efficient solutions from each cluster at 30, 40, and 50% were selected for consideration by the working group. Figure 6.10 shows the most efficient solutions, in which the SPEXAN objective function is minimized, from five different groups of solutions. Although these solutions differ in their geospatial coverage, all solutions include 30% of each conservation target.

The hundreds of solutions to the problem of reserve design were summarized by describing the number of times each planning unit was included in a final solution. The resulting map (Fig. 6.11) indicates the relative contribution (or "conservation value") of each planning unit

Figure 6.10. The model produced a large number of solutions to the complex problem of reserve design. Five solutions are shown for the goal of 30% representation of all conservation targets. All solutions meet the conservation goals, providing flexibility in the reserve design process.

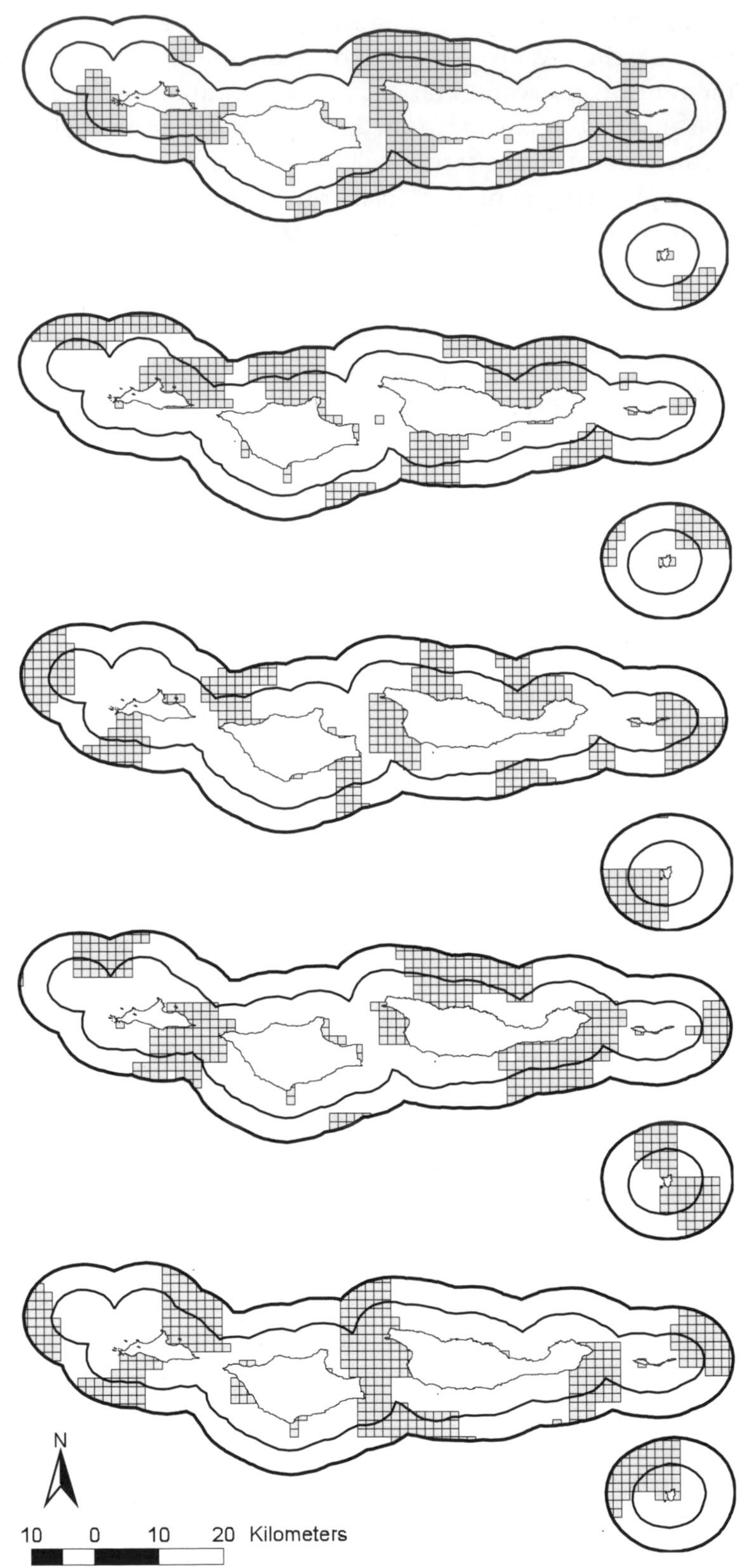

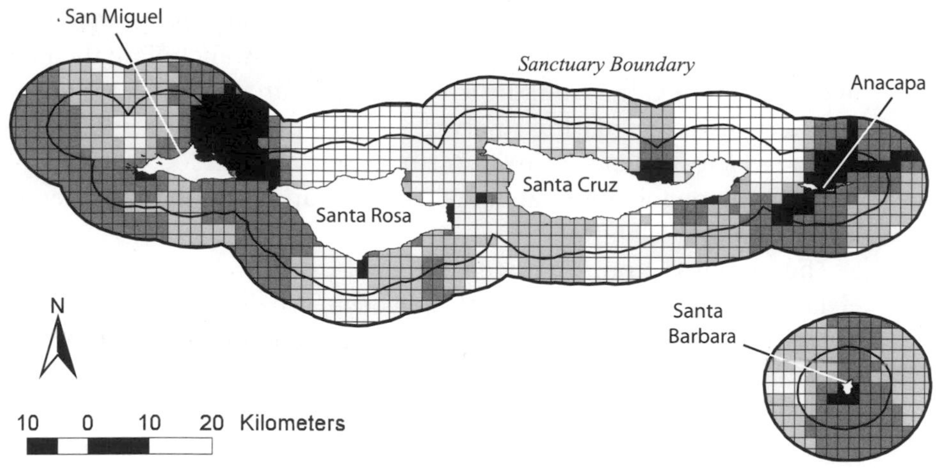

Figure 6.11. Conservation value of planning units in the Channel Islands National Marine Sanctuary. The conservation value is the number of times each planning unit was included in a solution. Black and dark gray areas were included in more than 75% of the solutions whereas light gray and white areas were included in less than 25% of the solutions.

to the conservation goals. Twenty-two planning units were chosen more than 75% of the time to meet goals of 30% of each conservation target. These sites were represented consistently in the solutions, suggesting that they may be priority sites for protection. In the Channel Islands case, the priority sites were located off the northeast coast of San Miguel Island, on the north side of Anacapa Island, and on the south side of Santa Barbara Island. A large number of planning units (576) were chosen less than 25% of the time, suggesting that these may not be suitable sites for reserves. Five planning units were never chosen during the runs. A planning unit that was not chosen in a final solution may not contain sufficient conservation targets to contribute to conservation goals. However, it is important to note that, even if a planning unit is not selected for a reserve, the planning unit may contain habitats and species of interest. Additional information may be gained by evaluating the raw data and consulting with experts who have information about the area in question. The map of conservation value (Fig. 6.11) was used as the ecological criterion data layer in CI-SSAT.

Results from CI-SSAT

In the Channel Islands process, the working group began their discussion about marine zoning with support of the CI-SSAT. The working group did not consider the locations of marine reserves during their first 12 months (July 1999 – Sept 2000). During the initial period, the working group considered the state of the marine ecosystem and goals for marine reserves. In May 2000, the science advisory panel delivered a suite of ecological goals for the design of marine reserves to meet goals of the working group. In August 2000, the science advisory panel delivered the initial results of the SPEXAN modeling process, including the map of conservation value from the irreplaceability analysis (Fig. 6.11) and 10 alternatives for marine reserves that achieved conservation goals (five of which are depicted in Fig. 6.10). In September 2000, the working group began deliberations about potential locations for marine reserves in the Channel Islands.

The working group met with the advisory panels to consider the ecological and socioeconomic data and begin the process of designing a network of marine reserves for the Channel Islands National Marine Sanctuary. After reviewing information provided by the advisory panels, the working group divided into four subgroups. Each subgroup worked with a technical facilitator to access and query data using CI-SSAT. Most subgroups used pre-set queries in CI-SSAT to understand the potential impacts of their decisions.

Nine scenarios were produced at the conclusion of the September 2000 workshop. The scenarios ranged from 11 to 38% of the Channel Islands National Marine Sanctuary. Figure 6.12 shows the number of times each planning unit was included in one of the nine scenarios developed by the working group. All groups agreed that a reserve should be established in the waters on the northwest side of San Miguel Island. All but one scenario included a reserve around Gull Island off the southwest coast of Santa Cruz Island. Seven of nine scenarios included reserves off the southwest side of Santa Barbara Island and at the Footprint (offshore between Santa Cruz and Anacapa Islands). Less than half of the working group agreed that reserves should be located on the north side of Santa Rosa Island, at Scorpion Rock on the northeast side of Santa Cruz Island, and off the northern coast of Anacapa Island.

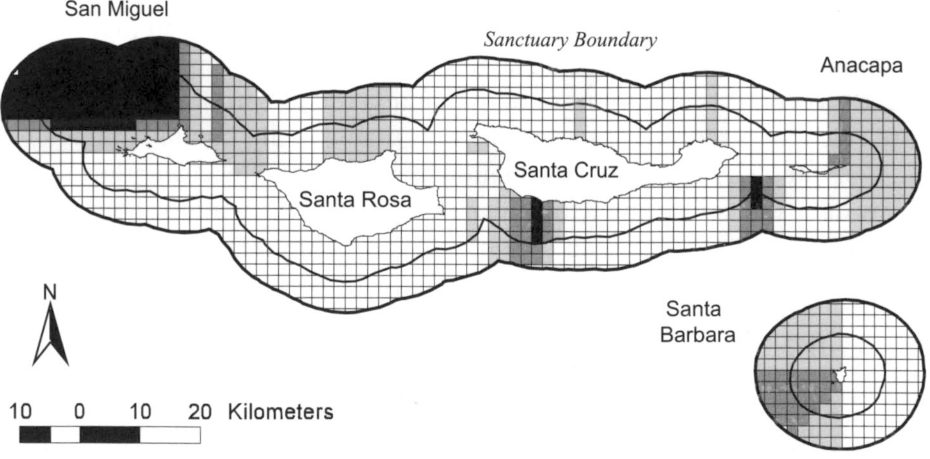

Figure 6.12. Results of the design workshop in September 2000. Nine scenarios were developed by the working group. Shading represents the number of scenarios that included each planning unit. Dark shaded areas were included in most of the scenarios whereas light shaded areas were not.

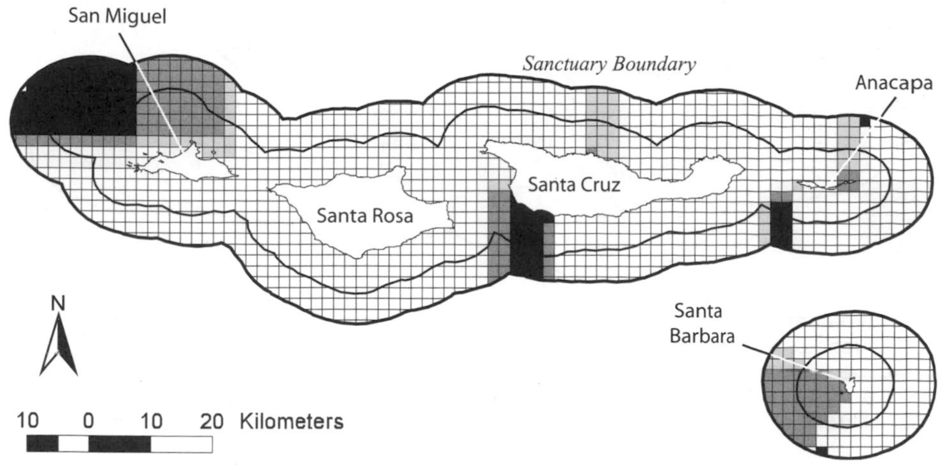

Figure 6.13. Results of the design workshop in October 2000. Five scenarios were developed by the working group. Shading represents the number of scenarios that included each planning unit. Dark shaded areas were included in most of the scenarios whereas light shaded areas were not.

Detailed ecological and economic analyses were provided to the working group, including the amount of each conservation target represented in each scenario and the potential impact of each scenario on commercial and recreational activities.

In October 2000, the working group convened to revise and refine the reserve scenarios. To vary the approach, the working group divided into five subgroups representing similar perspectives and each subgroup produced a single alternative reflecting the common views of the subgroup. From this exercise, the working group produced five alternatives representing the interests of commercial fishing, the environment, recreational users, government, and community. The working group achieved a greater degree of consistency through this exercise. Figure 6.13 shows the number of times each planning unit was included in one of the five scenarios developed by the working group in October 2000. Most groups agreed that reserves should be located in waters off the north side of San Miguel Island, around Gull Island off the southwest coast of Santa Cruz Island, at the Footprint, between Santa Cruz and Anacapa islands, and off the west coast of Santa Barbara Island. Three of five groups agreed that reserves should be located off the north coast of Anacapa Island whereas the other groups suggested placing reserves off the southeastern corner of the island. Detailed ecological and socioeconomic analyses were provided to the working group in November 2000.

With assistance from the advisory panels, the working group continued to refine the alternatives during subsequent meetings. By May 2001, the working group had developed over forty different designs for marine zoning and evaluated the ecological value and potential economic impact of each design. The working group selected two designs

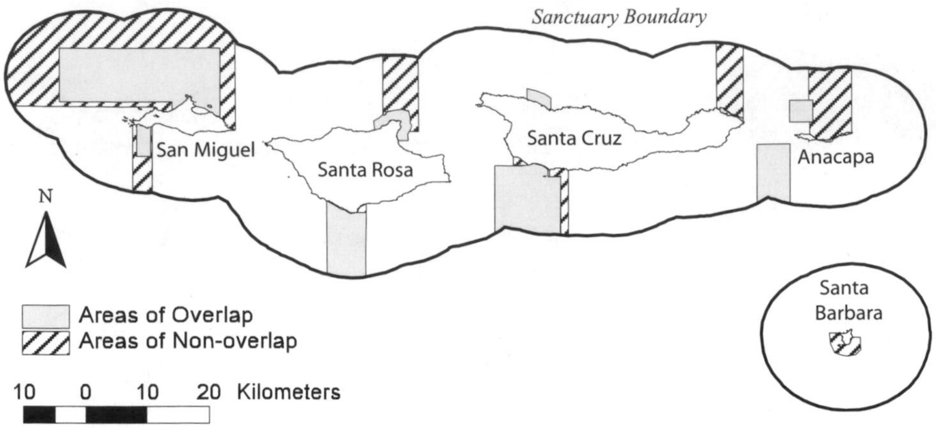

Figure 6.14. Two designs for networks of marine reserves developed by the working group using CI-SSAT. The solid polygons represent areas that all group members agreed to set aside in marine reserves. The hatched marks represent additional areas that some members of the group considered essential to meet conservation goals.

to represent the diverse views of the group (Fig. 6.14). All members of the working group agreed that some areas (known as the "areas of overlap") should be set aside in marine reserves. Some members of the working group did not agree that the areas of overlap, which totaled 12% of the Sanctuary, were sufficient to meet the conservation goals established by the working group. A second set of areas was proposed (known as "areas of non-overlap") to satisfy the concerns by some members of the working group that the areas of overlap were not sufficient to meet their goals. The combination of the areas of overlap and non-overlap, totaling 28% of the Sanctuary, formed the network of marine protected areas proposed by some members of the working group.

These maps, together with policy, scientific, and economic information, were provided to state and federal agencies for consideration in June 2001. During the summer of 2001, the California Department of Fish and Game and the Channel Islands National Marine Sanctuary worked together to develop a compromise between the two maps. The proposed reserve area on the north side of San Miguel Island was divided into two smaller areas to protect specific targets at Harris Point and Richardson Rock. The southern boundaries of proposed reserves on the south sides of Santa Rosa and San Miguel islands were moved two and three nautical miles, respectively, inshore to reduce impacts to trawling and trapping industries. A small patch of coastline and shallow subtidal waters were removed from the proposed reserves to accommodate consumptive recreational divers. The boundary of the reserve proposed at Carrington Point on the north side of Santa Rosa Island was constrained within three nautical miles of shore to avoid conflicts with offshore set and gill net fisheries. The boundaries of the

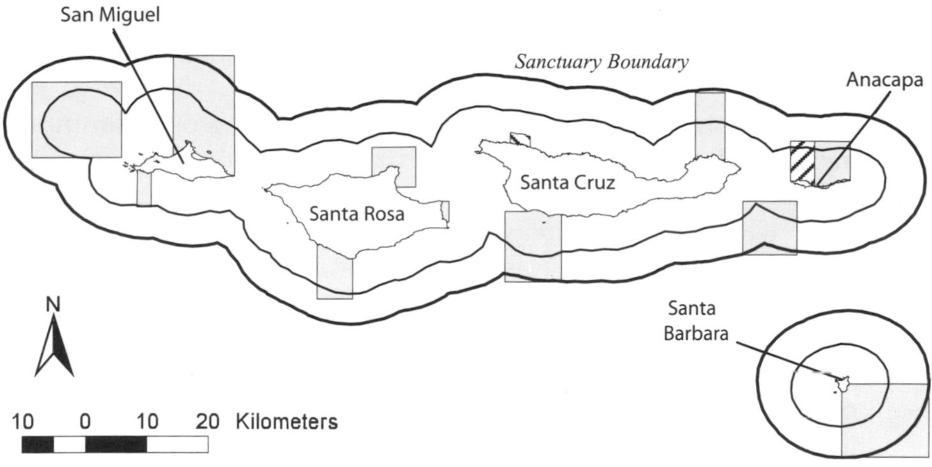

Figure 6.15. The design for a network of marine protected areas developed by the California Department of Fish and Game and the Channel Islands National Marine Sanctuary. The design is based on the two maps provided by the working group and the ecological and socioeconomic data gathered during the Channel Islands process. Solid shaded polygons represent no-take marine reserves; hatched polygons represent marine parks and conservation areas where limited commercial and recreational fishing is allowed.

proposed reserve at Scorpion Rock on the northeast side of Santa Cruz Island were moved west to alleviate pressure on recreational fishers and divers. The west side of the proposed reserves on north Anacapa Island and at Painted Cave on the northwest side of Santa Cruz Island were opened to limited commercial and recreational fishing to reduce potential impacts to recreational fishers and commercial lobster fishers. The resulting network of marine reserves (Fig. 6.15) was a compromise between the perspectives represented on the working group.

In October 2002, the California Fish and Game Commission took action to protect marine ecosystems at the Channel Islands using network design based on the extensive information generated by the working group and advisory panels. Because its jurisdiction is limited to waters between the mean high tide and three nautical miles offshore, the Commission was not able to implement the entire network of marine protected areas. The Commission considered the proposed areas within state waters only. In April 2003, a network of state marine protected areas was established in the Channel Islands National Marine Sanctuary (Fig. 6.1).

Discussion and Conclusion

Geospatial modeling tools contributed substantially to the process of designing "no-take" marine reserves for the Channel Islands National Marine Sanctuary. SPEXAN was used by the science advisory panel to evaluate ecological information and provide a set of solutions to meet conservation goals established by the working group. CI-SSAT was used by the working group to view and query ecological and socioeconomic information and to evaluate the solutions produced during the SPEXAN analysis.

Both tools are displayed within a GIS, providing opportunities to view, display, and manipulate spatial information. During public meetings of the working group and advisory panels, both geospatial tools were used to display data and solutions (or alternatives). Information in the tools was projected onto a screen and members of the working group or advisory panels were able to discuss the information and make changes while others observed the process. The science advisory panel used their public meetings to review the effects of changing the boundary length modifier and the conservation goals in SPEXAN. Consideration of the results led the science advisory panel to recommended use of a boundary length modifier of 1 and conservation goals between 30 and 50%. Additionally, the science advisory panel reviewed results from SPEXAN to refine the data classification system, providing flexibility and repeatability of the analysis. The working group used CI-SSAT to view and query data in public meetings. Additionally, the working group generated, reviewed, and modified alternative designs for marine reserves, while the designs were projected on a screen in a meeting room.

The use of geospatial modeling tools bolstered public confidence in the reserve design process for the Channel Islands. In the case of SPEXAN, the tool provided the flexibility needed to address policy concerns within the framework of an analytical process that was repeatable and rigorous. Using simulated annealing, the science advisory panel explored the process of designing a network of marine reserves by considering the level of detail of habitat classification, the type of data in the analysis, and the overall conservation goals. After the working group established the overarching goals for the Channel Islands process, the science advisory panel developed a corresponding set of conservation targets to achieve the working group's goals. The flexibility in the model allowed policy-makers and scientists to evaluate the effects of different types of data, classification schemes, and goals on the model output.

Simulated annealing produced many good solutions that achieved conservation goals and minimized the area and perimeter of the network. More compact solutions, in spite of their greater area, were selected for consideration because of the relative ease of implementation, enforcement, and monitoring. The alternative solutions were particularly useful in the Channel Islands process because of the flexibility introduced to the discussion about where reserves should be located. From the SPEXAN analysis, it was clear that various configurations of marine reserves could satisfy the conservation goals. Given the range of solutions, the working group was able to identify constructive alternatives to establishing reserves in areas of high conflict among working group members.

The map of conservation values, generated from the irreplaceability analysis in SPEXAN, was particularly valuable for advancing discussions about marine zoning. Biodiversity "hot spots" were identified as planning units selected for a large number of solutions. In the Channel Islands process, the map of conservation values (Fig. 6.11) provided the foundation for discussions about reserve design. The network of state marine protected areas, established in April 2003, includes many of the hot spots identified from the map of conservation values.

The ecological and socioeconomic information in CI-SSAT advanced the reserve design process in the Channel Islands. Working group members wanted each conservation target to be represented in the network of reserves with a minimum impact to commercial and recreational users. Working group members drew potential reserves in CI-SSAT and used the built-in queries to investigate the potential benefits and impacts of the reserves. Before a design concept became a feasible alternative, adjustments were made to reduce potential impacts to commercial and recreational users and incorporate conservation targets. CI-SSAT supported this process of exploration of the data, thus facilitating the development of reserve designs.

In spite of the vast amount of information provided, the working group was unable to come to consensus on the size and location of potential reserves for the Channel Islands. One of the goals of the Channel Islands process was to bring together a diverse group of stakeholders and work together to develop a consensus view of the management needs, based on shared information. This approach was used by the Florida Keys National Marine Sanctuary during discussions that led to the creation of the Tortugas Marine Ecological Reserve (Florida Keys National Marine Sanctuary, 2000). In the Florida Keys process, the working group agreed to set aside a reserve that was outside the study area. In the Channel Islands process, the working group agreed on the problem and the goals of the process, but was unable to come to consensus on the size and location of reserves within the study area.

From the numerous designs generated during the Channel Islands process, it is clear that there was some agreement on locations of reserves, particularly at San Miguel Island, at Gull Island off the southwest coast of Santa Cruz Island, and at the Footprint area between Santa Cruz and Anacapa islands. Consensus eluded the Channel Islands process, in part, because diverse views represented on the working group reached their limits of acceptable compromise before a final solution was developed. A majority of working group members was willing to accept the combined areas of overlap and non-overlap (Fig. 6.14) as a network of reserves. However, a few minority views on both sides of the debate prevented the group from reaching full consensus.

In such complex management problems, the lofty goal of consensus may not be a realistic target. To reach a solution, participants must either adjust their expectations to be satisfied with a compromise or make explicit policy decisions *a priori* to weight the contributions of different interests.

Acknowledgments

The work reported here was conducted as part of the marine reserves process supported by the Channel Islands National Marine Sanctuary (CINMS) and the California Department of Fish and Game. The use of SPEXAN was made possible with assistance from H. Possingham and I. Ball. The use of the Channel Islands Spatial Analysis and Support Tool was made possible by NOAA's Coastal Services Center with assistance from D. Killpack. Special thanks to the marine scientists who donated their time and expertise to advise the CINMS Marine Reserves Working Group: M. Cahn, M. Carr, E. Dever., S. Gaines, P. Haaker, B. Kendall, M. Love, S. Murray, D. Reed, D. Richards, J. Roughgarden, D. Schroeder, S. Schroeter, D. Siegel, A. Stewart-Oaten, L. Washburn, and R. Vetter. Socioeconomic data and analyses were provided by the socioeconomic advisory team lead by V. B. Leeworthy and P. Wiley, and supported by C. Barilotti, M. Hunter, C. Kolstad, M. Kronman, and C. Pomeroy. In addition, I would like to extend my thanks to B. Kinlan for his thoughtful comments on this manuscript.

Notes

1 Fisheries permitted within the state marine conservation areas include recreational fishing for pelagic species and recreational and commercial fishing for California spiny lobster (*Panulirus interruptus*).

2 In April 1998, the California Fish and Game Commission received a recommendation from the Channel Islands Marine Resources Restoration Committee to set aside 20% of the shoreline and waters out to 1 mile. The recommendation was criticized as the work of one stakeholder group representing a narrow range of perspectives. A public process was recommended by the California Fish and Game Commission to resolve the conflicts between stakeholders.

3 The geospatial tool SPEXAN (SPatially EXplicit ANnealing) was developed for the Nature Conservancy for the purpose of locating terrestrial reserves. Later, the geospatial tool was modified to more directly reflect biophysical principles and the new tool was named "MARXAN" (Ball and Possingham, 2000). A description and the program are available at the MARXAN Web site.

References

Airamé, S., Cassano, E., Pickett, M., Fangman, S., Hastings, S., Bingham, S., Walton, A., Waltenberger, B., Murray, M., Simon, M., Woodall, R., and Ugoretz, J., 2000. Species of Interest in the Channel Islands: For consideration by the Marine Reserves Working Group, Santa Barbara, California, Channel Islands National Marine Sanctuary.

Airamé, S., Dugan, J. E., Lafferty, K. D., Leslie, H., McArdle, D. A., and Warner, R. R., 2003. Applying ecological criteria to marine reserve design: a case study from the California Channel Islands, *Ecological Applications,* 13(1): Supplement S170-85.

Allison, G. W., Gaines, S. D., Lubchenco, J., and Possingham, H. P., 2003. Measuring persistence of marine reserves: Catastrophes require adopting an insurance factor, *Ecological Applications,* 13(1): Supplement 8-24.

Ardron, J. A., Lash, J., and Haggarty, D., 2002. Modelling a network of marine protected areas for the central coast of British Columbia. Version 3.1, Sointula, British Columbia, Canada, Living Oceans Society.

Ball, I. R., and Possingham, H. P., 2000. MARXAN Manual. Adelaide, South Australia. Adelaide University. http://www.ecology.uq.edu.au/marxan.htm.

Beck, M. W., and Odaya, M., 2001. Ecoregional planning in marine environments: identifying priority sites for conservation in the northern Gulf of Mexico, *Aquatic Conservation: Marine and Freshwater Ecosystems,* 11: 235-42.

California Department of Fish and Game, 2002. Final Environmental Document. Marine Protected Areas in NOAA's Channel Islands National Marine Sanctuary, Volume 1, October 2002, State of California, The Resources Agency, Department of Fish and Game.

Clarke, K. R., and Warwick, R. M., 2001. Change in Marine Communities: An Approach to Statistical Analysis and Interpretation, 2nd edition, Plymouth, PRIMER-E Ltd.

DeLong, R. L., and Melin, S. R., 1999. Thirty years of pinniped research at San Miguel Island. Alolkoy, The Publication of the Channel Islands Marine Sanctuary Foundation, 12(2): 3.

Florida Keys National Marine Sanctuary, 2000. Tortugas Ecological Reserve, Final Supplemental Environmental Impact Statement/Draft Supplemental Management Plan, U.S. Department of Commerce, NOAA/NOS/Office of Ocean and Coastal Resource Management/ Marine Sanctuaries Division.

Jostes, J. C., and Eng, M., 2001. Facilitator's Report: Regarding the Channel Islands National Marine Sanctuary Marine Reserves Working Group, Santa Barbara, California, Channel Islands National Marine Sanctuary Advisory Council, 27 pp.

Killpack, D., Waltenberger, B., and Fowler, C., 2000. Using the Channel Islands-Spatial Support and Analysis Tool to Support Group-based Decision Making for Marine Reserves, Charleston, South Carolina, NOAA Coastal Services Center, http://www.csc.noaa.gov/pagis/html/ web_cissat_paper.pdf

Kronman, M., Airamé, S., and Simon, M., 2000. Channel Islands National Marine Sanctuary Ethnographic Data Survey, Santa Barbara, California, Channel Islands National Marine Sanctuary.

Leeworthy, V. B., and Wiley, P., 2002. Socioeconomic Impact Analysis of Marine Reserve Alternatives for the Channel Islands National Marine Sanctuary, Silver Spring, MD, U.S. Department of Commerce, NOAA National Ocean Service, Special Projects.

Leet, W. S., Dewees, C. M., Klingbeil, R., and Larson, E. J. (eds.), 2001. California's Living Marine Resources: A Status Report, State of California. The Resources Agency, The California Department of Fish and Game.

Leslie, H., Ruckelshaus, M., Ball, I. R., Andelman, S., and Possingham, H. P., 2003. Using siting algorithms in the design of marine reserves, *Ecological Applications* , 13(1): Supplement 185-98.

Lewis, A., Slegers, S., Lowe, D., Muller, L., Fernandes, L., and Day, J., 2003. Use of Spatial Analysis and GIS techniques to Re-Zone the Great Barrier Reef Marine Park, Coastal GIS Workshop, July 7-8, 2003, University of Wollongong, Australia.

Margules, C. R., and Pressey, R. L., 2000. Systematic Conservation Planning. *Nature* 405: 243-53.

McDonnell, M. D., Possingham, H. P., Ball, I. R. and Cousins, E. A., 2002. Mathematical methods for spatially cohesive reserve design, *Environmental Modelling and Assessment*, 7:107-14.

Possingham, H., Ball, I. and Andleman, S., 2000. Mathematical models for identifying representative reserve networks, in Ferson, S., and Burgman, M. (eds.), *Quantitative Methods for Conservation Biology*, New York, Springer-Verlag, 291-306.

Pressey, R. L., Possingham, H. P., and Margules, C. R., 1996. Optimality in reserve selection algorithms: When does it matter and how much? *Biological Conservation*, 76:259-67.

Roberts, C. M., Andelman, S., Branch, G., Bustamante, R. H., Castilla, J. C., Dugan, J., Halpern, B. S., Lafferty, K. D., Leslie, H., Lubchenco, J., McArdle, D., Possingham, H. P., Ruckelshaus, M., and Warner, R. R., 2003a. Ecological criteria for evaluating candidate sites for marine reserves, *Ecological Applications*, 13(1): Supplement 199-214.

Roberts, C. M., Andelman, S., Branch, G., Bustamante, R. H., Castilla, J. C., Dugan, J., Halpern, B. S., Lafferty, K. D., Leslie, H., Lubchenco, J., McArdle, D., Possingham, H. P., Ruckelshaus, M., and Warner, R. R., 2003b. Application of ecological criteria in selecting marine reserves and developing reserve networks, *Ecological Applications*, 13(1): Supplement 215-28.

Sala, E., Aburto-Oropeza, O., Paredes, G., Parra, I., Barrera, J. C., and Dayton, P. K., 2002. A general model for designing networks of marine reserves, *Science*, 298: 1991-93.

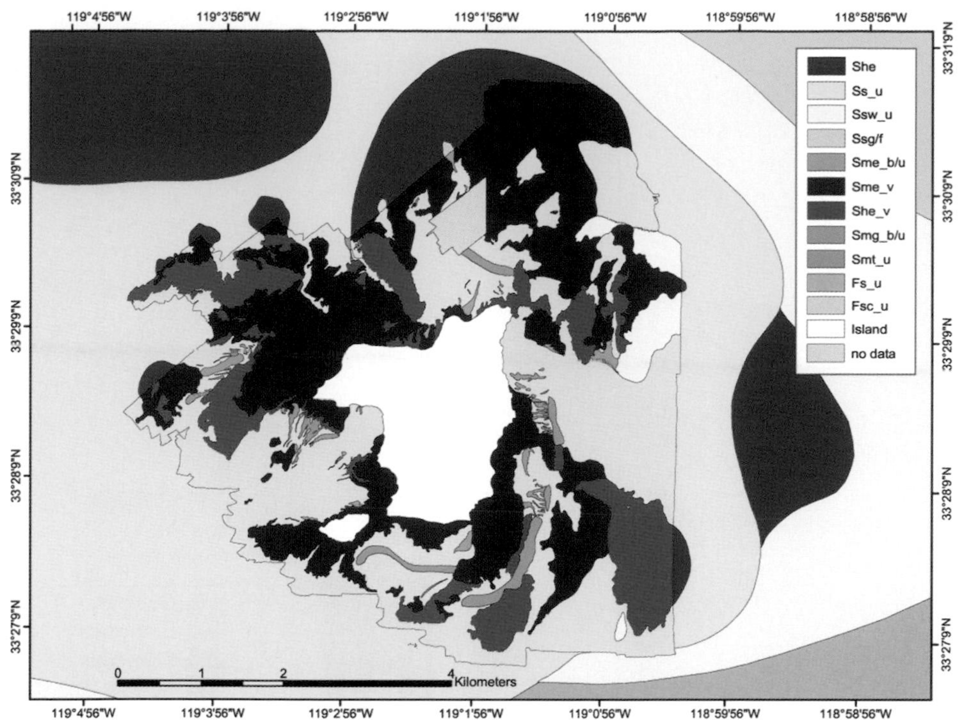

Figure 3.9. Merged map of potential habitats interpreted from Reson 8101 (240 kHz) multibeam data (Fig. 3.8) and previously mapped geologic data (Fig. 3.7) collected around Santa Barbara Island, southern California. Refer to Appendices 3.1 and 3.2 for explanation of habitat code.

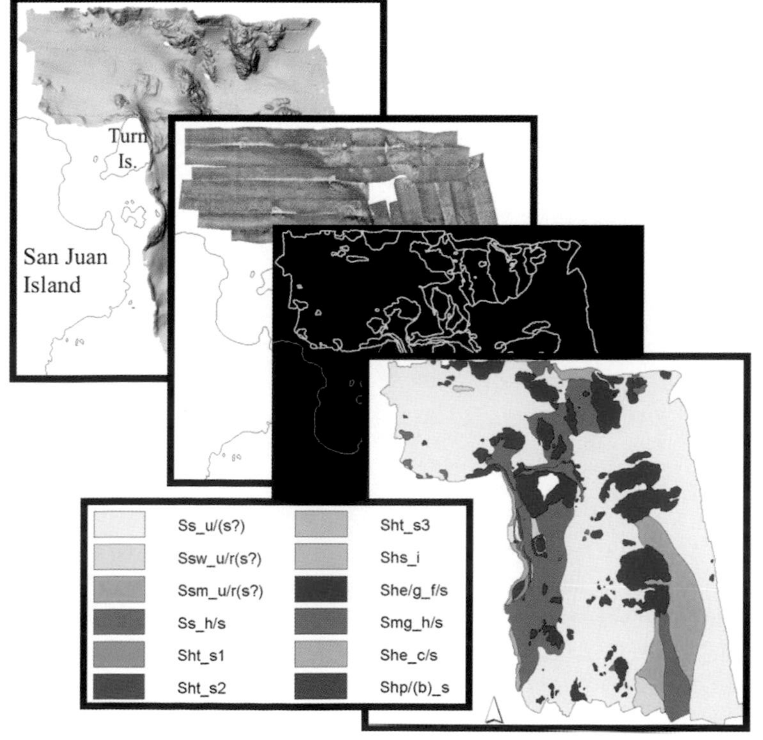

Figure 3.10. Example of nested data layers used in the compilation and construction of maps of potential marine benthic habitats. Locality is San Juan Channel offshore of San Juan Island, Washington. 3.10a = multibeam bathymetry; 3.10b = multibeam backscatter; 3.10c = line drawing of interpreted habitat types; 3.10d= potential habitat polygons. Refer to Appendix 3.1 for explanation of habitat code.

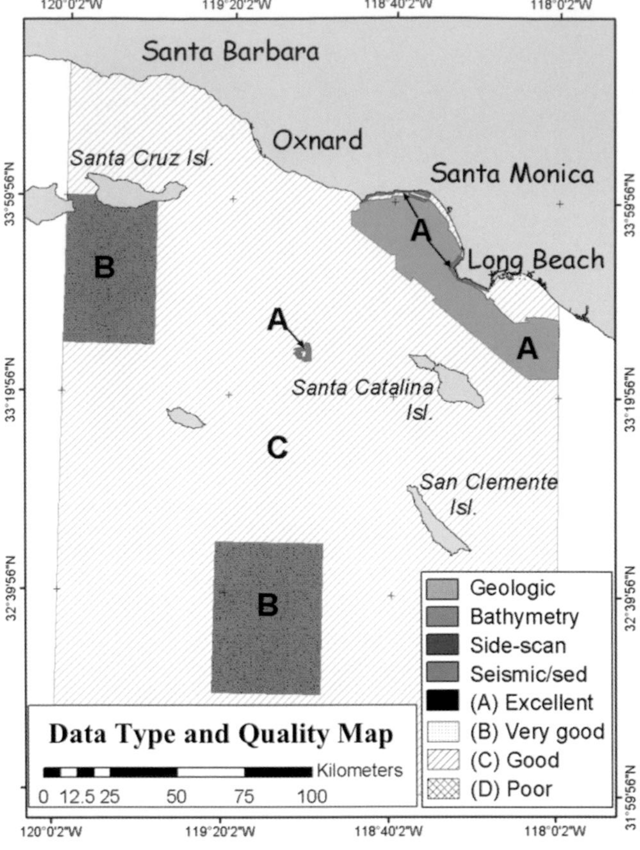

Figure 3.11. Example of data quality map showing type, quality, and coverage of data used in the compilation and construction of a map of potential habitats off southern California.

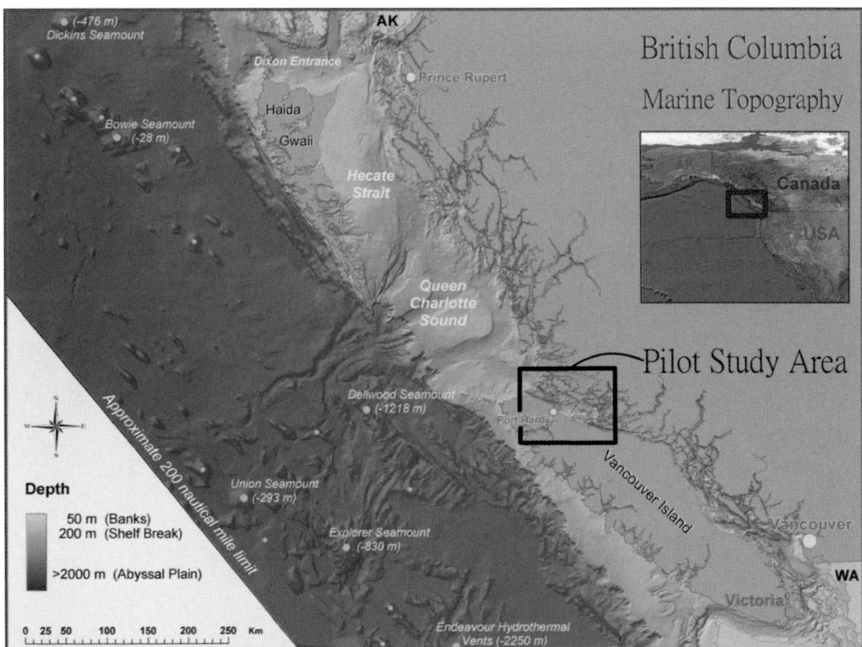

Figure 4.1. The study area is part of the southern Central Coast of British Columbia. It is approximately the extent of DFO Statistical Area 12. For our analysis, we excluded the inlets. The remaining sea area is approximately 2,400 square kilometres (931 sq. mi; 703 sq. nautical miles).

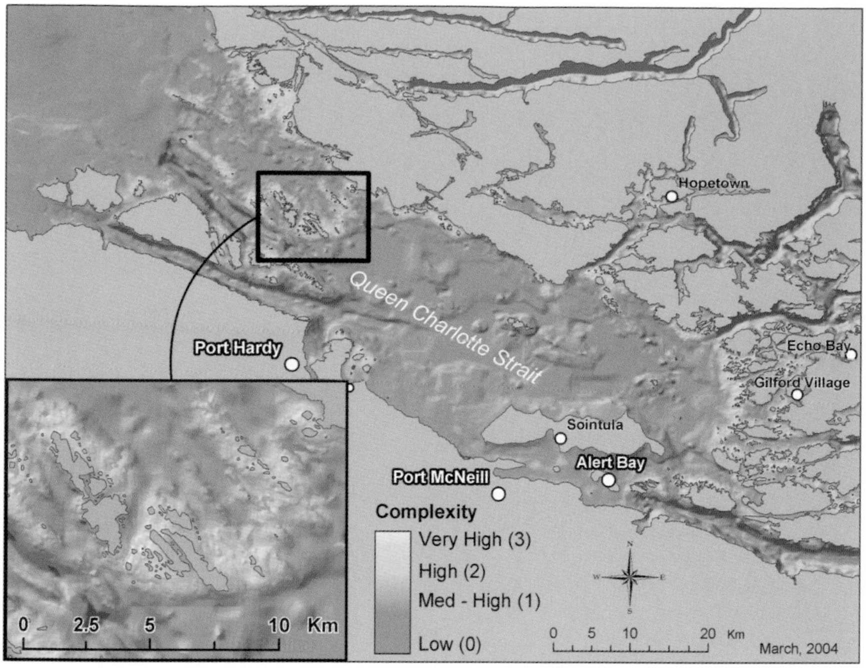

Figure 4.2. Benthic topographical complexity is a measure of how convoluted the sea floor is per given unit area. At this scale, it is a good method for identifying rocky reef habitat.

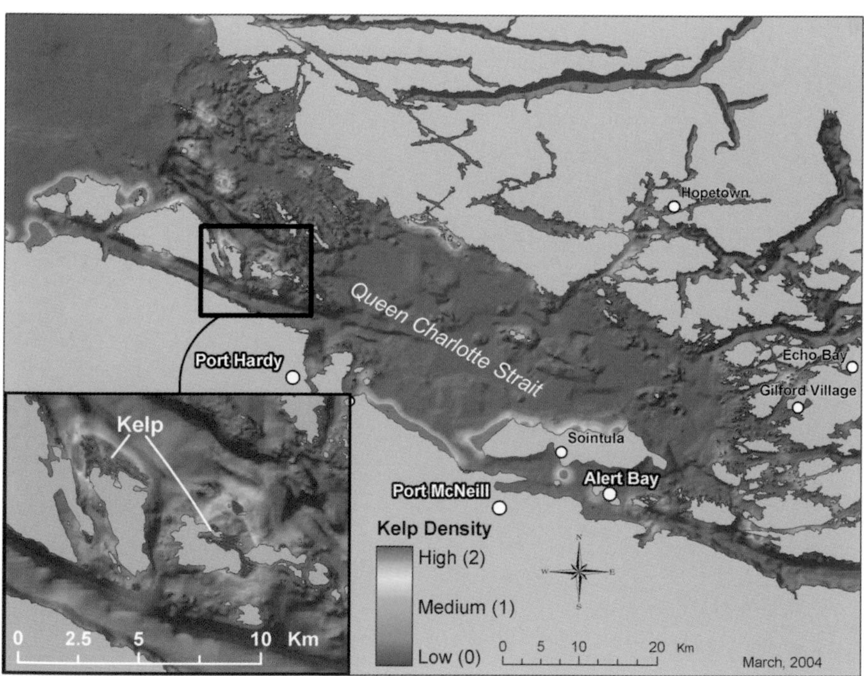

Figure 4.3. Kelp beds (*Nereocystis luetkeana* and *Macrocystis intergrifolia*) occur in varying sizes and distributions. We wanted to model proximity to complex areas, but realized that a buffer would exaggerate the influence of small beds. Thus, we performed a density analysis, and re-classed the results into either high (2) or medium (1).

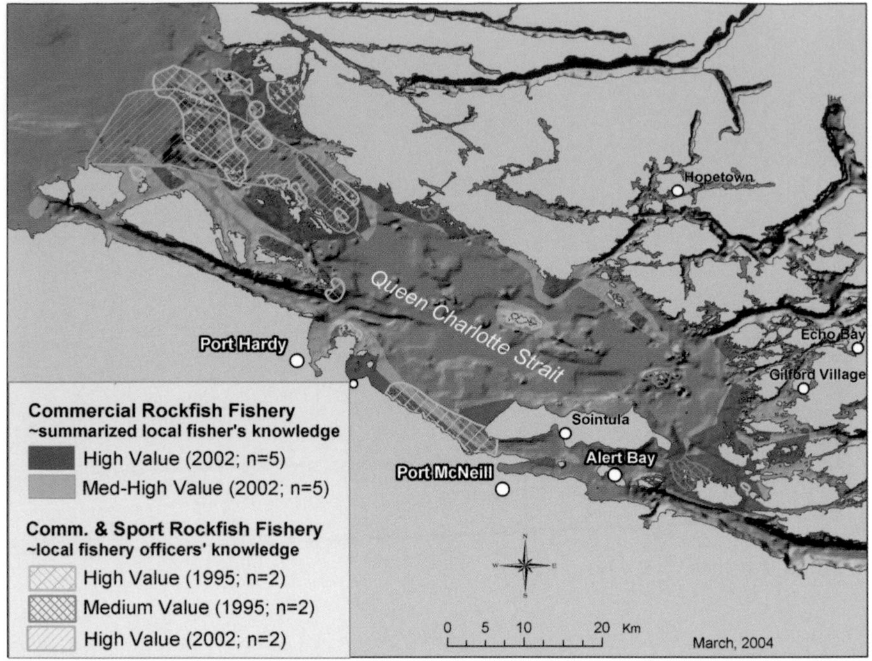

Figure 4.4. Known as "Area 12," Queen Charlotte Strait has several areas where commercial and recreational rockfish fishing occur. While there are areas of overlap, there are also discrepancies between the fishers' knowledge and that of the fishery officers and managers.

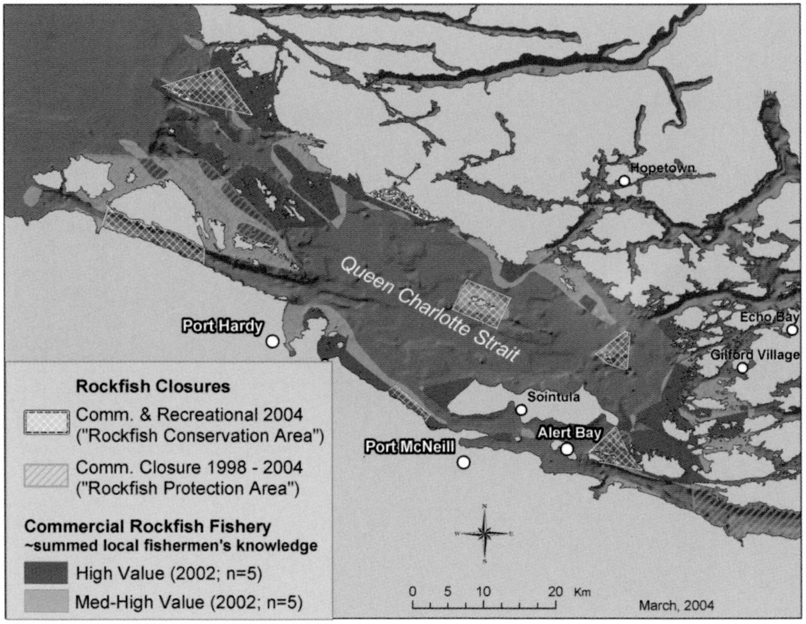

Figure 4.5. Previous experience with the commercial closures (orange hatching) had indicated that while some areas were known to harbour rockfish, others were not. When the commercial and recreational Rockfish Conservation Areas (white hatching) were tentatively first announced in 2003, it again appeared to be a portfolio of possibly mixed results. This prompted our analysis.

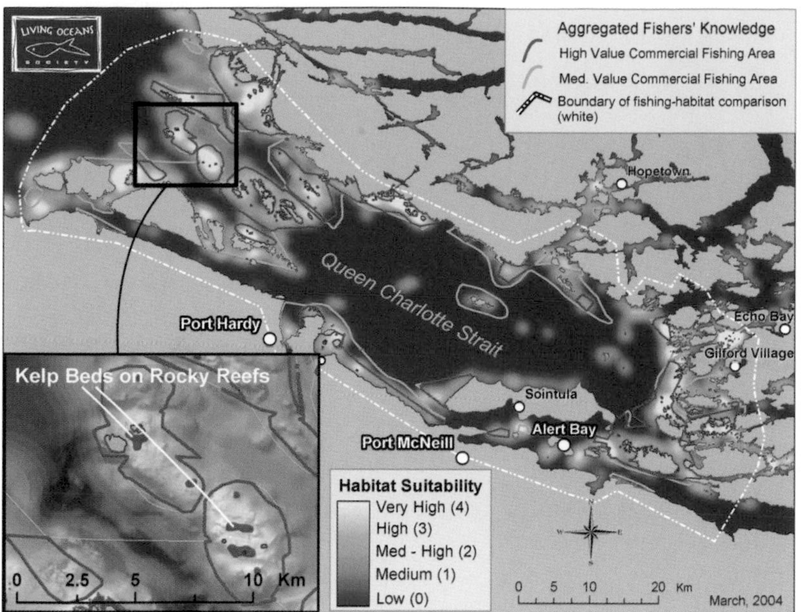

Figure 4.6. The model had a high degree of overlap with areas identified by commercial fishers as being important fishing areas. Note that the commercial fleet rarely fishes in the inlets. Thus to keep the comparison meaningful, we excluded the inlets (white dashed line). With only a few small exceptions, every high-value fishing area in the study area contains high or very high-value habitat identified by our model. While not every high-value habitat area is accompanied by a high-value fishing area, local anecdotal knowledge would indicate that most of these areas are also known fishing areas. (We are currently involved in expanding the data collection to include recreational fishers as well as more commercial fishers.)

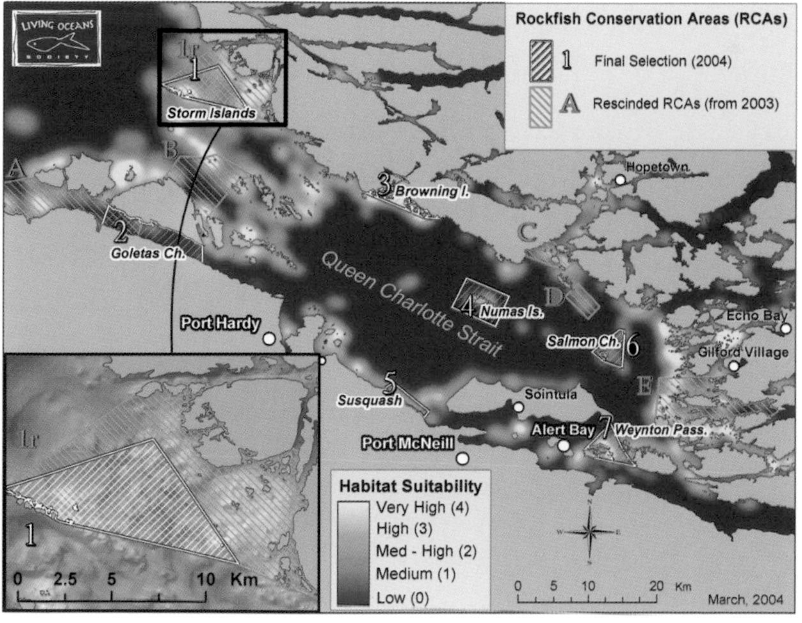

Figure 4.9. A number of rockfish conservation areas were initially proposed within the study area. Of those, some were finalized (black & white) and others were rescinded (orange). As can be seen, these RCAs overlap rockfish habitat to varying degrees, with the rescinded areas often being in what the model predicts to be quite good habitat (e.g., C, E, and 1r – inset).

Figure 5.6. Areas where trawl activity would likely not occur. Untrawlable tow paths (red arcs) are removed.

Figure 5.7. Removing tow paths in steep terrain (slopes greater than 1%). Red arcs are removed from analysis.

Figure 5.10. (a) Catch per area in the Monterey Bay National Marine Sanctuary.

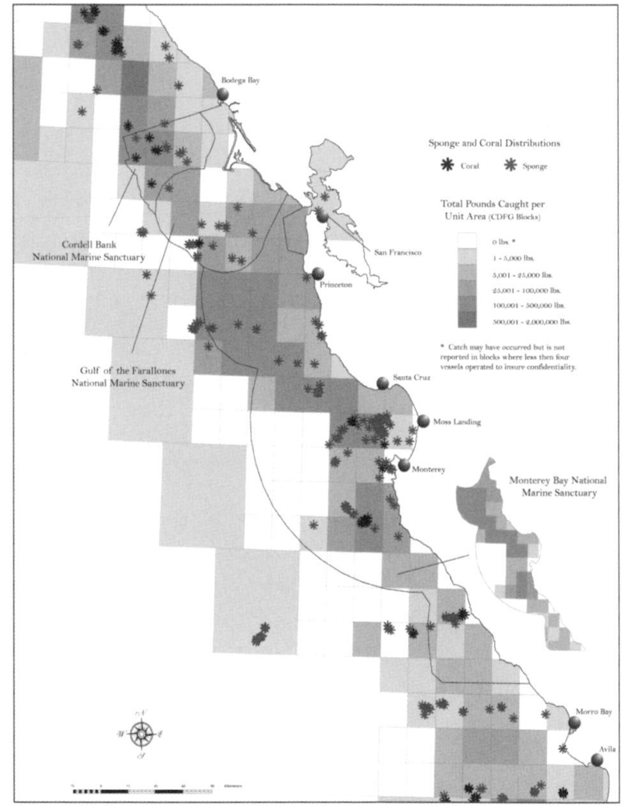

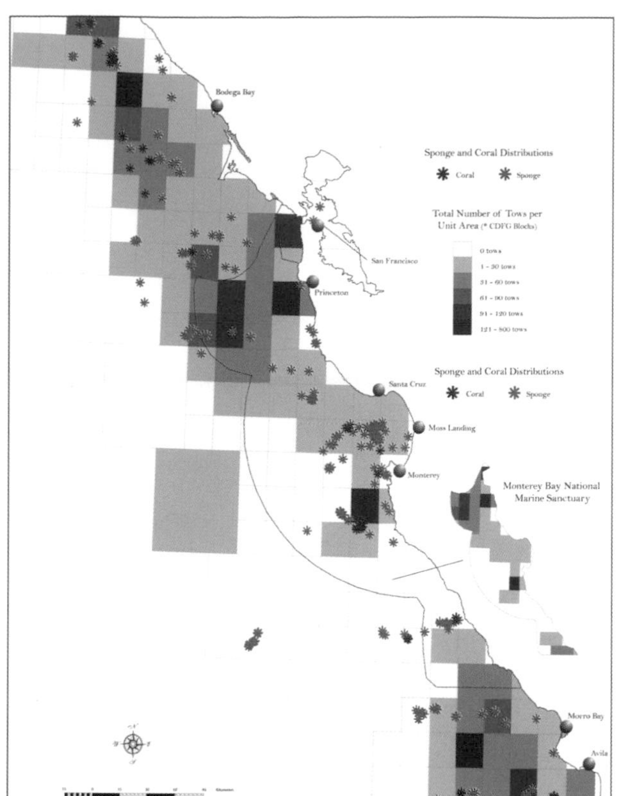

Figure 5.10. (b) Number of trawl tows in the Monterey Bay National Marine Sanctuary.

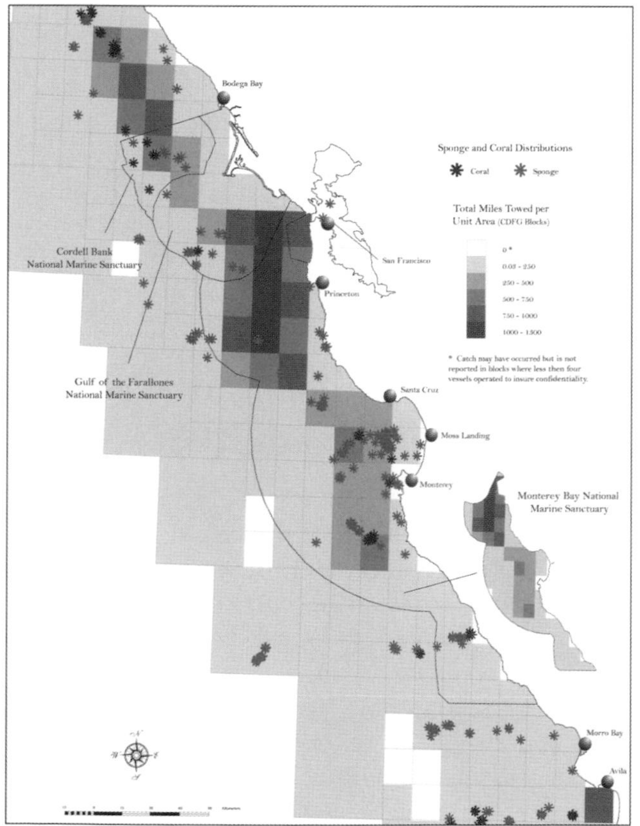

Figure 5.10. (c) Total miles towed in the Monterey Bay National Marine Sanctuary.

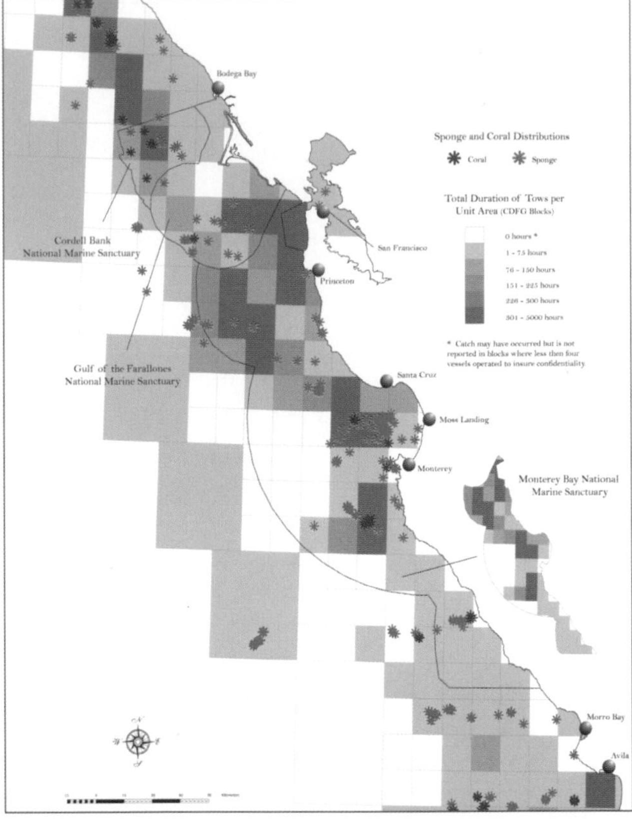

Figure 5.10. (d) Duration of tows in the Monterey Bay National Marine Sanctuary.

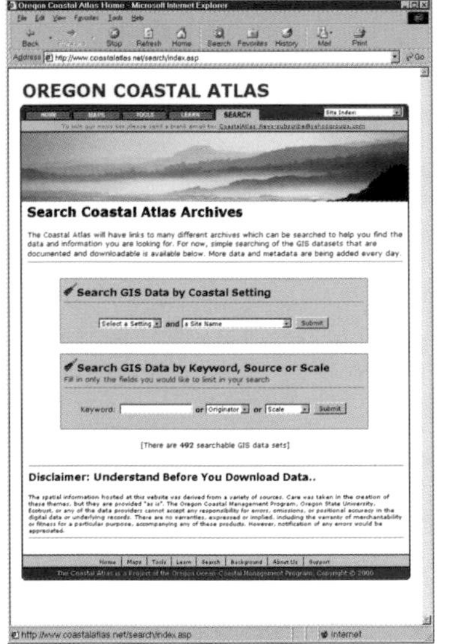

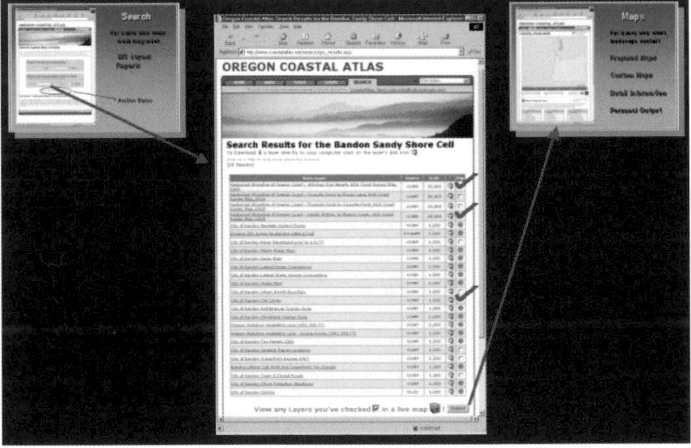

Figure 7.1. Searching for metadata and data in the Oregon Coastal Atlas. (a) Screen of the Search tool inviting the user to search either by coastal setting, keyword, source of the data (by agency), or map scale. At the time of publication, the Atlas held nearly 1,100 searchable datasets. (b) Search results are provided in list form with links to the respective metadata and download files. The user may also view the dataset in map form before downloading. For most datasets, a link to the Atlas interactive map interface is provided so that the data can be browsed via the built-in Internet map server.

Chapter 7

The Tools of the Oregon Coastal Atlas

Tanya Haddad, Dawn Wright, Michele Dailey, Paul Klarin, John Marra, Randy Dana, David Revell

Abstract

The Oregon Coastal Atlas (http://www.coastalatlas.net), a collaboration of the Oregon Coastal Management Program, Oregon State University, and Ecotrust, is an interactive map, data, and metadata portal targeted at coastal managers, scientists, and the general public. The site was developed to meet long-standing needs in the state for improving information retrieval, visualization and interpretation for decision-making relating to the coastal zone. It has the ambitious goal of being a useful resource for the various audiences that make up the management constituency of the Oregon coastal zone. The site provides background information for different coastal systems, as well as the expected access to interactive mapping, online geospatial analysis tools, and direct download access

Tanya Haddad and **Randy Dana**
Oregon Ocean-Coastal Management Program, Department of Land Conservation and Development, 800 NE Oregon St # 18, Suite 1145, Portland, OR 97232
Corresponding author: tanya.haddad@state.or.us

Dawn Wright
Department of Geosciences, Oregon State University, Corvallis, OR 97331-5506

Michele Dailey
Ecotrust, Jean Vollum Natural Capital Center, Suite 200, 721 NW Ninth Avenue, Portland, OR 97209; now at Portland Metro, 600 NE Grand Ave., Portland, OR 97232-2736

Paul Klarin
Oregon Ocean-Coastal Management Program, Department of Land Conservation and Development, 635 Capitol St NE, Suite 150, Salem, OR 97301

John Marra
Oregon Ocean-Coastal Management Program, Department of Land Conservation and Development, 800 NE Oregon St # 18, Suite 1145, Portland, OR 97232; now at NOAA Pacific Services Center, Honolulu, HI

David Revell
Oregon Ocean-Coastal Management Program, Department of Land Conservation and Development, 800 NE Oregon St # 18, Suite 1145, Portland, OR 97232; now at Earth Sciences Department, University of California, Santa Cruz, CA

to an array of natural resource datasets relating to coastal zone management. Therefore, as a portal, the Oregon Coastal Atlas enables users to search and find data, but also to understand its original context, and put it to use via online tools in order to solve a spatial problem. This chapter describes, in detail, the various online tools of the Atlas, most of which can be used online, some of which need to be downloaded to the user's desktop, some that are used for viewing data, some that are used for geospatial analysis; and tools for a diverse audience– some best suited for coastal planners and researchers, some for contributors to the Atlas, and some for the general public.

Background

The Oregon Coastal Atlas (OCA), a collaboration of the Oregon Ocean-Coastal Management Program, Oregon State University, and Ecotrust, is an interactive map, data, and metadata portal for coastal resources managers and scientists, with additional outreach sections for the general public. The portal enables users to obtain data, but also to understand their original context, and to use them for solving a spatial problem via online tools. The design of the Atlas draws from the reality that resource decision-making applications require much more than simple access to data. Resource managers commonly make decisions that involve modeling risk, assessing cumulative impacts, and weighing proposed alterations to ecosystem functions and values. These decisions involve pulling together knowledge from disparate disciplines such as biology, geology, oceanography, hydrology, chemistry, and engineering. Practitioners within each one of these disciplines are often vested in the technologies that dominate the market within their particular field. This presents significant data integration difficulties for investigators involved in management decisions that are as inherently interdisciplinary as those in the coastal zone. The goal of our proposed effort is to address these problems by incorporating a variety of geospatial data and "tools" within a common framework. The OCA was designed to meet long-standing needs in the state of Oregon for improving information retrieval, visualization and interpretation for decision-making relating to the coastal zone, but is also one of a handful of major information gateway portals across the country that contribute to the U.S. Coastal National Spatial Data Infrastructure, and therefore service users from many states and countries (Goss, 2003; Wright et al., 2003; Wright et al., 2004).

Without resorting to a dictionary definition, the OCA project team roughly defines a tool as "something used by someone to accomplish some task." In this loose equation, there are two variables for what ultimately constitutes the functionality of a given tool: the "someone" or audience that must use the tool, and the "task" or problem they are trying to solve or resolve. Of these two variables, the "someone" or audience is of particular importance. Different audiences have different problems, and also different levels of skill or resources available to resolve their problems. Tool providers need to be sensitive to the nature of their audiences, the specifics of the problems those audiences are faced with solving, and the resources and abilities available for resolving them.

Early on in the project, the audiences that make up the constituency of the OCA were characterized as falling into four distinct groups. In no particular order they are: the public, researchers, planners and contributors. Simply by listing the names of these groups one begins to get a sense for the different types of tasks each group might need to accomplish. Not so simply, it was also determined that there would be instances where individuals would be members of more than one audience (e.g., a member of the public who also serves on a volunteer watershed council in a planning role). The task at hand was to build the OCA as a resource that would be of some use to each of these groups. It was to be the "Swiss army knife" of information tools for the Oregon coast.

Open Access Tools and Audience Diversity

As tool providers faced with such a diverse audience, we found that there was no way to simplify or standardize our task. Tools would have to be built audience-specific, and as a result, every tool could not work for every group. In some cases, it would even be necessary to make two versions of the same tool for different audiences. In this chapter, we present a detailed explanation of our audience types and examples of tools created for each, in the process detailing how each tool resolves a common task characteristic of the audience it was created for.

Tools for the Public

Members of the public tend to be interested in or have a stake in learning about places on the coast: places they have visited or are going to visit, places they live or would like to live. Sometimes this interest is general, sometimes specific. For example, coastal residents, both full and part-time, have special knowledge of and vested interest in local resources and economies, and can benefit from having accurate, standardized inventories that raise awareness of local resources that provide community benefits. Coastal residents and homeowners are also at ground zero for risk from a host of hazards that can uniquely threaten life and property on the coast, and can thus also benefit from clear and concise information that clarifies levels of risk, and outlines measures that homeowners can take to minimize danger to people and property. We will present examples of OCA tools that serve both the need for standardized information presentation and for specific information disbursement (Coastal Atlas Maps Tool, Coastal Access Inventory Tool, Marine Visioning Tool).

Tools for Researchers

Researchers tend to be well informed about their particular field of expertise, and are often looking for specific datasets or pieces of information that will inform or be of use in their research activities. They may know or suspect that research has been done in a particular location or on a particular topic and would like easy access to prior data products, published and unpublished results, metadata regarding how, when, and why data were collected, etc. They tend not to require, be interested in, or even be impressed with informational layout per se, but rather with ease of access to the information they are seeking. Certainly the existence of adequate, consistently presented documentation or informational pedigree allows them to determine if what they have found is relevant to the inquiry they are pursuing at hand. We will present examples of two OCA tools that are built to serve up large archives of geospatial data and research products in ways that are quickly and easily searched (Coastal Atlas Geospatial Archive Search Tool and Catalogue of Marine and Coastal Information (COMCI), and one example of a specific research topic tool (Erosion Hazard Tool Suite).

Tools for Planners

Planners, characterized broadly as formal planners and other public managers whose responsibilities relate to planning and coordinating activities on the coast, are often faced with recurring management problems in different locations on the coast. They tend to need ways to help integrate the variability of issues and geographies into consistent decision-making processes. In many ways, this audience is the most challenging to serve, and presents the greatest potential for improvement in service. We will demonstrate two coastal atlas tools that attempt to take specific recurrent coastal management tasks (relative hazard vulnerability and watershed assessment) and present standardized processes for working through completing such tasks.

Tools for Contributors

Contributors to the Atlas consist of current and future coastal atlas project partners who are dispersed throughout a wide area and need ways to be able to add their own particular data/information/tools to the collective project in a way that is easy and consistent with the common interface familiar to the other audiences. While not specifically addressing the needs of the more public audiences of the coastal atlas project, the thought is that by making it easy to participate as a contributor to the Atlas, more contributors can be recruited over time, which in turn will increase the odds that the content of the site will serve the needs of a particular audience member. Two OCA tools that simplify a contributor task are the Glossary Linking and Bug Reporting

Tools. Future directions for such tools include tools that allow contributors to input geospatial, bibliography, and photo data, and even to build Internet map server (IMS) templates.

Tools, Tools, and More Tools...
Coastal Atlas Geospatial Archive Search Tool

http://www.coastalatlas.net/search/

AUDIENCE: PLANNERS, RESEARCHERS

The heart of the OCA is an archive of geospatial data collected over the years by various program partners of the Oregon Ocean-Coastal Management Program. The data range from point-based species data through shore-type delineation line data, offshore geology polygon data, to raster products such as digital nautical charts, orthophoto quadrangles, and Landsat TM (thematic mapper) imagery. Typically, the data were originally collected as part of a specific planning project involving a partner agency, and once that original project was completed, the data became part of the published projects products, sometimes never to be seen again. Rather than allow such data to gather dust on shelves and in storage boxes, the Atlas project team made a concerted effort to mine existing partner hardcopy archives for any digital data that could be preserved and brought into the future via the new OCA archive. The intent was to create a "one-stop shop" for finding the fruits of past data collection efforts.

To accomplish this, all archive data media formats (floppy disk, digital tape, CD, etc.) were acquired and opened, and any data contained within them were matched with documentation from accompanying hardcopy reports. Federal Geographic Data Committee (FGDC) compliant metadata were then created for individual datasets, and all datasets were brought into a common coordinate projection, so that all OCA archive products could easily be used together. Individually documented datasets were entered into a database and a Web search interface was constructed, which allowed users to search for geospatial data based on coastal locations, data originator, data scale, or even keywords. Search results are provided in list form with links to the respective metadata and download files, and should the user wish to view the data pre-download, a link to the Atlas interactive map interface is provided so that the data can be browsed via the built-in IMS (Fig. 7.1; see page 132).

Coastal Atlas Map Tool

http://www.coastalatlas.net/maps/

AUDIENCE: PUBLIC, PLANNERS

The OCA includes a stock IMS as a top-level tool, which can be used by visitors to view a variety of standard, pre-formatted and commonly requested data served in the OCA archives. The Map Tool is powered by the open source software known as Minnesota MapServer (see: http://mapserver.gis.umn.edu/), which offers unparalleled interface customization to tool providers. In the OCA interface, we chose to provide common IMS utilities such as Data Display, Quick Zoom, Zoom In, Zoom Out, Pan, Identify, and Print (Fig. 7.2; see page 197). The purpose of this tool is to allow users, such as coastal planners and members of the public who do not have access to a desktop geographic information system (GIS), to build simple, personalized maps using data relevant to the coast.

Commonly requested data, such as various base layers (orthophotos, digital topographic quads, shaded elevation, shaded bathymetry, and land cover), as well as common overlay groups (transportation networks, hydrography, jurisdiction/administrative boundary lines, and landmarks), and common coastal specific layers (coastal zone boundary, territorial sea line, locations of estuaries, rocky shore sites, sandy shore sites, and important ocean areas) are provided in easy-to-use switchboard lists, which allow a high degree of user customization in personalized map content design. In addition to the provided stock layers, users can also search the entire OCA geospatial data archive (see previous information) and add any layers of specialized interest to their maps. Maps can be given personalized titles and output to portable document format (PDF) format for use in printed reports, email, etc.

Of special interest in this tool is the "Info" utility provided in the main navigation bar. As in most IMS implementations, this tool will return basic identifying information for any selected feature. However, a special feature of this tool is that when used in conjunction with the coast-specific data layers provided in the "Info Layers" switchboard, the tool will return a pop-up window presenting a database template that displays a range of location specific information stored in the Atlas about that site. This one-page layout includes one or more site photos, a text-based informational narrative, a current weather report and a listing of any geospatial datasets relevant to that specific site. This feature allows the Map Tool to essentially provide an alternative route to an easy-to-read, compound data layout, which gives the user a summary of whatever the OCA has to offer for their site of concern.

Coastal Access Tool

http://www.coastalatlas.net/tools/public/coastal_access.asp

AUDIENCE: PUBLIC, PLANNERS

Both coastal planners and members of the public are interested in information about beach access points on the Oregon coast. Typically, the public consists of frequent practical users of the access points, and the planners are tasked with maintaining the sites, and preventing their loss. The Oregon Ocean-Coastal Management Program maintains a Coastal Access Information System containing location and inventory information about each of the access sites on the open ocean coast. This Oregon Coastal Access Information System contains valuable information pertaining to Oregon's public beach access locations, including characteristic information such as access availability, type, presence of human-made and natural features and services, and activities available at each location. Supplemental data have been provided including orthophotography, boat ramp data, line work indicating the Pacific Coast Trail, and other base data layers.

While this original Coastal Access Inventory is a wealth of good information, the sheer quantity of data itself is cumbersome to navigate, and in its original GIS form (ArcView project) the data were really only available to users with ArcView and access to one of the originally published inventory CDs. The Coastal Atlas Access Tool was therefore created to provide much more widespread access to the inventory (via the Web, and with no need for proprietary software), and was designed to present a simple way to search the data and return the results of searching the inventory in an easy-to-read, printable format (Fig. 7.3; see page 197).

COMCI: A Coastal and Marine Bibliography Tool

http://scallywag.science.oregonstate.edu/website/comci/viewer.htm

AUDIENCE: RESEARCHERS, PLANNERS

The Catalogue of Oregon Marine and Coastal Information (COMCI), was developed through the collaboration of groups at Oregon State University: the Partnership for Interdisciplinary Studies of Coastal Oceans, Oregon Sea Grant, the Department of Zoology, and the Northwest Alliance for Computational Science and Engineering. Originally intended as a stand-alone resource that would provide policy makers, resource managers, and interested members of the public (planners and researchers) with information about the coastal and marine environment, it became evident after its inception that COMCI contained a natural complement to the geospatial data archive of the OCA. As such, a partnership was formed, which allowed the COMCI searchable archive to be incorporated into the structure of the OCA, making it easy for users of the OCA to access results relevant to their area of interest from COMCI and vice versa. This work is ongoing.

When completed, COMCI will allow users to quickly and efficiently search for information about Oregon's coast and ocean using either a text- or map-based interface. Sources referenced in COMCI range from articles, proceedings, and other documents that provide background information about oceanography and ecology, to those that include specific details about particular regions of the Oregon coast, such as habitat maps and local economic statistics. The six different, detailed subject areas contained within COMCI are: coastal zone management, marine fisheries, marine species, marine habitats, oceanography, and nearshore and open-ocean activities. If and when the referenced information is available in an online form, users can access it directly through a hot link in the meta-database.

COMCI is intended to be a valuable tool for informing ocean and coastal decision-makers in Oregon as they strive to develop scientifically sound and integrated marine policy, management, and conservation measures. For this reason, the original architects included a description of how the data are or can be relevant to area-based management in Oregon's marine environment.

Watershed Assessment Tool

http://www.inforain.org/coastalatlas.net/tools/watershed/intro.asp

AUDIENCE: PLANNERS

The Coastal Atlas Watershed Assessment Tool is based on the Oregon Watershed Enhancement Board (OWEB) assessment manual, a very large paper document used by volunteer watershed councils as a "cookbook" of how to go about doing a watershed assessment. Due to the technical capacity of many Oregon watershed councils, the OWEB manual did not focus on providing instruction on assessment procedures using GIS when it was last published in 1999. However, GIS is critical to preserving field data, monitoring restoration success or failure, and modeling priorities and conditions, and is increasingly available to non-profit groups such as watershed councils. Therefore, the purpose of the Coastal Atlas Watershed Assessment Tool is to provide GIS data, instruction, and an IMS to facilitate the use of GIS in the watershed assessment process (Fig. 7.4; see page 198).

The Watershed Tool uses the components of the OWEB manual to develop parallel online chapters addressing certain aspects of an assessment for which GIS is appropriate. The manual has eleven modules, from start-up to monitoring. It also includes a chapter placeholder on estuary assessment, which is being developed by OWEB for inclusion in the manual in 2004. Each chapter has three sections: Overview, Methods, and GIS. "Overview" reviews the reason for that component in an assessment. Additionally, it asks the "critical questions" that are part of each manual component. "Methods" discusses the steps and final products identified in the manual. "GIS" focuses on those steps and products that can be accomplished through the use of GIS.

To further enhance the range of user abilities addressed by the Tool, there are three components of the GIS section: data, procedures and tools. These three components were developed to serve three tiers of GIS users: those with GIS software and skills who only need data, those with GIS software and few skills who need data and instruction in GIS software, and those users with no software who benefit from an IMS concentrating on each OWEB manual component.

With regard to data, it is not presumed that all the spatial data necessary for a complete watershed assessment will always be available. However, well-documented, standardized data, commonly needed for an assessment are being provided. The data download feature in the GIS section therefore links the user to standard, best-available data that have been prepared and clipped for each watershed. The procedures feature of the GIS section provides links to the GIS data as well as general instructions for how to use the GIS to accomplish assessment tasks, and instructions for how to create and incorporate new data into the assessment process. Last but not least, the IMS tools feature of the GIS section uses consistent, best-available data for the coastal watersheds of Oregon to create maps for those users without GIS software, allowing them to browse and print maps of their choosing

Marine Visioning Tool

http://www.inforain.org/coastalatlas.net/tools/marine/intro.asp
AUDIENCE: PUBLIC, PLANNERS

The Marine Visioning Tool was designed to give the user a better understanding of the complex nature of the marine landscape. In a way, it is the least "tool-like" of all the tools described in this chapter. Rather than being a step-by-step GIS decision-making tool, the Marine Visioning Tool is designed to illustrate the web of interconnections between various components of the marine landscape, thereby abandoning the "step-by-step" model. With the help of Bob Bailey, Oregon Coastal Management Program Manager and one of the original authors of *The Oregon Ocean Book,* a template of marine landscape subjects was developed that fell within the headings of time scale, fluid, physical, plankton, benthos, nekton, and human. With the help of a steering committee that included Bob Bailey; Hamilton Smillie, physical scientist, GIS analyst, and program manager at the NOAA Coastal Services Center in Charleston, South Carolina; and Peter Huhtala, acting executive director of the Pacific Marine Conservation Council in Astoria, Oregon, the Tool was further developed to include more categories under these headings.

The Tool functions by providing an overview of a main topic such as time scales of oceanographic processes, then goes on to describe in more detail the individual subjects within that, such as Pacific Decadal Oscillation, El Niño Southern Oscillation, and winter/summer effects (Fig. 7.5; see page 198). Once the user enters the description of a subject, they can then look at the connection between that subject and another subject within any of the headings. The content for these descriptions and interconnections was developed from an extensive literature review, during the course of which an expansive bibliography and glossary of specialized terms was created. The contents of the bibliography have been incorporated into COMCI and, by virtue of that tool's incorporation into the wider OCA project, are available to anyone using COMCI as a bibliographic search tool. Similarly, the glossary of scientific and technical terms was incorporated into the larger atlas wide glossary, and relevant phrases imbedded throughout the narrative text are linked to their definitions via pop-up windows.

Erosion Hazards Tool Suite

One of the most chronic hazards facing Oregon coastal communities is erosion, which occurs over relatively short time periods in certain geographic areas, greatly affecting shoreline stability. The Oregon Department of Land Conservation and Development created a dune and bluff hazards decision support tool based on existing foredune erosion models (Marra, 1998; Ruggiero et al., 2001) and traditional ground survey beach elevation data. Since its initial development, this tool has been updated once through a partnership with NOAA Coastal Services Center to take advantage of newly available light detection and ranging (LIDAR) data for the Oregon coast, creating a combination of a science-based decision support tool and a cutting-edge remote sensing technology. This union resulted in an initial product, the Dune Hazard Assessment Tool (DHAT) that can help coastal managers identify the relative risk to properties from coastal erosion.

The system being modeled by the Tool is quite complex. Dune Hazard Zones (DHZ) occur within segments of shoreline backed by a sandy beach and dune. They refer to the area landward of a reference feature, such as the existing line of vegetation that is subject to coastal hazards. Along dune-backed shorelines, the extent of short-term shoreline change attributable to episodic wave attack events is the primary factor affecting shoreline stability and thus determines the landward extent of the zone. Long-term trends of shoreline change, attributable to factors such as the sediment budget or relative sea level rise, may also need to be accounted for in this environment. A typical analysis will involve assigning a range of actual values to the individual terms in a formula that relates all the relevant environmental parameters. By, in effect, "bracketing the truth," the dune hazard assessment formula provides a rational basis for assessing the relative risk to life and property posed by coastal hazards. In creating a tool for managers, parameters such as significant wave height, wave period, and mean water level are embedded in a Storm Event Magnitude drop-down list, which can draw from actual geospatial datasets. Other parameters could be derived from choices or entries made by the user.

The DHAT is an example of an effort to aid local managers with a complex but not uncommon technical task encountered in the coastal zone on a routine basis. For the OCA, this Tool has essentially been ported into an Internet accessible form such that it is seamlessly available to users alongside the available data archives in OCA. The result is actually a suite of three erosion hazard tools for the OCA: Overtopping Hazards, Undercutting Hazards, and Bluff Recession Hazards. Each of the tools walks the users through a process that explains points of relevant science, and then asks the users to make a selection or input a value. All selected or entered values are stored and used to calculate and/or populate variable parameters in a mathematical model that is designed to estimate hazard recession based on the entered values and the appropriate model for the specified geomorphology of the relevant section of shoreline. All three tools are designed for users with a reasonably high level of understanding of coastal processes, and existing coastal geomorphology, and this is reflected in the tool steps. In the case of one tool (Overtopping Hazards), a simplified version of the tools was also constructed to cater to interest from more average lay users.

Fig. 7.6 (see page 198) shows the results of a GIS query to locate vacant parcels in private ownership within a high-risk zone. Assuming that the landward boundary of the zone of high relative risk represents an oceanfront construction setback line, then it is readily apparent how many property owners (i.e., tax lots) are affected along different segments of shoreline. For policy development purposes, this visualization tool makes it readily apparent that a construction setback may be effective along the central portion of the segment of shoreline depicted on the left, but unlikely to be effective in the segment depicted on the right. However, a review of additional attribute information provided by the tool suggests that other hazard avoidance measures may be viable. For example, an examination of the zoning designation of the affected lots indicates that they are zoned medium-density residential and high-density residential. Consideration may be given to downzoning as a means of reducing risk. An examination of the ownership of adjacent lots indicates that many are in public ownership. Land acquisition may be a viable, if not the preferred, alternative in this instance. By moving along segments of shoreline within a littoral cell and conducting the types of analyses illustrated above, a suite of preferred hazard avoidance measures can be identified for various segments of shoreline. When taken together they constitute a hazard avoidance strategy for a given littoral cell.

Overtopping Hazards Tool

http://www.inforain.org/coastalatlas.net/tools/overtopping/intro.asp
AUDIENCE: PLANNERS, RESEARCHERS

The primary determining factor as to whether a location on the coast will be inundated during a storm is the amount of water expected to hit the site as calculated from a variety of factors. These contributing factors range from water level values resulting from sea level, storm size, wave size, presence, absence or strength of El Niño, etc. The Coastal Atlas Overtopping Tool allows the user to input parameters to determine the final total water level, and apply it to a specific section of coastline (area of interest). The ultimate output of the Tool is a map that illustrates the extent of inland flooding.

This Tool was developed in both amateur and professional versions, in order to appeal to a wide variety of user abilities. Both versions begin by asking the user to select an area of interest from an IMS map or a dropdown list of site names. The amateur version performs the same analysis as the professional version, however, it eliminates many user choices and asks for input on only three parameters: sea level scenario, beach slope, and storm scenario. In contrast, the professional version asks the user to input a wide variety of parameters beginning with mean water levels associated with return intervals ranging from 5 to 100 years for a selected tide gauge, and progressing to those associated with elevated water levels resulting from storm surges, El Niño, and global sea level rise. Shorter-term variations in water levels resulting from storms are considered when the user is asked for storm scenarios associated with return intervals ranging from 5 to 100 years for a selected wave buoy, as well as an elevated wave height trend in winter. The user is finally asked to input beach slope as a parameter of beach morphology.

Regardless of which version of the Tool a user is engaged in, after all selections regarding locations and water level components are made, the choices made thus far are displayed on a summary page that shows the parameters the user has chosen, as well as the model calculation of total water level. This final water level, as well as the coordinates of the area of interest, is passed to an IMS that zooms into that area and queries the processed LIDAR contours for the relevant water level value. The LIDAR contours are then drawn on the map with all contours falling below the "flood level" being drawn in blue. This resultant map shows the tool user the approximate potential level of ocean flooding for that area under circumstances based on input parameters.

Undercutting Tool

http://www.inforain.org/coastalatlas.net/tools/undercutting/intro.asp
AUDIENCE: PLANNERS, RESEARCHERS

The Undercutting Tool allows the user to model foredune retreat—an indicator of the potential for ocean erosion or "wave-undercutting." There is only one version of this Tool and it assumes the user has a good understanding of the subject.

Similar to the Overtopping Tool, the Undercutting Tool has a step that asks the user to select an area of interest using an IMS or drop-down menu. Because undercutting is also partly a function of total water level, many of the initial steps in the Overtopping Tool are repeated here, such as choosing the mean water level value associated with 5 to 100 year intervals for the South Beach tide gauge, storm surge, El Niño, and global sea level rise, and short-term storm scenarios as defined by wave buoy values and elevated wave height trend. In addition, the Undercutting Tool asks for two beach slope parameters: mid-beach and upper-beach. The mid-beach slope is used in total water level calculations and the upper-beach slope is used in the wave undercutting calculation. A beach-dune junction value is also requested before the user is directed to the summary page that displays user input parameters and calculated total water levels and potential extent of landward foredune retreat due to wave undercutting. The coordinates for the area of interest and the potential retreat value are passed to the IMS where the LIDAR contour equal to the retreat value is highlighted.

Bluff Recession Tool

http://www.inforain.org/coastalatlas.net/tools/bluff/intro.asp
AUDIENCE: PLANNERS, RESEARCHERS

The Bluff Recession Tool departs from the approach explained above for overtopping and undercutting scenarios by offering users a choice between two methodologies for estimating potential bluff recession. These methods are the Recession Rate and Planning Period method, and the Slope Stability method, respectively. Because both methods require knowledge of hazard-related science, the overall Tool is written with the assumption that the user has a hazards background.

As with the Erosion Suite Tools explained above, the user is first asked to choose an area of interest via an IMS or drop-down list. The progression is then to the selection of either the recession rate method or the slope stability method. The Recession Rate method relies on an estimate of the average annual rate of bluff recession. This rate is often determined by analyzing changes in bluff crest position over time using aerial photographs. The Slope Stability method requires an understanding of the type of slope movement as expressed in terms of the "effective angle of internal friction."

If the Recession Rate method is selected, the Tool first asks the user for elevation of the bluff top. It then asks for the estimated recession rate and planning period to be either entered by hand or selected from provided lists of common rates and common planning periods. Users are then directed to the standard summary of selections page, and the Recession Buffer calculation result– (simply rate x planning period) is presented. The calculated value is the recession distance from the bluff top edge. The corresponding LIDAR contour and area of interest are reflected in the next page containing the IMS, which displays the Tools' resultant map product.

The Slope Stability method asks for the bluff top elevation as well as for the elevation of the bluff toe. It then asks for input of the "effective angle of internal friction" from either a list of common values or a user input text box. This angle value is used to represent the potential type of slope movement or mode of bluff failure, which is determined by material type, geologic structure, moisture content, and vegetative cover of the bluff itself. The next step directs the user to the summary page and the resulting calculated effective angle of internal friction –simply the tangent of bluff face slope. This value is multiplied by the bluff height to determine the recession distance inland from the bluff top, which is passed to the IMS with the area of interest coordinates to construct the resultant tool map product.

Coastal Atlas Glossary Tool

http://www.coastalatlas.net/glossary/

AUDIENCE: CONTRIBUTORS, PUBLIC, PLANNERS

The Coastal Atlas Glossary Tool is primarily a contributors' tool, but as with any tool that contributes value to the overall OCA, the glossary itself benefits any user of the Atlas who encounters an unfamiliar word or term of coastal science or policy.

The Tool is very simple in form, essentially presenting a contributor to the Atlas with two methods for linking any word in their content pages to a centralized coastal glossary. Contributors using the Tool simply follow instructions for how to embed their preferred type of glossary links into their HTML pages and ensure that the links point to the appropriate definition. The most common implementation is via a link that results in a small pop-up window containing the word, definition, and definition source citation. The central glossary itself was compiled from a variety of public sources, and contains over 1,000 definitions. Contributors can add new definitions to the central glossary simply by emailing the OCA Project Team.

Coastal Atlas Bug Reporting Tool

http://www.coastalatlas.net/glossary/

AUDIENCE: CONTRIBUTORS, ANYONE ELSE THAT SPOTS A BUG!

The Coastal Atlas Bug Reporting Tool is primarily a contributors' tool, originally implemented in the beta-testing phase of the Atlas as a way to centrally report all the quirky items that contributors and testers encountered while playing with the system as the project took form. As with any networked, Web-based project, the "under construction" light could be perennially lit above our URL, and hence the continued relevance of this Tool.

The Tool works as a simple database input form, requesting that users enter information about the offending page, the particulars of the problem they saw, and relevant information about their particular operating system environment in case this proves to be a contributing factor in the manifestations of the bug. Upon submitting the form, users are thanked and shown a list of the most recent bugs submitted to confirm that their finding has been successfully registered. On the back end, Atlas project partners can view the lists of bugs submitted and take appropriate action to remedy the identified problem. The database archive of submitted bugs becomes a record of improvements and progress in the overall OCA effort.

Conclusion

Technologies such as GIS could make it easier for resource managers, planners, scientists, and even elected officials to process and use different sources of information. However, the current process of integrating such technology into daily activities is typically carried out in a different manner from locality to locality, which results in a vast array of system choices and further complicates data sharing on a regional or coast-wide level. This makes the development and application of tools that can integrate datasets and increase efficiency more important.

Effectively sharing data means more than just the ability to download a file. Many datasets and analytical tools available on the Web are underused, and thus do not fulfill their potential for advancing research. One primary reason for this situation is the unfamiliarity of investigators with the methods required to use the Web-based tools and datasets. The research presented has addressed issues in the development and use of analytical tools for *Web*-based data, an area that has been underutilized. We have built a broad suite of decision-specific tools to facilitate both access to and use of data by local level resource managers and scientists, and the public, using the Web as the common gateway interface for access. These tools need to be continually adapted so that they serve different levels of needs according to the user community. For example, a hazards tool for the general public will be quite generic in its inputs and outputs, while one for planners is much more sophisticated (enabling queries such as in the Erosion Hazard Suite), while one serving the explicit needs of scientists must entail the highest levels of intricacy. All should draw from and operate off of the same data archives, however, but to varying levels of complexity in interaction. In the continued evolution and adaptation of tools, one user community that is extremely important is the contributor community, those with new, emerging tools and relevant datasets. The infrastructure of the OCA is built upon a foundation of tools that facilitates continued contribution and feedback from the Oregon coastal management and GIS communities.

Acknowledgements

The Oregon Coastal Atlas is currently funded by an Information Technology Research/Digital Government grant from the National Science Foundation (EIA-011359) and a Directors Grant from the NOAA Coastal Services Center (NA16OC2709), as well as supplemental funding from the Federal Geographic Data Committee. We are also indebted to our terrific team of student workers for programming support, the writing of metadata records, and general research and database assistance. These include undergraduate interns Mike Tavakoli-Shiraji at OSU, Ben Donaldson and Sarah Klain at Ecotrust, graduate students Ken Crouse, Chris Zanger, Danielle Pattison, Colin Cooper, and Peter Bower, and metadata maniacs Amythyst O'Brien, Christina Ryan, Anthea Fallen-Bailey, Jenny Allen, Jessica Adine, and Ryan Field. And we very much appreciate the guidance and support of Bob Bailey of the Oregon Ocean-Coastal Management Program and Mike Mertens of Ecotrust.

References

Goss, H., 2003. Stepping through Oregon's coastal information gateway. *Coastal Services* 6(6): 2-3.

Marra, J. J., 1998. *Chronic Coastal Natural Hazards Model Overlay Zone: Ordinance, Planners Guide, and Practitioners Guide.* Newport, OR, Shoreland Solutions, Report to the Oregon Department of Land Conservation and Development, January.

Ruggiero, P., Komar, P. D., McDougal, W. G., Marra, J. J., and Beach, R. A., 2001. Wave runup, extreme water levels and the erosion of properties backing beaches, *Journal of Coastal Research,* 17(2): 407–19.

Wright, D., Haddad, T., Klarin, P., Dailey, M., and Dana, R., 2003. The Oregon Coastal Atlas: A Pacific Northwest NSDI contribution, Proceedings of the 23rd Annual ESRI User Conference, San Diego, CA, Paper 105.

Wright, D. , Haddad, T., Klarin, P., Dana, R., and Dailey, M., 2004. Infrastructure for data sharing, spatial analysis, resource decision-making, and societal impact: The Oregon Coastal Atlas, *Proceedings of The National Conference on Digital Government Research,* Seattle, WA, pp. 131-132.

Related Web Sites

Ecotrust: http://www.ecotrust.org/

Inforain: Ecotrust's Bioregional Information System for North America's coastal temperate rain forest, http://www.inforain.org

Minnesota MapServer: http://mapserver.gis.umn.edu/

NOAA Coastal Services Center http://www.csc.noaa.gov/

Oregon Ocean-Coastal Management Program: http://www.lcd.state.or.us/coast/

OSU Geosciences, Davey Jones Locker Lab: http://dusk.geo.orst.edu/djl/

PISCO: Partnership for Interdisciplinary Studies of Coastal Oceans, http://www.piscoweb.org/

Writing for the Web, John Morkes and Jakob Nielsen, http://www.useit.com/papers/webwriting/

CHAPTER 8

Nearshore Marine Conservation Planning in the Pacific Northwest:

Exploring the Utility of a Siting Algorithm for Representing Marine Biodiversity

Zach Ferdaña

Abstract

Terrestrial conservation planning is well developed. In comparison, there are only a handful of published marine conservation plans and few of them are quantitative. The Nature Conservancy's strategic planning approach, called "Conservation by Design," employs an ecoregional planning methodology to construct conservation portfolios, or high-priority conservation areas. This chapter reports on the development of a regional nearshore marine analysis in the inland seas of Puget Sound, Washington, in the U.S., and the Strait of Georgia, British Columbia, in Canada. The purpose of marine planning within the Willamette Valley-Puget Trough-Georgia Basin ecoregion was to identify a set of conservation areas that capture the full array of representative shoreline ecosystems and a subset of the existing nearshore biodiversity. This chapter outlines the basic steps of the ecoregional planning process, including the identification of conservation targets, assigning conservation goals, assessing population viability and ecological integrity, and selecting conservation areas. A marine planning team identified 37 shoreline ecosystems, one for rocky reef habitat, and 72 marine species as conservation targets. In order to achieve our conservation goals in the most efficient manner possible, we concluded that at least 30% of the total shoreline (excluding humanmade shore units) in the ecoregion warrants an evaluation of ecosystem integrity in order to place these sites in some form of protected status or conservation management. Four terms were adopted to describe and examine representation: overrepresentation ($p > 1.3$), adequately captured ($p => 1.03$ and < 1.3), efficiency of representation ($p = 1.0 +/- .03$),

Zach Ferdaña
Global Marine Initiative, The Nature Conservancy, 217 Pine Street Suite 1100 Seattle WA 98101
zferdana@tnc.org

and missing values (p = < .97). SITES, a reserve site selection algorithm, was used to examine site selection. Marine analyses were developed both in combination with terrestrial information and as a separate, yet parallel, process. The first approach used a seamless 750-hectare hexagon assessment unit across both environments. Although integrative in nature, the seamless unit tended to over-represent shoreline targets (57%), thereby exceeding set conservation goals. However this analysis met most goals, with the missing values category containing only 4% of targets. The second approach used a linear shoreline and nearshore hexagon unit. A 4-tiered nearshore analytical framework was designed to analyze information according to our confidence in the spatial data, and systematically incorporate expert input. It was found that overrepresentation was not as much of a factor as when using the seamless hexagon approach, with only 18% of targets considered over-represented. The missing values category contained 31%, thereby not performing as efficiently as the seamless hexagon in meeting conservation goals. Where the uniformity of the seamless hexagon provided the means to include information across environments in the land/sea interface, the nearshore and shoreline units tended to be more spatially explicit and follow ecological boundaries. In total, there were 186 shoreline/nearshore sites comprising 2,910 km of shoreline in the final conservation portfolio.

Introduction

Coastal regions across the globe are under particular stress today as human populations concentrate along shoreline environments. Estuarine and marine environments bear the cumulative, negative impacts of land-use and resource-management decisions carried out in adjacent terrestrial and freshwater areas. At the same time, humans are exploiting marine fisheries with an efficiency that threatens to undermine trophic relationships and biodiversity. In response to these threats and challenges, more effort has focused on marine conservation planning in the last few years.

Setting priority areas for conservation often involves a strategic planning approach. The Nature Conservancy's approach is called "Conservation by Design." This is directing the organization to systematically identify the array of places around the globe that embrace the full spectrum of the Earth's natural diversity. It is also a framework for developing the most effective strategies to achieve tangible, lasting results, and to work collaboratively to catalyze action on a scale great enough to ensure the survival of entire ecosystems (The Nature Conservancy, 2001).

In order to protect diversity in a cost-effective manner, the field of conservation has developed general planning principles (Pressey et al., 1993; Margules et al., 2000; Groves et al., 2002). These widely applied planning approaches originated in terrestrial settings and have only recently been applied to marine environments. For example, the World Wildlife Fund recently completed plans for the Sula-Sulawesi Seas, the Meso-American Reef, and Nova Scotian Shelf (Day and Roff, 2000). The Nature Conservancy has completed ecoregional assessments for the central Caribbean (Sullivan-Sealey and Bustamante, 1999) and the northern Gulf of Mexico (Beck and Odaya, 2001), Chesapeake Bay, southern California, and Cook Inlet in Alaska. In addition, The Nature Conservancy and World Wildlife Fund have collaborated in a plan for the Bering Sea (Banks, 1999). Other examples exist outside the non-profit community (e.g., Ward et al., 1999; Airamé et al., 2003; Leslie et al., 2003). In response to the present human impacts and threats facing coastal ecosystems in the Pacific Northwest, methods have been developed for constructing a conservation portfolio across terrestrial and nearshore environments in the Willamette Valley-Puget Trough-Georgia Basin ecoregion (Fig. 8.1; see page 199). Reported here is a regional planning exercise to test methods of capturing representative nearshore marine biodiversity in a conservation portfolio for the waters of Puget Sound and Strait of Georgia.

Conceptual Framework for Marine Ecoregional Planning

Marine planning at the ecoregional scale provides a larger context for selecting high-priority conservation areas in estuarine, nearshore, and offshore environments. Ecoregions, not political boundaries, provide a framework for capturing ecological and genetic variation in biodiversity across a full range of environmental gradients. There are three key components to consider during the ecoregional planning process: conservation targets, conservation goals, and population viability and ecological integrity. They are briefly outlined in the following text. For a more in-depth treatment, see Beck (2003).

Conservation Targets

The first step in the Conservancy's regional planning approach is to select conservation targets. These are ecosystems, habitats, and species that represent a diversity of the biotic assemblages in a region. In marine environments, the most effective planning approach is to focus on marine ecosystems and the ecological processes that sustain them (Beck, 2003). This presumes that the conservation of a representation of all the ecosystems will also conserve a representation of the diversity of species found in these ecosystems. Examples include rock platforms that support tide pools, kelp forests, and seagrass meadows.

A robust classification scheme to identify the different types of ecosystems is critical for selecting conservation targets. The choice of a particular classification scheme can significantly influence siting algorithms and decision support tools that identify potential conservation areas. Where possible, classification schemes should be based on biological data, but in the marine environment, surrogate data are usually required. While coastal classifications generally rely on physical factors such as landform, slope, and wave energy, a combination of abiotic and biotic-based targets will likely be most effective in conserving the full array of biodiversity in any given planning region.

Marine species targets that are least likely to be represented by ecosystem level information are endangered, imperiled, or species considered keystone (see Power, 1996). Many of these species require individual attention because management of their habitats alone is necessary but insufficient for their conservation needs. In addition, aggregation sites usually associated with the physical convergence of water and land or of different water masses, are also used as a marine target type. Examples include the spawning aggregations of reef fish, or breeding congregations of seals and sea lions on haulout sites.

Conservation Goals

A conservation goal is characterized by the amount of the target that should be represented in conservation areas across the planning region. The objective is to assess how much representation is required to maintain its persistence over time. This should ideally be based on historical estimates of the abundance and distribution of the targets. Unfortunately, goals often have to be based on current distributions (Beck and Odaya, 2001). This being the case, different approaches have been adopted to test the representation question in marine environments (Leslie et al., 2003). One such approach is to conduct sensitivity analyses. This involves systematically varying the conservation goals to determine how they affect the overall size of the area selected.

Population Viability and Ecological Integrity

As data are gathered on the distribution of the targets and their locations, attempts are made to ensure that only populations of species and examples of ecosystems that are likely to persist into the future are included (Beck, 2003). However, formal analyses of viability are rare for marine species and similar analyses of integrity are virtually non-existent for ecosystems. While one may not have these sources of information, there are often factors that can be used to "screen" or filter out areas that are not likely to have the best or most viable examples of species and ecosystems. These factors are often built into a "suitability index." Examples include shoreline impacts such as bulkheading (seawalls, jetties) and coastal development (boat ramps, docks, marinas), adjacent impacts such as land-use designation (urban, agriculture), and freshwater impacts such as water quality. These factors generally guide site selection algorithms away from these impacts.

Selecting High Priority Conservation Areas

One of the primary tools being used by The Nature Conservancy in selecting areas that deserve conservation attention is the use of site selection algorithms. For this ecoregion, SITES was used, an optimal reserve selection algorithm (Andelman et al., 1999; Possingham et al., 2000). SITES, previously known as SPEXAN, and subsequently known as MARXAN, is becoming well established in conservation planning circles. SITES is best suited (but not restricted) to the situation where an ecoregion has been divided into a set of candidate sites, or planning units that completely fill the region. Examples include abstract units such as equally sized grids, and natural units of analysis, such as watersheds. These are the basic building blocks for assembling a conservation portfolio.

At the core of reserve selection problems is the overall objective of minimizing the area encompassed with the network of reserves (Pressey et al., 1993). SITES uses a simulated annealing algorithm to evaluate alternative site selection scenarios, comparing a very large number of alternatives to identify a good solution. The procedure begins with a random set of planning units, and then at each iteration, swaps planning units in and out of that set and measures the change in "cost." Cost here does not mean dollars for land purchase, but the amount of area selected in the alternative. The algorithm's objective function is to minimize total area while meeting the desired amount of target representation. This function is a nonlinear combination of the total area and the boundary length of perimeter of the site selection output (Leslie et al., 2003). A boundary length modifier setting in the algorithm's parameters determines the relative importance placed on minimizing the perimeter relative to minimizing area. When this modifier is set very small, the solution algorithm will concentrate on minimizing area, whereas when the modifier is set relatively larger, the solution method will put the highest priority on minimizing the boundary length of the feasible reserve system. In its iterative nature, if the change in cost tends to improve the selected set, the new set is carried forward to the next iteration until the maximum number of iterations is reached.

There are many methods for solving this nonlinear integer programming problem. Do we want fewer, larger sites, or smaller, more dispersed sites of nearshore marine ecosystems and habitats across the seascape? There is never just one "optimal" solution (i.e., the definitive set of conservation areas) in regional planning, but it is possible to identify those areas that are both essential and representative as part of a plan. Siting algorithms provide a context for objective representation that is both measurable and spatially explicit. This chapter explores these parameters in order to test methods for efficient representation of nearshore biodiversity.

Methods

The purpose of our efforts at The Nature Conservancy was to develop a conservation portfolio that, if conserved and properly managed, will in part protect a representative subset of the existing nearshore marine biodiversity in the Puget Sound and Strait of Georgia. Protecting this representative subset will help ensure the long-term survival of the region's marine resources. Our goal was to illustrate the efficiencies and overrepresentation issues between including terrestrial and nearshore information into a single planning unit, and optimizing for minimum area and target representation using separate planning units. The nearshore data and available information used was limited in several ways when compared with terrestrial inputs. To truly match terrestrial datasets and information would require (1) historic information on marine habitats and species; and (2) viability or population assessments for marine species in setting ecological conservation goals. Unlike terrestrial environments, where extensive surveys report on the condition of individual species, and in some cases, entire habitats, there is no nearshore marine analogue. Therefore, this analysis had to rely on available data that report on quantities of marine ecosystems, habitats, and species, but not on their condition or quality. This translates to the most striking difference between terrestrial and marine analyses, where one is based on landscape condition, the other on representation. This representation was largely based on current classification schemes and surveys along the shoreline. Given these limitations, our intent is to show how different planning approaches can be conducted to best represent the present arrangement of nearshore marine ecosystems, habitats, and species using available data.

The Planning Unit

The design and selection of the appropriate planning unit is heavily debated within and among conservation planning teams. Choosing the spatial configuration and size are the two main debatable components. Planning units can be categorized into two realms: abstract and natural units. Abstract units are generally equally sized areas that arbitrarily fall across the land and/or seascape. Examples are grids or hexagons. Natural units are generally of variable size that fit within ecological boundaries. Examples are watersheds determined by drainage area, and shoreline segments or reaches determined by the length of a dominant beach substrate. There are advantages and disadvantages with choosing either abstract or natural units for analysis.

Abstract units have the advantage of incorporating information across ecosystems, in this case, the terrestrial and nearshore environments. In addition, equally sized units equally weight the amount of boundary or perimeter per unit in the selection process. From a modeling standpoint, this may be the preferred option. However, abstract units arbitrarily cut across ecosystem lines and randomly bin information.

Along the coast, for example, hexagons may encompass shoreline reaches from both shores of narrow water bodies like inlets or fjords. This aggregation of shoreline units associated with different landmasses may be problematic in that one side of the water body may be more ecologically significant than the other. This may lead to an overrepresentation of ecosystems and habitats in the selection output. Often a unit is chosen for a specific target in order to fulfill its representation goal, but may also select less desirable types that are aggregated along with it. This leads to a less optimized solution. An objective of the selection process is to not only meet each conservation target goal, but also to minimize the amount of that target's representation above the stated goal. In addition, abstract unit size usually lacks ecological justification. If units are too large they may over-generalize the more spatially explicit target data. Likewise, if the units are too small, they may be misrepresenting the level of spatial detail of the data.

Natural units have the advantage in that they fall within ecological boundaries, and in the case of shoreline reaches, are more spatially explicit than generalizing them within abstract units. In addition, using the output from natural unit selection is more intuitive in delineating sites for a conservation portfolio. However, natural units are of variable size, with the algorithm often choosing larger units over smaller ones. This, too, can also lead to overrepresentation problems. Furthermore, using a linear planning unit like shorelines may be inappropriate in that the selection of segments are often randomly scattered and shorter segments may be insignificant at a regional scale. It is appropriate here to come back to the original set of questions being asked of the data and the selection process. Given the level of spatial detail required over a large study area. it may be adequate to generalize the more spatially explicit input data and retain its detail for more site-specific conservation planning. Therefore, several approaches to planning unit configuration were tested.

The first approach was to combine terrestrial and marine target information into a seamless, 750-ha hexagon. The second approach was to use two spatial planning units for the nearshore and shoreline environments; 750-ha hexagons and linear units respectively. This approach separated terrestrial and nearshore site selection. The linear shoreline reaches (determined by the length of the dominant beach substrate) were adopted as analysis units in their original form. Since linear planning units are not the native spatial configuration for SITES, the boundary length modifier had to be customized. SITES uses this modifier to aggregate planning units based on the real extent of shared boundary between planning units. In order to use this parameter, a linear boundary to calculate the adjacency of shoreline units was developed. For every shoreline reach, the adjacent units were identified and the boundary modifier could therefore be set higher to select more contiguous lengths of shoreline, or lower to disperse shoreline units into shorter sections. For this ecoregional assessment, the outputs of these different spatial configurations were compared in terms of the efficiency of target representation.

Nearshore Conservation Targets: Shoreline Ecosystems and Intertidal Habitats

Representative ecosystem types were derived from shoreline classifications and inventories first developed in British Columbia, and then modified in Washington State. The Province of British Columbia developed its physical and biological ShoreZone mapping system based on shore types after Howes et al. (1994) and Searing and Frith (1995). These shore types are biophysical types that describe the substrate, exposure, and vegetation across the tidal elevation, as well as the anthropogenic features and supratidal types. In Washington State, the ShoreZone mapping system was adopted and attributed to both the British Columbia classification scheme as well as the Dethier (1990) method that more precisely segregated intertidal communities. These shoreline inventories were assessed by helicopter over the entire region, then interpreted, classified, and digitized into a geographic information system, or GIS.

Eight thousand and seventy kilometers have been flown and interpreted over the ecoregion, or nearly fifteen thousand linear reaches of shoreline classified according to their landform, substrate, and slope (see Howes et al., 1993; Berry et al., 2001). These data, and the underlying British Columbia summary classification (34 coastal classes and 18 representative shore types), served as the basis for constructing shoreline ecosystem conservation targets.

Eighteen representative shore types were examined within the original classification and aggregated further into 8 shoreline substrate types. This process generalized shoreline ecosystems into discernable coastal communities for planning purposes, distributed evenly across the ecoregion. We then augmented the representative or physical component of the classification with a biological one. To do this, biological information was extracted from the dataset to identify chosen intertidal vegetation types. Saltmarshes (high and low tidal marshes, sedges), seagrasses (eelgrass, surfgrass), and kelps (giant and bull kelps) were chosen as the three vegetation categories to capture the major biological communities of the nearshore zone from the supratidal to shallow subtidal. Although these categories alone do not represent the entire range of intertidal habitats or the most diverse habitat types, they are biologically productive and the most sensitive to man-made alteration. These categories are protected by policy, recognized to be ecologically important, and thus served as the best surrogates to represent a wide range of nearshore habitats.

Table 8.1. Shoreline conservation targets

Shoreline ecosystems	*Intertidal habitats*								
	K	*K+S*	*Sm*	*Sm+stv*	*Sg*	*Stv*	*Uv*	*V*	*Grand totals*
Mud Flat	0	0	618,206	136,656	0	47,821	195,630	0	998,313
Rock cliff	0	0	0	0	0	0	1,027,027	424,614	1,451,641
Rock platform	0	0	0	0	0	0	129,360	155,460	284,820
Rock with sand and/or gravel beach	254,981	109,765	0	18,365	80,514	0	1,042,151	0	1,505,777
Sand and gravel beach	182,046	179,744	37,128	36,695	262,747	0	431,813	0	1,130,172
Sand and gravel flat	74,565	121,036	41,104	56,599	207,792	0	320,206	0	821,301
Sand beach	135,651	63,885	66,841	57,266	194,266	0	162,832	0	680,741
Sand flat	33,337	60,181	67,378	113,689	232,900	0	152,712	0	660,198
Totals	680,580	534,611	830,656	419,270	978,219	47,821	3,461,732	580,073	7,532,962
Goals	30%	40%	30%	40%	30%	40%	25%	30%	30% (portfolio)
Goals in meters	204,174	213,845	249,197	167,708	293,466	19,128	865,433	174,022	2,186,972

Lengths in meters
K = Kelp; K+S = Kelp and seagrass; Sm = Saltmarsh; Sm+stv = Saltmarsh and subtidal vegetation; Sg = Seagrass; Stv = Subtidal vegetation; Uv = Unvegetated; V = vegetated

Eight vegetation combinations were derived from the three categories listed above. These combinations were assembled for each of the eight shoreline substrate types, creating 64 potential physical and biological combinations. These unique types were distilled down to a more manageable list of 37 shoreline ecosystem and intertidal habitat targets (Table 8.1). These distillations were based on expert opinion, with each shoreline substrate type considered separately. When there were not clear dominant vegetation types based on percentages available per substrate category, all vegetation was lumped into a "vegetated" class. In addition, the "saltmarsh and subtidal vegetation" class indicated that all types were present across the intertidal zone.

Additional Nearshore Datasets

Habitat and species information were collected in the nearshore to augment the data captured within ShoreZone. We calculated 469,461 ha of nearshore marine waters, defined in this study as the area extending from the supratidal zone above the ordinary or mean high water line (i.e., the top of a bluff or the extent of a saltmarsh in the upper intertidal) to the 40-m depth below mean lower low water. This represents 31% of all marine waters in the ecoregion, covering 1,509,733 ha. In addition, we divided the waters of the ecoregion into two distinct sections based on freshwater and oceanographic characteristics (Fig. 8.2; see page 199). This was done to ensure the selection of occurrences across the natural range of the target.

Fishery-independent video surveys were conducted in Puget Sound by the Washington Department of Fish and Wildlife between 1993 and 1997. These surveys collected data on rocky reef habitat and marine fishes. The rocky reef data contained important attributes including relief (elevation) and complexity (roughness of rock structures, crevices). Transformation of these data into conservation targets are discussed in Ferdaña (2002). This information was used to analyze areas deeper than the intertidal zone and associate species data on nearshore marine fishes with a subtidal benthic habitat. No comparable data were available to develop a portfolio of any subtidal habitats deeper than 40 m. Because habitat categorizations were lacking beyond this depth, planning efforts were not extended to deeper waters except in a few shoal areas away from the coast. In addition, rocky reef data were not available for the Strait of Georgia and therefore, only the ShoreZone data could be relied upon to interpret down to the shallow subtidal in British Columbia.

Additional marine species were included in the assessment, including various rockfishes and other marine fishes, seabirds, marine mammals, and invertebrates (Table 8.2). The final list included 72 species: 9 fish, 3 marine mammals, 50 seabirds and 10 invertebrates. For seabirds, individual species either served as surrogates for groups of species (e.g., American wigeons were used to represent 4 species of dabbling ducks),

Table 8.2. Marine species conservation targets included in the data analysis

Target Scientific Name	*Target Common Name*	*Taxa*
Branta bernicla	Black Brant	Bird
Various species	Dabbling ducks	Bird
Various species	Diving ducks/bay ducks	Bird
Histrionicus histrionicus	Harlequin duck	Bird
Gavia spp.	Loons	Bird
Brachyramphus marmoratus	Marbled murrelet	Bird
Aechmophorus spp.	Red necked grebes	Bird
Melanitta spp.	Scoters	Bird
Various species	Seabird [nesting colonies]	Bird
Various species	Shorebirds-mud/ aggregated	Bird
Aechmophorus occidentalis	Western grebe	Bird
Sebastes melanops	Black rockfish	Fish
Sebastes caurinus	Copper rockfish	Fish
Ophiodon elongatus	Lingcod	Fish
Clupea pallasi	Pacific herring [spawning]	Fish
Ammodytes hexapterus	Pacific sandlance	Fish
Sebastes maliger	Quillback rockfish	Fish
Hypomesus pretiosus	Surf smelt [spawning]	Fish
Sebastes nigrocinctus	Tiger rockfish	Fish
Sebastes ruberrimus	Yelloweye rockfish	Fish
Gorgonocephalus eucnemis	Basket star	Invertebrate
Lopholithodes spp.	Box crabs	Invertebrate
Pollicipes polymerus	Gooseneck barnacles	Invertebrate
Ptilosarcus gurneyi	Orange sea pens	Invertebrate
Haliotis kamtschatkana	Pinto (northern) abalone	Invertebrate
Polyorchis penicillatus	Polyorchis jellyfish	Invertebrate
Crassedoma giganteum	Rock scallop	Invertebrate
Tritonia diomedea	Rosy tritonia	Invertebrate
Virgularia spp.	Seawhips	Invertebrate
Various species	Spiny vermilion star	Invertebrate
Phoca vitulina	Harbor seal [pupping]	Mammal
Eumetopias jubatus	Steller sea lion [haul out and rafting]	Mammal

or individual targets represented multiple species (e.g., seabird nesting colonies represented 13 different species).

Sensitivity Analyses

There were three critical components of the SITES algorithm to construct before the strengths and weaknesses of a seamless planning unit and multiple, separate units could be tested: a suitability index, setting conservation goals, and a species penalty factor.

The suitability index is an assemblage of different costs or impacts that are scaled relative to each other. This relative cost value is assigned to every planning unit. Cost can be many different things, but here serves as an index of values that either adversely affect the health of an ecosystem (human impacts) or make conserving a particular area less feasible (designation of land use and socioeconomic values). This index tends to reduce representation in places where human uses or modifications restrict conservation options. It is user defined, usually incorporating a variety of impacts to the environment, but may also include land status. These costs are generally seen as either more (e.g., lands already in some protected status) or less (e.g., lands devoted to resource extraction) suitable for conservation action. For our purposes, costs to the nearshore were primarily impacts, making particular places less suitable for conservation. There were also land use designations as factors in the intertidal and shallow subtidal zones.

Direct impacts to the shoreline were defined, as well as offshore factors. Shoreline impacts included the amount of armoring along the shoreline, the presence of railroad beds in the higher intertidal zone, and the number of public and private boat ramps. Offshore factors included ferry and commercial shipping routes. Land use and designation factors included ownership of tidelands (public versus private), fisheries closures, marine reserves, and other protected areas. Using the same nearshore costs within both hexagons and shoreline planning units resulted in similar indices. This provided consistency in running the site-selection analysis using both spatial formats. Since there was no way to assess viability for individual marine targets, the suitability index was built as a means of driving the algorithm towards the least disturbed examples of habitats.

Since good, comprehensive historic records were not available, population viability assessments with which to set goals have not been conducted regionally, and there was a lack of survey work reported on the condition of nearshoreprocesses, conservative (low) goals were set to help the algorithm assemble an efficient portfolio of sites important to multiple targets. Using this approach, we attempted to answer the question, 'Where do we start?' in evaluating places for nearshore biodiversity, as opposed to "How much (area) is enough?" to conserve that biodiversity. ShoreZone data were the most uniform across the ecoregion, providing the best data for describing a portfolio representative of the ecoregion's nearshore. We began by setting a "portfolio goal" for how much of the total shoreline length should be included in the nearshore marine portfolio. We examined a variety of portfolio goal levels from 20% to 40%. Likewise, we examined multiple goals for individual shoreline ecosystem targets ranging from 15% to 50% of the target's current extent.

Goals for individual shoreline ecosystem targets were determined by the relative diversity of habitats across the tidal range. The biologically simplest targets—unvegetated shorelines—received a goal of 25% of current extent. The biologically richest targets—shorelines with saltmarshes, seagrasses, and kelps—received the highest goal of 40%. The other targets received goals within this range,corresponding to their presence across the intertidal zone. In this way, the site selection algorithm chose more occurrences of the biologically richest sites to ensure representation of the wider range of species that occupy them. This approach to goal setting attempted to integrate intertidal habitats. The final goals chosen for the analysis are shown in Table 8.1.

For rocky reef habitats we set a goal of 30% of known existing occurrences, attempting to capture places with the highest levels of relief and complexity. Although data were not comprehensive across the ecoregion, and represented only one subtidal habitat, rocky reefs with and without the confirmed co-occurrence of rockfish species were identified by the algorithm. In setting goals for species targets we considered the relative abundance, distribution, and number of occurrences as well as our confidence in the data. Datasets that were more comprehensive across the ecoregion, recently compiled, represented a specific life stage (e.g., spawning) as opposed to observational or behavioral (e.g., swimming), or represented a series of observations over time, received higher goals. With these factors in mind, goals ranged from 20% to 60% of known occurrences.

By setting goal ranges we tested the sensitivity of conservation target representation. The importance of sensitivity analyses is to evaluate the efficiency of representation during the site selection process when goals and other pararmeters are varied. In addition to establishing goal ranges to test sensitivity, we experimented with a range of boundary modifiers for clumping hexagons that represented nearshore species and rocky reef habitat, and connecting adjacent linear units of representative shoreline ecosystem types.

Another key component is the penalty factor assigned to each target. A penalty is weighed against the cost factors and applied to targets for not meeting conservation goals. In other words, the penalty means that the user defines the importance of the target in meeting its goals, and this weighting is factored against cost. Therefore, if a target is assigned a relatively high penalty factor, then the algorithm will try harder to meet its goal, even if it has to select planning units with high cost values. We used our confidence in the data to help set this parameter. Targets represented by data with higher confidence received a higher penalty factor.

Multiple Scenarios Explored

Optimized, efficient, and spatially explicit representation of conservation targets across the ecoregion is an important aspect of defining a marine conservation portfolio. There are challenges in over-representing certain target elements, and not meeting conservation goals for others. Our objective was to explore two site selection scenarios to test these levels of representation for shoreline ecosystems, intertidal habitats, and marine species. We evaluated site selection output using definitions of "overrepresentation" and "efficiency of representation" as defined in Leslie et al. (2003). In addition, we added two categories in examining representation—"missing values" and "adequately captured."

Overrepresentation was defined as a target exceeding its assigned goal by 30% or more ($p > 1.3$). Adequately captured were determined to be targets that exceeded their goals but were not considered overrepresented ($p = > 1.03$ and < 1.3). Efficiency of representation was defined as a target meeting its goals at or close to 100% ($p = 1.0$ +/- .03). The last category, missing values, were targets that did not meet their goals ($p = < .97$).

Our first approach was to adopt the 750-hectare hexagon as the seamless planning unit across terrestrial and marine environments. All spatial datasets were intersected by the over 8,000 hexagons from the foothills of the Cascade Range to the waters of Puget Sound and the Strait of Georgia. We examined a variety of SITES parameters in configuring planning unit selection as described earlier. Variations of each parameter were tested until an optimal setting was found. With these optimal settings in place, we ran a single scenario and evaluated the amounts of representation across all nearshore target types. We chose the "best" output from SITES, which reflects the least overall cost (minimizing area and perimeter of planning units) across all iterations within the scenario in meeting conservation goals.

We also tested representation levels by conducting a nearshore-only analysis using two spatial planning units (shoreline units and nearshore hexagons). We used an analytical and expert review framework to order targets into "tiers" for portfolio assembly, capturing sites where multiple targets occur, where suitability for conservation is highest, and where our confidence in individual datasets is sufficiently high. Taxon groups (e.g., marine fishes) were analyzed at different stages of portfolio assembly. This was an attempt to build stepwise analyses, control data biases, and minimize overrepresentation of individual targets. The approach for building a nearshore marine portfolio combined spatial analysis and expert review into a tiered system. At each tier we analyzed ecosystem, habitat, and species data, then called upon experts to choose a select number of sites for that stage from the overall representation. These sites then became the "locked-in" areas for subsequent SITES runs. This systematic approach of varying the goals and building expert designation into the framework was used to test each tier against the other while refining the portfolio.

Results

We concluded that an overarching portfolio goal of 30% of the entire shoreline ecosystem (not including human-made shore units) was appropriate to identify priorities in evaluating the conservation of the diverse coastal environment. Reviewers consistently indicated that a 20% goal omitted some critical sites, especially where extensive dikes have been built but ecological processes were still intact (e.g., adequate freshwater and tidal flow regimes in estuaries for juvenile fish rearing habitat). Further, reviewers indicated that a goal of 40% identified too many sites that were often felt to be low in potential quality. Given that the algorithm attempts to filter a large amount of information into a representative subset, we felt that 30% was the appropriate level to test efficiency and overrepresentation of targets within a selection arrangement.

Influence of the Seamless Hexagon

In an attempt to have a consistent planning unit across the entire planning region, all target information was input into hexagons. SITES parameters were varied until optimal settings were established that best met conservation goals and minimized overrepresentation as much as possible. We have focused our results for this report on the nearshore portion only.

We input 119 data elements representing all of our shoreline ecosystems, habitats, and species targets. We stratified the 37 shoreline ecosystem targets into two sub-regions (Fig. 8.2) to make 74 data elements. For marine species, we generated 44 data elements from 72 targets. Some of these elements served as surrogates for more than one target (e.g., seabird colonies), while others had two data elements for one conservation target (e.g., some seabird targets were represented by two different datasets because of their spatial differences in British Columbia and Washington). We used an additional data element for rocky reef habitat, bringing the total discrete data elements to 45. For this exercise, we analyzed representation by data elements as they were input into SITES, and not by conservation target.

The best scenario output tables from SITES were used to evaluate the results of the seamless hexagon planning unit. We examined the stratified shoreline ecosystems and intertidal habitats both combined with and separately from the marine species and rocky reef elements (Table 8.3). The overrepresentation ($p > 1.3$) of all data elements was 57%, or 68 out of 119. Within the 74 stratified shoreline elements, 52, or 70%, were considered over-represented. For the 45 marine species and rocky reef data elements, only 16, or 36%, were considered over-represented. The adequately captured category ($p = > 1.03$ and < 1.3) contained 23% of all data elements, or 16% of the stratified shoreline elements and 33% of the marine species and reef elements. The efficiency of representation ($p = 1.0 +/- .03$) for the seamless hexagon analysis revealed 19 total data elements, or 16% (7% shoreline stratified targets and 31% marine species/reef elements). Finally, the missing

Table 8.3. Results of the seamless hexagon and separate nearshore analyses.

Data elements	*1A & B*		*2A & B*		*3A & B*		*4A & B*	
	SU1	*SU2*	*SU1*	*SU2*	*SU1*	*SU2*	*SU1*	*SU2*
Stratified shoreline	52	12	12	18	5	9	5	35
	70%	16%	24%	16%	7%	12%	7%	47%
Species/rocky reef	16	10	15	27	14	6	0	2
	36%	22%	33%	60%	31%	13%	0%	4%
Total	68	22	27	45	19	15	5	37
	57%	18%	23%	38%	16%	13%	4%	31%

Columns 1A & B: Overrepresentation (p> 1.3); Columns 2A & B: Adequately captured (p = > 1.03 and < 1.3); Columns 3 A & B: Efficiency (p = 1.0 +/- .03); Columns 4A & B: Missing values (p = < .97); SU1 = Seamless units; SU2 = Separate units. Amounts indicate the number of data elements in each category and their associated percentages. The total number of data elements is 119, or 74 stratified shoreline and 45 rocky reef/marine species elements

values category (p = < .97) contained 4% of all elements, or 7% of shoreline and 0% species/reef designations (Fig. 8.3; see page 199).

The optimal reserve program attempts to represent target elements at their assigned levels and minimize the total area selected. Since aggregating shoreline reaches within hexagons tended to generalize the spatial data into arbitrary groups, the seamless hexagon analysis over-represented ecosystem and habitat conservation targets. Additionally, terrestrial data was input into the coastal hexagons, further influencing the selection process. On the other hand, this scenario either adequately captured or efficiently represented most of the remaining data elements, leaving only 4% of the elements that did not meet their goals. This is an important component when evaluating overall efficiency of site selection in comparison to elements that either met or exceeded goals (96%).

This result facilitated the need to extract shoreline information from the hexagons and apply a distinct linear planning unit. The nearshore species and rocky reef habitat information, however, could continue to be run within hexagons but was also separated from the influence of terrestrial target data. Going from one to two planning unit configurations meant that we had to construct an analytical framework for tracking the selection process while keeping a close eye on the various categories of representation.

Influence of the Shoreline Unit and Nearshore Hexagon

We designed an analytical framework for constructing shoreline and nearshore site selection separate from the terrestrial environment. This

-tiered framework was an attempt to control data biases and overrepresentation that we found in running SITES using the seamless hexagon approach.

Tier 1 involved the experimentation of different goals and expert evaluation to come up with initial seascape sites. A stepwise analysis was performed to identify initial seascapes based on various SITES scenarios applied to hexagons. The objective was to use multiple goals to evaluate which planning units would get chosen most often. This form of an "irreplaceability analysis," or the selection of core sites, included the evaluation of the importance of the conservation target, data confidence and co-occurrence of species. From this we combined taxonomic groups into four categories. First we input data on the forage fish targets (e.g., herring and sand lance) into SITES, with areas chosen over a range of goals for each target (20% to 40%) being locked into the algorithm for subsequent data computations. Next, data for lingcod, rockfish, and the rocky reef habitats were input, and again the most selected areas were locked in. This procedure was repeated for seabird, marine mammal and invertebrate targets. We then evaluated this initial analysis and nominated the first portfolio seascape sites based on regional importance throughout the ecoregion. This identified 9% of the shoreline toward our 30% portfolio goal.

Tier 2 added sites most important to nearshore marine fish targets outside of Tier 1 sites. The marine team selected forage fish, rockfish, and lingcod as primary conservation targets for portfolio assembly based on their regional significance and international recognition as keystone species ecoregion-wide. This time SITES re-evaluated these selected marine fish species and rocky reef habitat data with goals set between 30% and 60%. Tier 2 identified an additional 14% of the shoreline, bringing the total to 23%.

Tier 3 added the rest of the target information, including seabirds, marine mammals and invertebrates, with goals for these targets set between 30% to 60%. In addition, two quality assessment workshops were conducted after the Tier 2 analysis to review the selected sites. We reviewed the portfolio to ensure best quality shorelines were included, especially any unique, diverse, or pristine sites known from field surveys. This tier added 4% of the shoreline, bringing the total to 27%.

Throughout the ecoregional planning process, we assembled coastal ecologists and biologists to fill data gaps by identifying places of known nearshore diversity or individual species significance. The information we learned from experts provided us with a means to compare to the spatial analysis. The purpose of Tier 4 was to evaluate all previously nominated expert sites that were not selected from the analysis up to this point, and incorporate a subset of them into the portfolio. Our primary measure of including an expert-derived site was to verify whether the site was of importance at the scale of the ecoregion and captured the targeted nearshore biodiversity. We determined that many sites originally nominated were not significant for inclusion in the portfolio because they were insufficiently important to warrant removing other sites to make room for them within our 30% portfolio goal. If we were to have added the shoreline associated with all the expert-nominated sites, Tier 4 would have inflated by 11%, or 832.5 km of shoreline, thereby over-representing shoreline types without analytical discrimination. By applying scrutiny to the expert-nominated process, Tier 4 added only 1% of the shoreline, bringing the final draft nearshore marine portfolio to 28%. This represented approximately 2,095 km of shoreline out of the total natural (excluding human-made) shoreline of 7,533 km (Fig. 8.4; see page 200). We elected not to search for the additional 2% of shoreline to meet our portfolio goal of 30%, anticipating the addition of some shoreline when integrated with a separate terrestrial analysis.

In looking at representation of the marine targets through this 4-tiered process, we found that overrepresentation was not as much of a factor as when using the seamless hexagon approach (Table 8.3). The final draft nearshore-only analysis over-represented 22, or 18% of the 119 data elements. In analyzing the shoreline planning unit results, only 12 of the 74, or 16% of the stratified shoreline elements were considered over-represented.. For the 45 marine species and rocky reef data elements, only 10, or 22 % were considered over-represented. For the adequately captured category, 38% of all data elements were represented, or 24% of the stratified shoreline and 60% of the marine species and reef elements. The efficiency of representation for the separate analysis units revealed 15 total data elements, or 13% (12% shoreline stratified targets and 13% marine species/reef elements). The missing values category contained 31% of all elements, or 47% of shoreline and 4% species/reef designations.

Utilizing the results from both the seamless hexagon and nearshore-only analysis using two planning units, we then conducted a site delineation process that refined the portfolio into priority conservation areas (Fig. 8.5; see page 200). The final nearshore marine component of the integrated ecoregional assessment identified 186 shoreline/nearshore sites within the Puget Trough and Georgia Basin. In combining the separate analyses and refining boundaries, all nearshore conservation goals were met.

Discussion

The nearshore is subject to forces both oceanic and terrestrial, producing ecosystems that are dynamic and "open" in nature. This openness of marine populations, communities, and ecosystems probably has marked influences on their spatial, genetic, and trophic structures and dynamics in ways experienced by only some terrestrial species (Carr 2003). The nearshore is not easily defined and mapped, thus conservation planning is more difficult than on land. Conducting this regional analysis of nearshore biodiversity, however, was a step forward in objective representation and quantitative marine site selection.

Comparing the utility of an abstract spatial planning unit (e.g., hexagons) with a natural one (e.g., shoreline reaches) was an important step in evaluating site selection. Although uniform abstract units, such as hexagons, decrease the accuracy of ecological data because of the arbitrary boundaries imposed on the land or seascape (see Fotheringham, 1989; Stoms, 1994), they provide a simple method for mapping data and evaluating effects of selection unit variation on reserve selection (Warman et. al., 2004). This uniformity also provides the means to include information across environments such as the land/sea interface. Alternatively, natural units of analysis tend to be more spatially explicit and follow ecological boundaries. This may reduce additional refinement of site boundaries after the selection process when determining a priority conservation area. However, they are often quite variable in size, and some may be considered too small for effective implementation of conservation areas. In either case, size is an essential component to consider prior to building a conservation portfolio. Typically, the larger the planning unit size, the more generalized and therefore over-represented the conservation targets. Morphology of the study area (e.g., long, straight coastal environments versus highly convoluted coasts with many estuaries or fjords), scales of data input, and spatial unit configuration are all factors to consider. It is essential to examine both the habitat complexity of the region and the scale of the data available for reserve selection before deciding the size and shape of the selection unit (Warman et. al., 2004). We therefore attempted to match the size of the abstract planning unit roughly in relation to the scales of data input, and used both an abstract and natural unit for comparing target representation.

The objective of this planning process was to examine multiple approaches to spatial planning unit configuration in order to achieve the most efficient suite of sites representing nearshore ecosystems, habitats, and species. We established a set of site selection definitions to help us determine what conservation targets contained missing values, which were efficient in their representation, those that were adequately captured, and others that were over-represented. The marine planning team decided that overrepresentation was a key issue to consider because we wanted to be conservative in our approach to selecting shoreline as potential conservation areas. We concluded that the 750-hectare hexagon unit was too large and therefore generalized the shoreline data. This drove the cost of selecting those sites up, decreasing optimization. This decrease in optimization yielded an overrepresentation of shoreline targets. The development of a separate nearshore-only analysis using two spatial planning units therefore gave us the opportunity to increase the overall efficiency of the portfolio.

We determined that the shoreline planning units tended to minimize overrepresentation, thereby increasing overall efficiency (i.e., combining efficiency of representation and adequately captured categories). The under-representation of many elements in the nearshore-only analysis was not a big concern for us given that we still had to combine terrestrial analyses along the coast, thereby adding shoreline to the overall portfolio through a site delineation process.

Other planning teams may not consider over-representation as a significant factor, even though it drives up the cost of the reserve system. For instance, if efficiency of representation is considered to be the key issue, then overrepresentation of conservation targets may be viewed as beneficial for the long-term conservation of species (see Warman et. al., 2004). The seamless hexagon analysis reduced the number of missing values. This is essential if there is not any further refinement of the portfolio, either through integration with other analyses (e.g., combining separate terrestrial and freshwater site selection) or a site delineation process that ensures that all conservation targets have met their goals. If this is the emphasis of the planning team, then using the nearshore-only approach may not be preferred. The results of our separate unit analysis illustrated that 31% of all data elements did not meet goals in relation to the 69% that met or exceeded them. This could be seen as a less overall efficient portfolio in comparison to the seamless hexgon analysis that met or exceeded 96% of its goals. Depending on the teams' definition and expectations of the optimal reserve program (e.g., ecological accuracy versus spatial efficiency), there may always be trade-offs to make when examining issues of representation.

Our nearshore analysis framework combined both spatial data and expert opinion in building a conservation portfolio. ShoreZone data provided an excellent baseline on coastal characteristics, painting a regional picture over thousands of kilometers of shoreline. Through this lens, we were able to capture concentrations of multiple nearshore targets within the conservation portfolio. This was a starting point, however, and not a complete picture. As it can only represent the places to begin evaluating the current condition of nearshore marine ecosystems, incorporating expert review into the site selection process provided a critical assessment of quality. However, extensive ground-truthing of the selected areas is necessary to further assess condition and ecological integrity. Varying goals or representation levels did allow us to consider the "irreplaceability" of selected sites. Irreplaceability here means the number of times a single planning unit is chosen over mutiple scenarios. The varying of goals to determine irreplaceable sites is based on where conservation targets are found, and not on the attributes of a single best scenario and the particular rules used to select sites (see Pressey et al. 1994; Hopkinson et al., 2001). The nearshore-only analysis helped identify both representative and high-quality places because it combined irreplaceability, optimization, and expert review. Consequently, the nearshore portfolio captured an array of representative habitats and species in addition to providing an indication of priority conservation areas.

The Puget Sound and Strait of Georgia region is heavily impacted by human development, and this development is expected to only increase. Point and nonpoint source pollution, invasive species distributions, aquaculture, and coastal development continue to threaten the integrity of the region's marine life. "Conservation by Design" sets forth The Nature Conservancy's vision for abating these threats and calls to action the implementation of conservation strategies. It is anticipated that marine ecoregional planning methods will be improved upon as advances in our understanding of marine biodiversity and the threats they face are made, and the datasets that underlie these analyses, improve. Until then, utilizing automated site selection algorithms and expert opinion provide a foundation for initially comparing ecosystem and habitat representation across ecoregional land and seascapes. We therefore put forth these methods and results as the first iteration with the expectation that future iterations will improve our confidence in conservation portfolios intended to preserve nearshore marine biodiversity.

Acknowledgements

The author would like to thank the following individuals, agencies, and organizations for participating on the marine technical teams and providing helpful assistance: Curtis Tanner (U.S. Fish & Wildlife Service), Helen Berry and Betty Bookheim (Washington Department of Natural Resources), Jacques White and Phil Bloch (People for Puget Sound), Megan Dethier (Friday Harbor Laboratories, University of Washington), Pierre Iachetti and Gary Kaiser (Nature Conservancy Canada), Mary Lou Mills, Dave Nysewander, Bob Pacunski, Brian MacDonald, and Mary Lou Mills (Washington Department of Fish & Wildlife), Paul Dye (The Nature Conservancy), Mark Zacharias and Carol Ogborne (Ministry of Sustainable Resource Management), and Mary Morris (Archipelago Research LTD). Additional thanks goes to Mike Beck, Dan Dorfman John Floberg, and Peter Kareiva of The Nature Conservancy for reviewing this manuscript.

The Nature Conservancy is an international non-profit conservation organization that seeks to preserve the plants, animals, and natural communities that represent the diversity of life on Earth by protecting the lands and waters they need to survive.

References

Andelman, S. A., Ball, I., and Stomms, D., 1999. SITES 1.0: *An Analytical Toolbox for Designing Ecoregional Conservation Portfolios,* Arlington, VA, The Nature Conservancy. 55 pp.

Airamé, S., Dugan, J. E., Lafferty, K. D., Leslie, H., McArdle, D. A., and Warner, R. R., 2003. Applying ecological criteria to marine reserve design: a case study from the California Channel Islands, *Ecological Applications,* 13(1) Supplement: S170-S1854.

Banks, D., Williams, M., Pearce, J., Springer, A., Hagenstein, R., and Olson, D., 1999. *Ecoregion-Based Conservation in the Bering Sea,* Washington, DC, World Wildlife Fund, and Anchorage, AK, The Nature Conservancy. 72 pp.

Beck, M. W., and Odaya, M., 2001. Ecoregional planning in marine environments: identifying priority sites for conservation in the northern Gulf of Mexico, *Aquatic Conservation: Marine and Freshwater Ecosystems,* 11: 235-42.

Beck, M. W., 2003. The Sea Around – Planning in marine regions, in Groves, C. (ed.), *Drafting a Conservation Blueprint,* Washington DC, Covelo and London, Island Press, pp. 319-44.

Berry, H. D., Harper, J. R., Mumford, T. F., Jr., Bookheim, B. E., Sewell, A. T., and Tamayo, L. J., 2001. *The Washington State ShoreZone Inventory User's Manual. Nearshore Habitat Program,* Washington State Department of Natural Resources, Olympia, WA. 29 pp.

Carr, M. H., Neigel, J. E., Estes, J. A., Andelman, S., Warner, R. R., and Largier, J. L., 2003. Comparing marine and terrestrial ecosystems: implications for the design of coastal marine reserves, *Ecological Applications,* 13(1) Supplement: S90-S107.

Day, J., and Roff, J. C., 2000. *Planning for Representative Marine Protected Areas: A Framework for Canada's Oceans.* Report prepared for World Wildlife Fund Canada, Toronto. 147 pp.

Dethier, M. N., 1990. *A Marine and Estuarine Habitat Classification System for Washington State.* Washington Natural Heritage Program, Department of Natural Resources, Olympia, WA. 56 pp.

Evenson, J. R., and Buchanan, J. B., 1997. Seasonal abundance of shorebirds at Puget Sound estuaries, *Washington Birds,* 6:34-62.

Ferdaña, Z. A., 2002. Approaches to integrating a marine GIS into The Nature Conservancy's ecoregional planning process, in Breman, J. (ed.), *Marine Geography: GIS for the Oceans and Seas,* Redlands, CA, ESRI Press, pp. 151-58.

Fotheringham, A. S., 1989. Scale-independent spatial analysis, in Goodchild, M., and Gopal, S. (eds.), *The Accuracy of Spatial Databases,* London, Taylor and Francis, pp. 221-28.

Groves, C. R., Jensen, D. B., Valutis, L. L., Redford, K. H., Shaffer, M. L., Scott, J. M., Baumgartner, J. V., Higgins, J. V., Beck, M. W., and Anderson, M. G., 2002. Planning for biodiversity conservation: putting conservation science into practice, *BioScience* 52(6): 499-512.

Hopkinson, P. J., Travis, M. J., Evans, J., Gregory, R. D., Telfer, M. G., and Williams, P. H., 2001. Flexibility and the use of indicator taxa in the selection of sites for nature reserves, *Biodiversity and Conservation,* 10:271-85.

Howes, D. E., Harper, J. R., and Owens, E. H., 1994. *Physical shore-zone mapping system for British Columbia,* Technical Report, Victoria, British Columbia, Canada, Coastal Task Force of the Resource Inventory Committee (RIC), RIC Secretariat. 71 pp.

Howes, D. E., Wainwright, P., Haggarty, J., Harper, J., Owens, E., Reimer, D., Summers, K., Cooper, J., Berg, L., and Baird, R., 1993. *Coastal Resources and Oil Spill Response Atlas for the Southern Strait of Georgia, Victoria, British Columbia*, Canada, BC Ministry of Environment, Lands and Parks, Environmental Emergencies Coordination Office. 317 pp.

Leslie, H., Ruckelshaus, M., Ball, I. R., Andelman, S., and Possingham, H. P., 2003. Using siting algorithms in the design of marine reserves, *Ecological Applications*, 13(1) Supplement: S185-S198.

Margules, C. R., and Pressey, R. L., 2000. Systematic conservation planning, *Nature*, 405: 243–53.

The Nature Conservancy, 2001. *Conservation by Design: A Framework for Mission Success*, Arlington, VA, The Nature Conservancy. 16 pp.

Possingham, H., Ball, I., and Andleman, S., 2000. Mathematical models for identifying representative reserve networks, in Ferson, S., and Burgman, M. (eds.), *Quantitative Methods for conservation Biology*, New York, Springer-Verlag. pp. 291-306.

Power, M. E., Tilman, D., Estes, J. A., Menge, B. A., Bond, W. J., Mills, L. S., Daily, G., Castilla, J. C., Lubchenco, J., and Paine, R. T., 1996. Challenges in the quest for keystones, *BioScience*, 46(8): 609-20.

Pressey, R. L., Humphries, C. J., Margules, C. R., Vane-Wright, R. I., and Williams, P. H., 1993. Beyond opportunism:key principles for systematic reserve selection, *Trends in Ecology and Evolution*, 8: 124–28.

Pressey, R. L., Johnson, I. R., and Wilson, P. D., 1994. Shades of irreplaceability: towards a measure of the contribution of sites to a reservation goal, *Biodiversity and Conservation* 3: 242-62.

Searing, G. F., and Frith, H. R., 1995. *British Columbia Shore-zone Mapping System*, Contract Report by LGL Ltd., Sidney, BC for the Land Use Coordination Office, Victoria, British Columbia, Canada, BC Ministry of Environment. 46 pp.

Stoms, D. M., 1994. Scale dependence on species richness maps, *Professional Geographer* 46: 346-58.

Sullivan-Sealey, K. M., and Bustamante, G., 1999. *Setting Geographic Priorities for Marine Conservation in Latin America and the Caribbean*, Arlington, VA, The Nature Conservancy. 125 pp.

Ward, T. J., Vanderklift, J., Nicholls, M. A., and Kenchington, R. A., 1999. Selecting marine reserves using habitats and species assemblages as surrogates for biological diversity, *Ecological Applications*, 9: 691-98.

Warman, L. D., Sinclair, A. R. E., Scudder, G. G. E., Klinkenberg, B., and Pressey, R. L., 2004. Sensitivity of systematic reserve selection to decisions about scale, biological data, and targets: case study from Southern British Columbia, *Conservation Biology*, 18: 655-66.

PART III

Democratizing Data: Civil Society Groups' Usage of Marine GIS

Chapter 9

Continental-scale Conservation Planning

The Baja California to Bering Sea Region

Lance E. Morgan, Peter J. Etnoyer, and Elliott A. Norse

Abstract

The North American Commission for Environmental Cooperation (CEC) provides a facilitative body for Canada, Mexico, and the United States to address common environmental concerns and promote biodiversity conservation. The CEC is implementing a North American Marine Protected Areas Network (NAMPAN) with a wide array of partners in governmental and nongovernmental sectors in order to protect marine species throughout their ranges and across these three Exclusive Economic Zones. Identifying priority conservation areas for the Baja California to Bering Sea Region (B2B), is one initiative of this network. Marine Conservation Biology Institute (MCBI) worked jointly with the CEC in a multi-year process of consultation, data gathering, data analysis and GIS development culminating in an experts' workshop to define priority conservation areas. In this chapter we describe this effort, the workshop to identify priority conservation areas and briefly describe the final output of this process. A total of 28 sites were identified as priority conservation areas (PCAs), totaling 8% of the total Exclusive Economic Zone (EEZ) area of the three nations. PCAs vary by threat and protection status, but they represent a shared vision of critical places for North America's marine biological diversity. This portfolio

Lance E. Morgan
Marine Conservation Biology Institute, 4878 Warm Springs Road, Glen Ellen, CA, 95442
Corresponding author: lance@mcbi.org, phone 707-996-3437

Peter J. Etnoyer
Aquanautix Consulting, 3777 Griffith View Drive, Los Angeles, CA, 90039
peter@aquanautix.com, phone 323-666-3399

Elliott A. Norse
Marine Conservation Biology Institute, 15805 NE 47th Court, Redmond, WA, 98052
elliott@mcbi.org, phone 425-883-8914

of sites is a first step towards building a continental community to foster the development of cooperation and stewardship of the B2B region.

Introduction

In recent years, conservation strategies, noting past failures to stem the tide of extinctions, have focused to a greater degree on large-scale ecosystem approaches (e.g., Wildlands Strategy, World Wildlife Fund's Global 2000). Conservation efforts traditionally focused on individual populations, often protecting small areas that can safeguard only a small portion of the total population (Soulé et al., 2003). But outside small, isolated reserves, contamination, fragmentation, and the death of individuals occur daily. Thus, we must look to maintain ecological processes across an entire seascape.

Landscape ecology provides a new conceptual basis for continental conservation plans (Soulé and Terborgh, 1999). Sherman's "Large Marine Ecosystems" (LME) (Sherman et al., 1990; Sherman and Duda, 1999) and the work on "Biogeochemical Provinces" (BGCP) by Longhurst and colleagues (Longhurst, 1998) have conceptually helped bring marine ecosystems into a management context. Several multinational planning exercises have been carried out in the marine realm in recent years, suggesting the importance of a broader ecosystem focus (e.g., Banks et al., 1999; Sullivan-Sealey and Bustamante, 1999). Marine conservation planning, similar to terrestrial conservation planning, should recognize four critical aspects necessary to conserve species and processes: (1) conserving species and processes that require the greatest area to persist; (2) conserving widespread species and continental phenomena; (3) quantifying patterns of beta diversity and endemism; and (4) predicting the location and intensity of threats to biodiversity (Olson et al., 2002). Conservation planning must also map important areas, such as biodiversity hot spots and other conservation priorities, in order to set priorities for action (e.g., Hixon et al., 2001; Roberts et al., 2002).

Although many conservation efforts and sustainable development initiatives exist at different scales along the Pacific Coast of North America, they generally work independently of each other. Unless these efforts are coordinated, species numbers will continue to decline and ecosystem integrity will continue to be at risk. For example, gray whales have rebounded, thanks to an international agreement to stop whaling, and local efforts to protect calving lagoons in Mexico. The successful conservation of the North American seascapes requires cooperative action from all three countries and from diverse sectors of society. The CEC was created by the governments of these three countries—Canada, Mexico, and the United States—to address common environmental concerns under the North American Agreement for Environmental Cooperation, a side agreement to the North American Free Trade

Agreement (NAFTA). The North American Marine Protected Areas Network (NAMPAN) represents one initiative to facilitate collaboration to safeguard ecological linkages, and conserve marine biodiversity and productivity throughout the exclusive economic zones (EEZs) of the three nations. This initiative also complements the conclusions of the World Summit for Sustainable Development, where participating governments, including the NAFTA signatories committed to implementing marine protected area networks by 2012.

This chapter describes the process of developing appropriate datasets and analyses for identifying priority conservation areas (PCAs) in the Baja California to Bering Sea Region (B2B). Iteratively over the course of this project, the definition of PCAs was refined to reflect the mandate of the CEC, the variable nature of data available in the three nations, and the spatial scale of the region. Other initiatives advance a common framework by mapping marine ecoregions (Wilkinson et al., 2004b), identifying species of common conservation concern, and working to provide an understanding of the institutions in each country through which an integrated network of linked organizations can implement the NAMPAN. This PCA initiative seeks to detail where conservation action is immediately necessary, and charts a course for future conservation alliances and action in the B2B region.

Methods

The methodology for identifying PCAs relied on teaming experts' knowledge with the development of a geographic information system (GIS). GIS systems are ideally suited to conservation planning across large ecosystems because they can combine physical, biological, and social data into a single spatial frame of reference. GIS can scale spatially from the continental to the regional, and temporally from the annual to the daily. Recent advances in GIS technology allow visualization of the seafloor and water column in three dimensions, a critical aspect of conservation initiatives such as this one, with benthic and pelagic components.

The large geographic extent of the B2B region limits the viability of an entirely data-driven analysis at this scale. Comprehensive data and dependable proxies do not exist. It was independently concluded several times that the most likely approach for the entire B2B region is a site nomination Delphic approach that combined specific datasets and analyses, and captured the range of habitat diversity based on expert judgment related to biodiversity, threat and opportunity.

The GIS included appropriate spatial datasets of physical, biological, and social information. Analyses focused on translating several of these datasets to highlight regions where physical processes lead to unique features or high abundances of species. At the final PCA identification workshop, experts reviewed the aggregated datasets and analyses to inform their judgments of ecological value and conservation priority.

Priority was identified based on the ecological significance of the areas to North America, threats to the area, and opportunities to advance conservation.

This initiative spanned the course of three years, and tapped the knowledge of over 200 marine and social scientists from three different countries and 75 different organizations. The process can be described in the following manner: outline of work plan, data needs assessment, data needs ranking, data gathering and distribution, analysis, and Delphic selection of priority conservation areas. Here we present a detailed description of the process, in the hope that this effort may inspire cooperation and greater openness in conservation planning in other multinational waters.

Goal of the Priority Conservation Areas Project

In 2000, the CEC identified the Baja California to Bering Sea region as one of its Priority Regions for Biodiversity Conservation of North America[1]—this region is defined as the EEZ of Mexico, the United States, and Canada from 22°N latitude to 65°N latitude. The B2B region was advanced as the first test case for the CEC to implement its strategic plan in the marine environment[2].

In May 2001, MCBI and the CEC convened a workshop in Monterey, California, in the United States, where scientific experts, resource users, and marine conservationists from the three countries addressed the goals and identified the types of baseline data that are required for conservation in the B2B region. They agreed on the need to identify PCAs as a step in a larger continental-scale conservation effort. They also reached consensus that the overarching goal of a PCA is to conserve biodiversity, and should also include benefits to fisheries, cultural values, recreation, and scientific research. These experts agreed on the development of a GIS, based on common physical data for the entire region, to serve as a framework for integrating other information. The GIS included biological, physical, and social data layers. Experts also addressed issues of size and spatial scale, incorporating previous priority setting efforts and anthropogenic threats (Morgan and Etnoyer, 2002).

What is a PCA?

The first and most challenging aspect of this project was the definition of a priority conservation area. The definition was iteratively defined and refined through out the scope of this project. Consensus was achieved in a statement that defined PCAs on the basis of significant biodiversity and continental uniqueness, and incorporating three factors: (1) ecological value; (2) anthropogenic threat; and (3) opportunity for conservation (government or local support, existing designations, and conservation initiatives). Since no comprehensive measure of biodiversity exists for the B2B region, experts were asked to assess biodiversity indirectly, relying on their accumulated knowledge of species, habitats and ecological processes. Several factors were to be considered in their assessment: (1) continental scale physiographic and oceanographic features (features on the order of 100–1,000 km^2); (2) high beta-level biological diversity (between-habitat diversity); (3) continental endemism; (4) key habitats—concentration areas such as breeding and feeding sites or migration routes—for marine species of common conservation concern (Appendix 9.1); (5) critical habitats of umbrella and charismatic species that require large areas to persist; (6) areas that provide whole region benefits, e.g., seasonally productive, migration corridors; (7) areas of high biomass and/or productivity, e.g., coastal upwelling centers; and (8) emphasis on transboundary areas.

These criteria are consistent with other approaches that suggest capturing areas that contain regional representation of major habitats, diverse types of habitats, rare and threatened species and habitats, and endemic species, is a viable conservation strategy for defining priorities. At the same time, it is important to capture oceanographic processes and ecological linkages that interconnect these habitats. The geographic scope of the project (EEZ from 22°N latitude to 65°N latitude) included estuaries and islands, but not upland areas of freshwater environments.

Data Compilation and Distribution

Conservation planning exercises include physical, biological, and social components, and the data that goes into GIS analyses must reflect each of these. The physical oceanography community has a long history of basin scale data collection efforts, but biological and social datasets are more limited in scope. Therefore, we generated relevant datasets through synthesis of country-level information such as EEZ boundaries, population, ports, national parks, and local priorities. Few of these datasets had ever been observed in the context of their partners, nor in the same software, or projection. All data were projected to a uniform Lambert Azimuthal equal area projection with a central longitude that bisects North America. This compromise permits easy comparison with forthcoming North Atlantic datasets.

Digital assets from the physical oceanography community range from ships of opportunity to moored buoy arrays to satellite-derived measurements for gravity, topography, surface temperature, surface height, and chlorophyll. The physical oceanography community also has a long history of data sharing and distribution, and these types of information are readily available on the internet from NASA's Jet Propulsion Laboratory (JPL) Physical Oceanography Distributed Active Archive Center (PODAAC).

Biological data were both the most difficult to obtain and the most revealing types of information. No single biological survey encompasses the entire B2B latitudinal extent. Satellite derived estimates of chlorophyll from the SeaWIFS project, which were converted to primary productivity using methods described by Behrenfeld and Falkowski

(1997), come the closest. Neither the NOAA triennial trawl surveys, nor the Partnership for Interdisciplinary Studies of Coastal Oceans (PISCO) consortium, nor the bi-national California Cooperative Oceanic Fisheries Investigation (CalCOFI) survey programs encompass a uniform spatial and temporal extent over the entire Pacific Coast of North America out to 200 nautical miles. We compiled several datasets, including almost 2,700 records of habitat forming deep-sea coral occurrences from 10 different record-keeping institutions, to represent the benthic component of the region (Etnoyer and Morgan, 2003), and blue whale tracks from the Department of Fisheries and Wildlife of Oregon State University. Future data gathering efforts should focus on and support large-scale efforts and seek to integrate national fisheries statistics and survey programs.

All the datasets described in Table 9.1 are included on the B2B 1.1 CD-ROM as either attributed points (e.g., deep-sea corals and ports and harbors), lines (e.g., blue whale tracks), and polygon (e.g., EEZ) coverages, shapefiles (federal MPAs), or rasters (4 km-resolution altimetry, 9 km-resolution sea surface temperature). The oceanographic datasets are bi- weekly or monthly over a four-year El Niño Southern Oscillation (ENSO) cycle (1996-1999). Postscript maps and animations of sea surface temperature and sea surface height from the US Navy Layered Ocean model were included in an extras folder on the CD-ROM.

All datasets on the B2B CD-ROM include federally compliant metadata regarding the data's origin, and most are groundtruthed for data quality, as well as vertical and horizontal accuracy. This is an important step in the process of analysis to insure the legitimacy of results, and to anticipate criticism from federal or commercial interests opposed to conservation measures. Smith and Sandwell's (1997) global satellite altimetry-derived bathymetry was compared to multi-beam data for peak heights on 12 seamounts in the Gulf of Alaska, and found to vary within a remarkable range of 39 to 504 m (Etnoyer, in press; Fig. 9.1; see page 201). This quality control exercise indicates our continuing need for finer scale coastal bathymetry data. All bathymetry values were upgraded from the satellite-derived resolution of 4 km to

Table 9.1. All of these datasets are uniformly projected and included on the B2B 1.1 CD-ROM with FGDC compliant metadata for use in a marine conservation GIS.

Physical	*Biological*	*Social*
Surface Currents	Chlorophyll	Local priorities
World Vector Shoreline	Mammals	Population
Sea Surface Temperature	Turtles	Ports and Harbors
ETOPO2 Bathymetry	Deep-Sea Corals	MPAs
Seamounts	EEZ	GTOPO30 topography

a 100 m tri-national synthesis product that covered approximately 40% of the non-Alaska study area. Maps of deep-sea corals were found to reflect research effort, and to likely underestimate the abundance and distribution of deep-sea corals within their cosmopolitan range and their 20 m–4,000 m vertical extent. The Hawaiian Undersea Research Laboratory regularly groundtruths NASA's Advanced Very High Resolution Radar from Pathfinder, finding that data to vary in accuracy from day to night, falling generally within one tenth of a degree.

It is important to note that different countries have different data standards and different policies regarding the freedom of information. The United States passed the Freedom of Information Act in 1967, but Mexico passed its own in 2003. Before B2B, medium-resolution seafloor bathymetry in Canada was proprietary, and some government agencies had to pay for access to bathymetry data off British Columbia. In keeping with the spirit of transparency and cooperation, the Canadian Department of Fisheries and Oceans granted MCBI distribution rights to that information. Mexico also met the challenge of cooperation, with ready access to medium-resolution bathymetry from the Gulf of California. These datasets were not made available as part of the B2B 1.1 CD-ROM, but they were made available to researchers participating in the B2B analyses.

Spatial scale is a critical aspect of GIS analyses that range, for example, over three different countries with three different data standards and three different levels of investment in data development. One must "draw the line" somewhere to avoid an overabundance of spatially or temporally irrelevant data. We considered two spatial approaches: first, selecting datasets with a common resolution over the B2B extent, and second, building a "patchwork" of variable resolution datasets. We agreed that it was "unfair" to ultimately discern smaller PCAs from countries with higher-resolution datasets (e.g., bathymetry), so we settled on common resolution data derived largely from satellites. In the end, we incorporated medium-resolution (~100 m) bathymetry in our benthic complexity analysis (Ardron, 2002). For temporal consistency, we identified a contemporary four-year time frame (1996-1999) that captures the extremes of ENSO variability.

In June 2002, MCBI, in collaboration with the CEC, Ecotrust, and Surfrider Foundation, organized a "Data Potluck" workshop in Portland, Oregon. In this workshop, nearly 80 representatives from 30 organizations offered and exchanged datasets that appeared relevant to the spatial scale of the B2B region. This information and advice was incorporated into the B2B CD-ROM. The Data Potluck was the second in a series of technical meetings designed to build consensus on spatial methods of analysis for priority conservation areas.

Whereas experts at the first meeting held in Monterey placed emphasis on data types such as sea surface temperature and surface currents, the Data Potluck presentations revealed a new emphasis on socioeconomic information that was not evident in the previous Monterey Workshop. This difference in emphasis may be a result of the different backgrounds of the attendees at the two workshops, or may reflect the evolving nature of marine protected area (MPA) science. Socioeconomic data that were included on the B2B CD-ROM included locations of fishing ports and landings information, cities, population, MPAs, and previous efforts to identify conservation priorities for different regions within the B2B realm. The mandate given to the project was to use existing sources. Thus, no new data were collected, although significant efforts to digitize certain datasets did occur. In several cases, we included previous exercises to define priorities at regional scales. The spatial data generated by this effort are available on CD-ROM in GIS format (Etnoyer et al., 2002).

Workshop participants identified many "parallel projects" within the continental B2B region that have strong sub-regional potential, and identified common data needs for a more evenly distributed workload, and a potential for these disparate organizations to begin to speak with one voice, without sacrificing their individual institutional goals. The Data Potluck idea was very well received. Ed Backus of Ecotrust mentioned that the Potluck idea seemed awkward at first, but his staff eventually came to terms with the idea that a Potluck methodology provides an incentive to contribute, lowers expectations, and levels the playing field by providing all participants with the same information, which they then might use to address their own concerns.

MCBI staff also took advantage of the assembled expertise to conduct a survey of marine GIS users to understand their opinions on the data needed for successful identification of priority conservation areas. Twenty respondents from the three countries, all familiar with data-driven priority-setting exercises for conservation goals, completed a survey distributed by MCBI. Respondents ranked data types for their potential contribution to a priority habitat analysis at the continental scale, commented on methods, and future data needs.

Bathymetry, primary productivity, existing MPAs, and fishing pressure data were ranked highest (>4.5) for their ability to strengthen a GIS for a priority habitat analysis. All listed data (Table 9.2) save LIDAR and NGO Activity ranked above 3.5 on a scale of 5. Respondents generally valued their personal contributions ("Other") very highly, with substrate type, spawning aggregations, submarine cables, political climate, pollution and "community will" each receiving unsolicited votes from 20% of the respondents.

The general response to the qualitative question, "How would you like to see these data layers used in a GIS to generate a list of priority habitats for the B2B region?" suggested that most viewed "priorities" as a qualification of either the degree of threat or the opportunity for action. That is, most felt that the highest conservation priorities were those sites in great danger, or those sites where conservation actions

Table 9.2. Responses to survey asking participants to rank the data types they believed had the greatest ability to strengthen a GIS priority areas analysis at the continental scale (1= will not add much to the analysis, 5 = will strengthen the analysis considerably).

Physical Data Type	*Rank*	*Biological Data Type*	*Rank*	*Social Data Type*	*Rank*
Bathymetry	4.67	Primary productivity	4.67	Other	4.78
Other	4.66	Other	4.66	MPA	4.53
Seamounts	3.93	Mammal tracks /dist.	4.23	Fishing pressure	4.49
Sea surface temperature	3.71	Submerged aquatic vegetation	4.19	Jurisdictions	4.00
Altimetry (surface currents)	3.59	Seabird tracks/dist.	4.10	Ports and harbors	3.87
LIDAR	2.67	Turtle tracks/dist.	4.05	Previous priorities	3.66
		Deep-sea corals	3.93	NGO activity	2.87
		NOAA Atlas	3.67		

Other	*Count*	*Other*	*Count*	*Other*	*Count*
Substrate	4	Spawning aggregations	4	Cables	4
Sediment transport	3	Fish spp. distributions	3	Political climate	4
Lagoons	3	Nurseries	2	Community will	4
Upwelling	2	Feeding aggregations	2	Pollution/ Dump sites	4
				Recreational uses	3

Other	*Count*
Consistent high resolution shoreline, upwelling, salinity, shelf, ocean features	2
Outfalls, shipping channels, economic impact, fishing grounds, indigenous use	2
Large predators, benthic spp. assemblages, historical abundance/ distribution, migration corridors, kelp/ mangrove	1
Pipelines, current litigation, shipwrecks, ongoing efforts, oil leases, population, enforcement, access, mega-development projects	1

Comments: Bathymetry and primary productivity were ranked highest for their ability to strengthen a priority area analysis. All data save LIDAR and NGO Activity ranked above 3.5 on a scale of 5. Respondents generally valued their contributions ("Other") highly, with substrate type, spawning aggregations, submarine cables, political climate, pollution and community will each receiving unsolicited votes from 20% of the respondents.

were in progress, but incomplete. Individuals varied widely in their opinion of PCA goals, but the above model stands out as the most common view, with the most applicability. Comments by Mexican participants strongly suggested that some kind of "governance index" or measurement of "community will" was important. This was reinforced by some U.S. respondents, who had previously thought existing MPAs were the strongest candidates for enhanced protection. This reflects the general opinion that existing legislation rarely translates into effective management, and that boundaries are poor indicators of protection. Respondents also felt that submarine features and upwelling indices could benefit any PCA algorithm. Survey results indicated a large amount of variability and opinion regarding critical data.

In the 12 months anticipating the Priority Conservation Areas Workshop, MCBI distributed nearly 200 copies of the B2B CD-ROM to more than 50 different organizations and more than a dozen different countries. We charged a nominal fee to offset the costs of data packaging and distribution, and encouraged researchers to perform and submit their own priority setting exercises for consideration. Two submissions of outside analyses were received, benthic complexity by Jeff Ardron of the Living Oceans Society in British Columbia, and primary productivity analyses by Chuanmin Hu and Frank Muller-Karger at University of Southern Florida. The B2B 1.1 CD-ROM provides the foundation for half a dozen graduate theses to date, and background data for many regional investigations. The B2B CD-ROM provides a model for future "democratic" priority-setting exercises in marine conservation by providing information to all as an open data source.

Data Analyses

Several data analyses were conducted in order to highlight the significance of selected datasets to the conservation priority setting exercise. These analyses include: (1) benthic complexity—a measure similar to rugosity; (2) sea surface temperature fronts—areas known to aggregate a wide-variety of pelagic sea life, including fishes, sea turtles, birds and mammals; (3) primary productivity—chlorophyll measures modified by relative factors like day length and water temperature; and (4) sea-surface height—a measure that discerns currents and eddies, which transport nutrients and aggregate ocean life.

BENTHIC COMPLEXITY

Benthic complexity is a unique measure related to both slope and roughness. Generally speaking, it is a measure of the intricacy of the seafloor, that is, how much it changes in a given unit of area. This is, in many ways, similar to "rugosity." However, unlike rugosity, complexity is not greatly affected by large, unidirectional changes in depth, such as cliffs. The benthic complexity methodology, described by Ardron (2002)[3], is used to capture regions of high seafloor irregularity that previous methods, such as slope and relief, had not. It differs from slope and relief by differentiating between uniformly steep features, such as fjords, and those features that display more complexity, such as rocky reefs, seamounts, and archipelagos. The latter are especially known for their ecological significance.

For the purposes of our analysis, bathymetry with sufficiently high resolution (roughly 1:250,000) was not uniformly available throughout the B2B region. Bathymetry was available for three large regional areas: (1) British Columbia; (2) coastal California, Oregon, and Washington; and (3) Baja California. This analysis selected areas of highest benthic complexity such as the shelf slope, canyons, gullies, island archipelagos, and seamounts.

SEA SURFACE TEMPERATURE FRONTAL DENSITY

Oceanographic fronts can be some of the most persistent features in the pelagic realm, and they are known to perform vital habitat functions for fishes (Seki et al., 2002), sea turtles (Polovina et al., 2000), seabirds (Decker and Hunt, 1996), and marine mammals (Davis et al., 2002). Fronts are characterized by the interaction of two dissimilar water masses, such as cold water and warm water, fresh water and salt water, or nutrient-rich water with nutrient-poor water. This interaction can bring deep-water nutrients to the surface, where sunlight and warm water stimulate a phytoplankton bloom, often followed by a zooplankton bloom, producing a pulse of resources to species at higher levels.

The multi-channel sea surface temperature (MCSST) data are derived from the five-channel advanced very-high-resolution radiometers (AVHRR) on board NOAA polar-orbiting satellites. Clouds hinder frontal detection by radiometry. Cloud-free, interpolated sea surface temperature (SST) data are available at coarse scales. We tested satellite-derived SST data at three different resolutions to examine the effect of scale upon edge-detection algorithms. We found that the coarse-scale, cloud-free MCSST interpolated data underestimated the total linelength of frontal features from finer scale raw AVHRR at nine-kilometer resolution, and Coastwatch data at two-kilometer resolution. However, MCSST data can reliably detect the strongest, most persistent temperature fronts within the B2B extent. We examined monthly MCSST data over a four-year period, from 1996–1999. This "cloudless" temporal window captured a strong El Niño, a La Niña and two "normal" years.

Using new analysis methods to detect temporal variation in SST frontal concentrations (Etnoyer et al., 2004), we found less than 1% of the Northeast Pacific is active for temperature fronts across seasons and between years. We identified three of these large features—offshore Los Cabos (Mexico), Point Conception (United States), and the southern California Channel Islands (United States). The frontal density signature off northern Baja California (Ensenada Front) appeared weaker and closer to shore in an El Niño year, and stronger and more offshore during a normal year. Satellite telemetry data and fisheries statistics demonstrate these pelagic habitats are important to blue whales (*Balaenoptera musculus*) (Fig. 9.2; see page 201), swordfish (*Xiphias gladius*), and striped marlin (*Tetrapturus audax*). B2B is the first marine conservation initiative to identify and quantify "persistence" for pelagic marine habitat.

SEA SURFACE HEIGHT: CURRENTS, GYRES AND EDDIES

At the scale of an ocean basin, the sea surface is not flat. Warm water expands, producing higher than average surface heights (hills), while cool water contracts, registering lower than average surface heights (valleys). Orbital satellites such as TOPEX/Poseidon use pulses of radar to measure minute differences in sea surface height. This is known as "altimetry." In an altimetry map, wind and waves are averaged, and sea surface height is expressed as an "anomaly"—a negative or a positive difference from the mean sea surface height.

These small differences in water height translate into current movement. Warm-core eddies, areas with higher than average sea surface heights, spin clockwise or anti-cyclonically. Lower than average sea surface heights, or cold-core eddies, spin counterclockwise or cyclonically. Furthermore, cold-core eddies create upwelling conditions that bring nutrients to the surface, and may result in trophic cascades and plankton blooms. Eddies can form when large freshwater flows from terrestrial rivers spill into the saline waters of the sea. The Haida Eddy (Pacific Canada) is a three-dimensional "swirling freshwater tornado," about the size of Lake Michigan, that transports coastal nutrients (such as iron) to nutrient-poor offshore waters, fertilizing the environment and creating a plankton bloom (Crawford and Whitney, 1999). The Haida Eddy appears most strongly in El Niño winters off British Columbia. The footprint of the Haida Eddy varies within El Niño Southern Oscillation cycles, and appears weakest in La Niña years.

For this analysis, we used altimetry to study surface current patterns in the Gulf of Alaska. The Colorado Center for Atmospheric Research provided four years of bi-weekly averaged surface current magnitude and velocity, derived from a blended product of TOPEX/Poseidon, and ERS-1 and ERS-2 satellites. We masked all but the highest waters or the greatest slope and sequenced the data to reveal the location and trajectory of warm core rings in the Gulf of Alaska.

We identified the 1998 Haida Eddy and tracked it from Gwaii Haanas (Queen Charlotte Islands) in a southwesterly direction to beyond the Canadian EEZ. The feature persisted for more than a year, originally 100 kilometers in diameter, then dissipating down to 75 km for much of the year. We identified an equally impressive anti-cyclonic feature that seemed to originate in Shelikof Strait and to propagate westward along the Aleutian Archipelago, gaining strength as it passed. This feature traveled more than 400 km in the course of six months. Several Sitka eddies came and went in the Gulf of Alaska throughout the four-year investigation period. These eddies represent a trans-boundary export of nutrients and larvae between British Columbia, Canada, and Alaska, United States. It is equally possible that these retentive eddies could concentrate and transport inorganic pollutants and contaminants to rare and delicate seamount ecosystems.

PRIMARY PRODUCTIVITY

Obtaining synoptic chlorophyll distribution in the global ocean is only possible with satellite ocean color sensors. Sea-viewing wide field-of-view sensor (SeaWiFS) and moderate-resolution imaging spectroradiometer (MODIS) satellites provide one- to two-day coverage of the entire Earth, allowing study of regional and global ocean color patterns. The primary data product from the sensors is the surface chlorophyll concentration (in mg/m^3). Combined with the SST data obtained from satellites with an AVHRR, primary production can also be estimated from empirical models.

Net primary productivity (NPP) can be estimated from three parameters: chlorophyll, photosynthetically available radiation (PAR), and SST. We estimate the NPP in g C m^{-2} $month^{-1}$ (Behrenfeld and Falkowski, 1997). Monthly chlorophyll data for the region bounded by 12°N–72°N and 180°W–100°W between September 1997 and June 2002 were obtained from the US National Aeronautics and Space Administration (NASA) Distributed Active Archive Center[4], PAR data from SeaWiFS[5], and monthly SST data from NASA's Jet Propulsion Laboratory[6].

Briefly, atmospheric effects were removed and chlorophyll concentration was estimated. To estimate primary production, the model takes into account the depth-dependent chlorophyll and light profile, and estimates the primary production per unit chlorophyll from SST, using an empirical relationship. Based on the NPP monthly results for each location, we estimated the number of occurrences (frequency) in a year when NPP exceeded a pre-defined number (10 g C m^{-2} $month^{-1}$). The number was chosen according to visual examination of the difference between oligotrophic and productive waters, but is somewhat arbitrary. The results serve as an index to describe how long enhanced productivity exists.

PCA Workshop

This process culminated in April of 2003 with an experts' workshop, held at Simon Fraser University in Burnaby, British Columbia, Canada, to map North American PCAs, summarized herein. The methodology selected for identifying PCAs relied on teaming experts' knowledge with the development of a geographic information system. The use of expert knowledge in such an interactive team approach to decision-making is referred to as a Delphic approach or poll. It is characterized by experts informed of current consensus but not harassed by arguments, with both majority and minority opinions maintained. Subsequent review and refinement based on these opinions results in consensus.

The GIS included appropriate spatial datasets and selected analyses available for the B2B region at a common resolution, as well as smaller subsets of regional information. Analyses focused on translating several of these datasets in order to highlight regions where physical processes lead to unique features or concentrations of species. At the final PCA identification workshop, experts reviewed the aggregated datasets and analyses to inform their judgments of ecological value and conservation priority.

Throughout all consultations, this process attempted to interact with the appropriate federal agencies in each of the CEC countries, rather than directly involving state, provincial, or regional governing bodies (though these offices were involved to differing degrees). This led to a number of significant restrictions on this project. For example, the use of local ecological knowledge was discussed and considered. During our consultative process it was agreed that this type of information was clearly an important component of local conservation efforts, but at the continental scale, it should be left to additional regional and local efforts. This constraint of top-down efforts highlights the necessity of eventually matching this project with a community-based action plan involving members of the communities within the PCA regions.

IDENTIFYING PRIORITY CONSERVATION AREAS WORKSHOP/CONSENSUS MAPPER

The final aspect of this work was an experts' workshop to select priority conservation areas. This workshop involved a series of interactive mapping exercises (detailed later in this chapter). In April 2003, MCBI and the CEC led a three-day workshop at Simon Fraser University, British Columbia, where marine experts from government agencies, nongovernmental organizations, academia, and regional organizations in Canada, Mexico, and the United States met to identify PCAs in the B2B region. These experts represented interests from resource use, science, management, and conservation. The experts were supported by a team of GIS experts from MCBI and the geography departments of Simon Fraser University and McGill University, Montreal, Canada, to provide technical support for the mapping workshop.

At the identification workshop, we reviewed for the experts the appropriate rationale for continental scale PCAs in accordance with the goals of the project and the consultations at previous meetings. Experts were asked to identify those regions that offered high diversity of all criteria ("the most bang for your buck"). We briefed experts on the history of the B2B initiative, goals of the workshop, definitions of key terms, and criteria for selecting priority conservation areas. The organizers also informed the participants that the end product of this workshop would guide the three nations' governments in their joint conservation collaborations, as well as provide a framework for regional conservation efforts and programs.

Workshop organizers presented the assembled data and analyses, and individual experts made presentations on a range of species and areas of concern. These presentations were on topics such as the natural history of seamounts, benthic complexity, sea surface temperature frontal regions, species hot spots, fisheries, threats from human activities, and ongoing conservation activities in each of the three nations.

Next, experts participated in a round-table mapping exercise. Consensus Mapper is a software program and methodology that allows exploration of spatial data, discussion of decision priorities and mapping of selected regions. Individual maps are overlaid to show areas of overlap, or consensus, between different working groups. The round table permits experts from different fields of expertise to uncover their commonalities, while those with divergent interests can clarify their points of disagreement and work towards compromise. This system was developed by Community-Based Environmental Decision Support at McGill University (Faber, 1996; Balram and Dragicevic, 2002). The advantages of collaborative mapping include the following (Balram et al., 2003; Balram et al., 2004):

- facilitating collaboration and consensus building within a dynamic social setting;
- providing structure and documenting the stakeholder participation process;
- incorporating inputs and policies at various levels of spatial aggregation;
- encouraging spatial thinking and exploration of environmental issues;
- providing feedback into the decision-making process;
- integrating data from expert sources;
- managing the technical and social network of the participation process; and
- facilitating collaborative monitoring of decision actions.

Following an overview of the available data and instructions from the workshop facilitators, experts learned how to use Consensus Mapper software, a simplified version of ArcView software. Finally, participants were assembled into expert working groups. The workshop was conducted as a series of break-out sessions for mapping and plenary discussions to review progress.

During the workshop, experts engaged in several exercises to identify PCAs. In order to do so, the experts first identified ecologically significant regions (ESRs) in the B2B extent. The experts were asked to base ecological significance on the data available, and on their personal knowledge of species, habitats, and physical and oceanographic features in the B2B region. Experts reached consensus on ESRs by overlaying individual team maps to show areas of agreement between expert working groups. In subsequent exercises, experts were asked to review the specific criteria for each ESR and rate it according to their knowledge of regional threats (e.g., resource extraction, pollution, coastal development) and opportunities for collaboration (e.g., previous designation as a priority or site of conservation interest, existing protected status, sustainable practices, local support) relative to the other ecologically significant regions. The resulting map of ESRs served to highlight places of high ecological significance. PCAs are a subset of ecologically significant regions that become priorities based on significant threats and/or opportunities.

MAPPING EXERCISES

Exercise One: Thematically Identify Ecologically Significant Regions. The participants were divided into six groups according to their expertise: one benthic environment group, two pelagic environment groups, and three planning and management groups. Within each group, there were six to ten participants and at least one representative from each of the four B2B subregions: (1) Mexico, (2) California, Oregon, and Washington, USA, (3) Canada, and (4) Alaska, USA. Each group identified areas that they knew to be ecologically significant, and discussed and debated these with others in their group. These areas were drawn on a digital map using the Consensus Mapper program. For each place identified, they noted the rationales in a spreadsheet, stating the physiographic, oceanographic, and biological features, species diversity, endemism, or other criteria they believed relevant to the site's ecological significance. Pelagic groups were also asked to focus on migratory species (including the CEC's list of marine species of common conservation concern, Appendix 9.1). In this exercise, each group was allowed to select up to 40% of each nation's EEZ within the B2B extent. They were also asked to refrain from selecting areas smaller than 1° square. At the end of this exercise, all the groups' selections were superimposed onto one consensus map, with areas shaded in accordance with the degree of overlap among the six groups. In a plenary session workshop, participants were able to review and comment on the overlaid map of ESRs.

Exercise Two: Review and Refine Ecologically Significant Regions. We divided experts into four groups by region: Mexico; California, Oregon, and Washington, USA; Canada; and Alaska, USA. Within each group, members had differing expertise. They reviewed the results of the previous exercise, seeking to refine the coarser-scale analysis. They either modified the boundaries of those high-consensus regions from Exercise One, adopted them as ESRs, or added new selections. In this exercise, the groups also documented the rationales for each ESR they identified. Each group was allowed to identify up to 40% of its respective EEZ as ecologically significant. At the end of this exercise, all the groups' selections were combined and shown on a map in a plenary session. The participants saw the final ESRs from Baja California to the Bering Sea. Each group had an opportunity to explain their selections to the other groups

Exercise Three: Identify Threats and Opportunities. In addition to ecological significance, threats and opportunities are crucial factors in assigning priority. In this exercise, the participants were again divided into regional groups to rate the relative level of threats and opportunities in each of the ESRs previously identified. The workshop organizers categorized threats into the following types: (1) non-renewable resource extraction; (2) exploitation of renewable resources; (3) coastal land use change; (4) pollution; (5) damaging recreational use; and (6) physical alteration of coastlines. Opportunities were categorized as: (1) existing legal protection; (2) available management; (3) local and/or regional support; (4) funding available for information management and/or conservation; and (5) sustainable business practices. Each group of experts received a list of these categories. Group members discussed the relative significance of the types of threats and opportunities existing in their ESRs. Where applicable, experts provided additional details pertaining to the threats, ranked their relative intensity (high, medium, or low), and assessed the current trend (getting better, the same, or getting worse). The description, intensity, and trend were all recorded in a spreadsheet.

Exercise Four: Identify Priority Conservation Areas. The final step in the workshop was to identify PCAs. The participants were divided into six tri-national teams with at least one expert from each of the four B2B geographic regions. In this exercise, the goal was to select not more than 20% of the area within the ESRs from Baja California to the Bering Sea as PCAs. The group members used Consensus Mapper to digitally map their selections, and specified their rationales for every PCA. At the end of this exercise, the six sets of PCAs selected by the six groups were overlaid and shown to all workshop participants in a plenary session. The selected areas were colored according to the degree of overlay. The participants saw the level of consistency across the groups. Each group had the opportunity to explain the reasoning behind their selection and to point out unique features they had taken into consideration.

Results and Discussion

Defining priority conservation areas (PCAs) is the fulfillment of a workplan by the three nations to identify opportunities to work collaboratively, at the North American level, on marine conservation. A total of 28 sites were identified as PCAs, totaling 8% of the total EEZ area of the 3 nations (Fig. 9.3; see page 202). By country, these areas represent approximately 7% of the B2B region within Mexico, 10% of the area in Canada's Pacific EEZ, and 8% of the US EEZ (within the B2B defined region). The full discussion of the expert criteria and descriptions of the PCAs are included in the final report (Morgan et al. 2005). Boundaries of these PCAs were purposely left fuzzy to reflect the inappropriateness of human delineation to ecological phenomenon.

The CEC, through convening and coordinating NAMPAN, is developing capacity for a network of MPAs to span the jurisdictions of the three CEC member countries. The aim of NAMPAN is to enhance and strengthen the conservation of biodiversity in critical marine habitats throughout North America by creating a functional system of ecologically based MPA networks that cross political borders and depend on broad cooperation. The identification of these PCAs is not intended as the MPA network design, but is rather a portfolio of continentally significant sites that can serve as nodes around which a network of reserves can be built. Networks of reserves are an important tool for conserving biological diversity (Lubchenco et al., 2003) and these PCAs should be viewed as places to begin building a more comprehensive, effective MPA network for North America. Although these PCAs are science-based and anchored in a continental perspective, they are not intended to be a marine reserve network design as envisioned by others (Margules and Pressey, 2000; Possingham et al., 2000). Rather, the workshop organizers and participants clearly intend this report to be a first step towards a continental conservation strategy for B2B species and ecosystems. We hope that these priority areas for conservation will be used in formulating MPA networks based on broad input from all interested sectors.

Participants' attitudes towards the concept of a priority conservation area designation at the North America continental scale ranged from enthusiastic to confused, to doubtful, concerning the challenges of data integration, international cooperation, and synthesis. Despite this, the final workshop to identify PCAs was successful, though subject to the bias of available data and experts involved. The most prevalent concern was how to incorporate existing MPA designations, previous priority designations, and local projects. This concern is mostly alleviated by noting the consistency of these PCAs with past works on identifying important biodiversity sites (Ford and Bonnell, 1996; Banks et al., 1999; Ardron et al., 2001; Sala et al., 2002). In part, the process was designed to address this issue by asking various groups to come forward with their data, priorities, and projects in a sense of greater community. In general, participants benefited from exposure to relevant projects and avenues for collaboration throughout the B2B region and the success of the process itself may ultimately rest with initiating such exchanges.

An alternative approach to a vision of "the map" that represents a unified, multi-institutional perspective on priority areas for conservation over this 6,000-mi extent, is a process that allows individuals open access to baseline data and analyses. To this end, North Americans should view the identified PCAs as expert advice; information to include in their own regional or local planning efforts. It is our hope that the identification of these areas will generate discussion, catalyze action, inform opinion, and foster future cooperation. MCBI, the CEC, and Ecotrust distribute the available information on a CD-ROM (B2B 1.1). Hopefully, this dataset will serve as a foundation for future regional analyses. Although it is tempting to incorporate all relevant data in order to produce a high-quality dataset with simple baseline information, the first release, B2B 1.0, delimits the data at the highest common denominator resolution across the entire B2B extent, e.g., ~4 km Smith and Sandwell (1997) bathymetry, 9 km AVHRR SST, 7 km surface currents. It is likely that finer-scale resolution will be necessary for regional analyses.

While biodiversity protection is the ultimate goal of this priority conservation area assessment, no such datasets are available. Comprehensive biogeographic datasets of species diversity will need to be researched and built if they are to be incorporated into future analyses. Continental scale biodiversity could be captured by protection of representative areas and endemic species at the regional scale.

Conclusion

From the Gulf of California, with its deep canyons, nutrient-rich upwellings, and high levels of endemism, to the 20,000 km of bays, inlets and inland drainage systems of the Pacific Northwest, to the high productivity of the Bering Sea, the west coast of North America is home to unique and important shared marine environments. It is also home to a great number of shared marine species—such as Pacific gray and blue whales, leatherback sea turtles, bluefin tunas, black brant geese and Heermann's gulls—that migrate thousands of kilometers, moving across national borders without hesitation. Hence, be it through shared species or ecosystems, the marine environments of Canada, Mexico, and the United States are intimately linked. Accordingly, action or inaction on one side of a border will have consequences for the shared living organisms occupying ecosystems with no such definite boundaries. The process of identifying PCAs attempted to provide individuals throughout the B2B region with the same information as well as incorporate processes already finished or underway. Future efforts in the B2B region have a place to start.

Notes

1 <http://cec.org/trio/stories/index.cfm?ed=2&ID=18&varlan=english>
2 <http://cec.org/pubs_docs/documents/index.cfm?varlan=english&ID=1088>
3 See also http://srmwww.gov.bc.ca/risc for further information on complexity.
4 <http://daac.gsfc.nasa.gov>
5 Frouin, R., B. Franz, and M. Wang. Algorithm to estimate PAR from SeaWiFS data <http://daac.gsfc.nasa.gov/CAMPAIGN_DOCS/OCDST/PDFs/seawifs_par_algorithm.pdf>
6 <http://podaac.jpl.nasa.gov/sst>
7 Later expanded to include species that were affected by actions of two or more countries, and were not necessarily migratory or transboundary, such as the endemic vaquita.

References

Ardron, J. A., 2002. A recipe for determining benthic complexity: An indicator of species richness, in Breman, J. (ed.), *Marine Geography: GIS for the Oceans and Seas*, Redlands, CA, ESRI Press, 169-75.

Ardron, J. A., Lash, J., and Haggarty, D., 2002. Designing a Science Based Network of Marine Protected Areas for the Central Coast of British Columbia, Living Oceans Society Technical Report, Version 3.1, Sointula, British Columbia, Canada, Living Oceans Society, pp.120.

Balram, S., and Dragicevic, S., 2002. Integrating complex societal problems theory in a GIS framework: the collaborative spatial Delphi methodology, in Egenhofer, M., and Mark, D. (eds.), *GIScience 2002, Lecture Notes in Computer Science 2478* , Berlin: Springer-Verlag, 221–24.

Balram S., Dragicevic, S., Meredith, T., 2003. Achieving effectiveness in stakeholder participation using the GIS-based collaborative spatial Delphi methodology, *Journal of Environmental Assessment Policy and Management*, 5(2): 365–94.

Balram S., Dragicevic, S., Meredith, T., 2004. A collaborative GIS method for integrating local and technical knowledge in establishing biodiversity conservation priorities, *Biodiversity and Conservation*, 13(6): 1195-1208.

Banks, D., Williams, M., Pearce, J., Springer, A., Hagenstein, R., and Olson, D., 1999. *Ecoregion-Based Conservation in the Bering Sea*, Washington, DC, World Wildlife Fund, and Anchorage, AK, The Nature Conservancy, 72 pp.

Behrenfeld, M. J., and Falkowski, P. G., 1997. Photosynthetic rates derived from satellite-based chlorophyll concentration, *Limnology and Oceanography*, 42: 1-20.

Crawford, W. R., and Whitney, F. A., 1999. Mesoscale eddies in the Gulf of Alaska, *Eos, Transactions of the American Geophysical Union*, 80(33): 365–70.

Davis, R. W., Ortega-Ortiz, J. G., Ribic, C. A., Evans, W. E., Biggs, D. C., Ressler, P. H., Cady, R. B., Leben, R. R., Mullinm, K. D., and Wursig, B., 2002. Cetacean habitat in the northern oceanic Gulf of Mexico, *Deep-Sea Research* I, 49: 121–42.

Decker, M. B., Hunt, G. L., Jr., 1996. Foraging by murres (*Uria* spp.) at tidal fronts surrounding the Pribilof Islands, Alaska, USA, *Marine Ecological Progress Series*,139: 1-10.

Etnoyer, P., in press. Seamount resolution in satellite derived bathymetry. *Geochemistry, Geophysics, Geosystems (G3).*

Etnoyer, P. Canny, D., and Morgan, L. E., 2002. *B2B 1.0 CD-ROM. Information for Conservation Planning—Baja California to the Bering Sea,* Redmond: Marine Conservation Biology Institute, http:// www.mcbi.org.

Etnoyer, P., Canny, D., Mate, B., and Morgan, L. E., 2004. Persistent pelagic habitats in the Baja California to Bering Sea (B2B) ecoregion, *Oceanography,*17: 90-101.

Etnoyer, P., and Morgan, L., 2003. *Occurrences of habitat-forming deep sea corals in the northeast Pacific Ocean: A report to NOAA's Office of Habitat Conservation,* Washington DC, National Oceanic and Atmospheric Administration, Department of Commerce, 31 pp.

Faber, B. G., 1996. A group-ware enabled GIS, in Heit, M. (ed.), second edition, *GIS Applications in Natural Resources,* Fort Collins, CO, GIS World Books, 3-13.

Ford, G. R., and Bonnell, M. L., 1996. *Developing a Methodology for Defining Marine Bioregions: The Pacific Coast of the Continental USA,* Report to the World Wildlife Fund. Portland, OR, Ecological Consulting, Inc, 26 pp.

Hixon, M. A., Boersma, P. D., Hunter, M. L., Jr. Micheli, F., Norse, E. A., Possingham, H. P., and Snelgrove, P. V. R. 2001. Oceans at risk: research priorities in marine conservation biology, in Soulé, M., and Orians, G. H. (eds.), *Conservation Biology: Research Priorities for the Next Decade,* Washington DC, Island Press, 125–54.

Longhurst, A. R., 1998. *Ecological Geography of the Sea,* San Diego, CA, Academic Press, 398 pp.

Lubchenco, J., Palumbi, S. R., Gaines, S. D., and Andelman, S., 2003. Plugging a hole in the ocean: The emerging science of marine reserves, *Ecological Applications,* 13: Supplement 3-7.

Margules, C. R., and Pressey, R. L., 2000. Systematic conservation planning, *Nature,* 405: 243-53.

Morgan, L., and Etnoyer, P., 2002. The Baja California to Bering Sea priority areas mapping initiative and the role of GIS in protecting places in the sea, in Breman, J. (ed.), *Marine Geography: GIS for the Oceans and Seas,* Redlands, CA, ESRI Press, 137-42.

Morgan L. E., Maxwell, S., Tsao, F., Wilkinson, T., Etnoyer P., 2005. *Priority conservation areas: Baja California to the Bering Sea.* Montreal: Commission for Environmental Cooperation.

Olson, D. M., Dinerstein, E., Powell, G. V. N., and Wikramanayake, E. d., 2002. Conservation biology for the biodiversity crisis, *Conservation Biology,* 16: 1–3.

Polovina, J. J., Kobayashi, D. R., Parker, D. M., Seki, M. P., Balazs, G. H., 2000. Turtles on the edge: Movement of loggerhead turtles (*Caretta caretta*) along oceanic fronts in the Central North Pacific, 1997-1998, *Fisheries Oceanography,* 9: 71-82.

Possingham, H., Ball, I., and Andelman, S., 2000. Mathematical models for identifying representative reserve networks, in Ferson, S., and Burgman, M. (eds.), *Quantitative Methods for Conservation Biology,* New York, Springer-Verlag, 291-306.

Roberts, C. M., McClean, D. J., Veron, J. E. N., Hawkins, J. P., Allen, G. R., McAllister, D. E., Mittermeier, C. G., Schueler, F. W., Spalding, M., Vynne, C., and Warner, T. B., 2002. Marine biodiversity hotspots and conservation priorities for tropical reefs, *Science,* 295: 1280–84.

Sala, E., Aburto-Oropeza, O., Paredes, G., Parra, I., Barrera, J. C., and Dayton, P. K., 2002. A general model for designing networks of marine reserves, *Science,* 298: 1991-93.

Seki, M. P., Polovina, J. L., Kobayashi, D. R., Bidigare, R. R. and Mitchum, G. T., 2002. An oceanographic characterization of swordfish (*Xiphias gladius*) longline fishing grounds in the springtime subtropical North Pacific, *Fisheries Oceanography* 11(5): 251-66.

Sherman, K., Alexander, L. M., and Gold, B. D. (eds.), 1990. *Large Marine Ecosystems: Patterns, Processes and Yields, Washington,* DC, American Association for the Advancement of Science, 242 pp.

Sherman, K., and Duda, A. M., 1999. An ecosystem approach to global assessment and management of coastal waters, *Marine Ecology Progress Series*, 190:271-87.

Smith, W. H. F., and Sandwell, D. T., 1997. Global seafloor topography from satellite altimetry and ship depth soundings, *Science*, 277: 1957-62.

Soulé, M. E., and Terborgh, J., 1999. *Continental conservation: scientific foundations of regional reserve networks,* Washington DC, Island Press, 227 pp.

Soulé, M. E., Estes, J. A., Berger, J., and Martinez del Rio, C., 2003. Ecological effectiveness: Conservation goals for interactive species, *Conservation Biology*, 17(5): 1238-50.

Sullivan-Sealey, K. M., and Bustamante, G., 1999. Setting Geographic Priorities for Marine Conservation in Latin America and the Caribbean, Arlington, VA, The Nature Conservancy, 125 pp.

Wilkinson, T., Agardy, T., Perry, S., Rojas, L., Hyrenbach, D., Morgan, K., Fraser, D., Janishevski, L., Herrmann, H., and De la Cueva, H., 2004a. Marine Species of Common Conservation Concern: Protecting Species at Risk across International Boundaries. In: Munro, N. W. P., Dearden, P., Herman, T. B., Beazley, K., and Bondrup-Nielsen, S., editors. *Proceedings of the Fifth International Conference on Science and Management of Protected Areas*. Wolfville, Nova Scotia: Science and Management of Protected Areas Association.

Wilkinson, T., Wiken, E., Hourigan, T., Madden, C., Bezaury, J., Padilla, M., Janishevski, L., Valdes, C., Herrmann, H., Gutierrez, F., and Sanders, N., 2004b. Mapping marine ecological regions of North America: Laying the foundation for cooperative ecosystem-based conservation in North America. In: Munro, N. W. P., Dearden, P., Herman, T. B., Beazley, K., and Bondrup-Nielsen, S., editors. *Proceedings of the Fifth International Conference on Science and Management of Protected Areas*. Wolfville, Nova Scotia: Science and Management of Protected Areas Association.

Appendix 9.1.

Marine Species of Common Conservation Concern

In an activity parallel to the identification of PCAs, the CEC convened an advisory group to identify the first list of marine species of common conservation concern, (Wilkinson et al. 2004a). The goal of the present project was to focus on key conservation actions and protected areas needed to support these populations. These umbrella species captured a different conservation perspective by shifting the focus to processes that affect species as well as the places they inhabit. Compulsory criteria focused the initiative towards species that were: (1) transboundary or migratory; and (2) at high risk of extinction, given their current status or trends, their inherent natural vulnerability and their susceptibility to anthropogenic threats. Using secondary or recommended criteria, priority was then given to species: (1) deemed ecologically significant, e.g., umbrella, keystone, or indicator taxa; (2) officially listed as being of conservation concern by one of the three North American countries, by the World Conservation Union (IUCN), or by the Convention on International Trade in Endangered Species of Wild Fauna and Flora (CITES); (3) whose recovery or management was feasible, including re-establishment potential, as well as the opportunity to strengthen management and learn from successes; and (4) which had a high potential for public engagement (flagship species). To this end, key habitats for these species, as identified in this report (Wilkinson et al., 2004a), were included as criteria for PCAs.

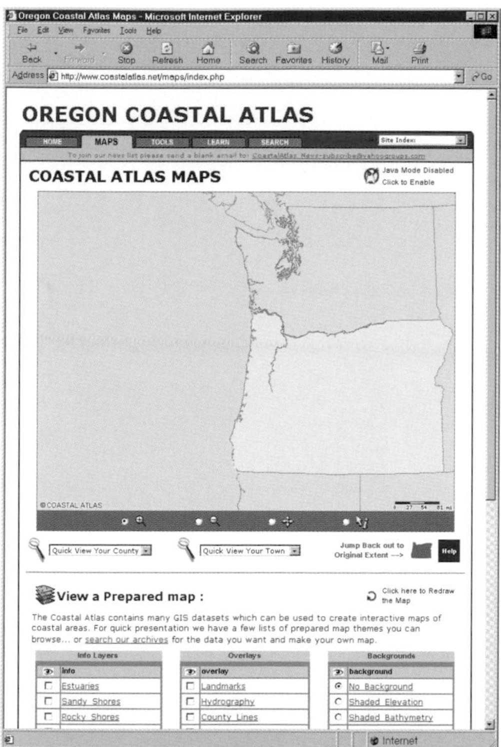

Fig. 7.2. Map Tool for users who want to see a landscape/seascape context. They may choose from prepared maps, make their own custom maps, add detailed information from the OCA archives, and create their own resulting PDF.

Fig. 7.3. Users may browse through a coastal encyclopedia within the Learn section of the OCA, and then focus on a coastal setting or topic by way of the online Coastal Access Information Tool containing location and inventory information about each of the access sites on the Oregon coast.

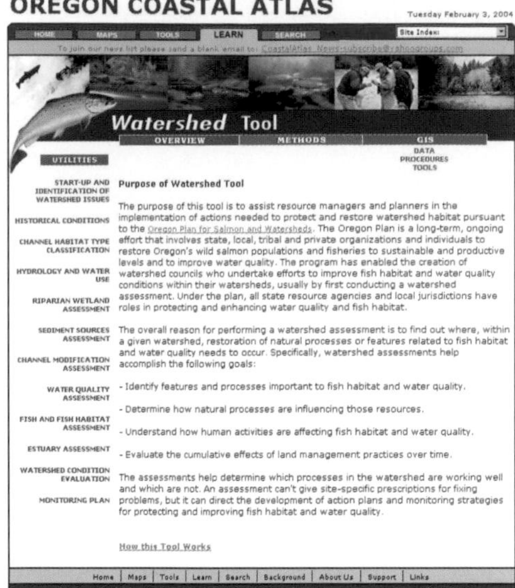

Figure 7.4. Opening page of the online Watershed Assessment Tool, a step-by-step GIS decision-making tool that provides guidelines, instructions, and then access to GIS data, and an Internet map service to aid the user in a watershed assessment process.

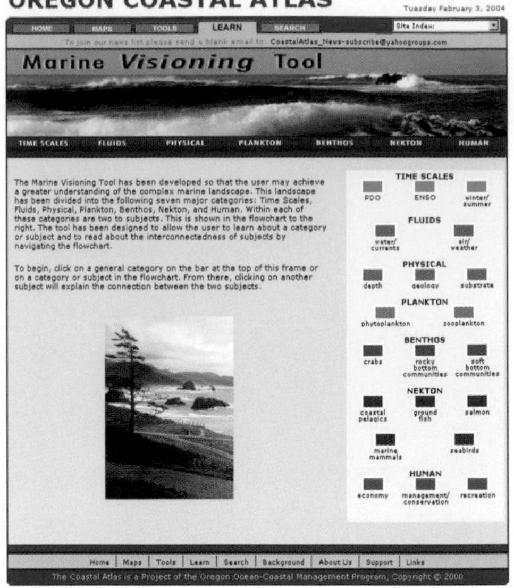

Figure 7.5. Opening page of the online Marine Visioning Tool, which allows the user to explore various topics about oceanographic processes, including time scales, fluids, physical parameters, benthic communities, pelagic communities, and human activities.

Figure 7.6. Two views that show the results of a GIS query to locate vacant parcels in private ownership within a high-risk zone.

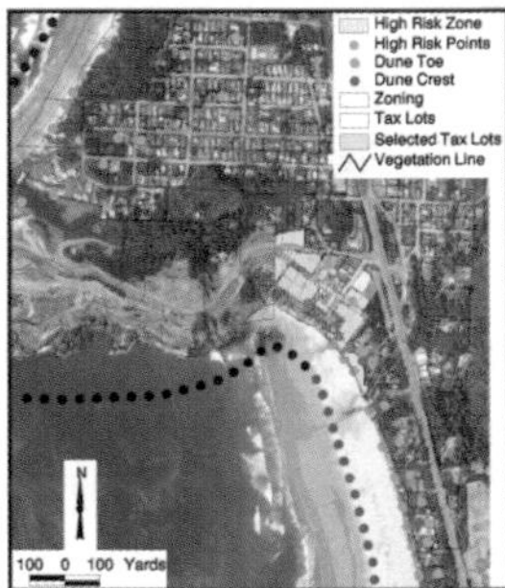

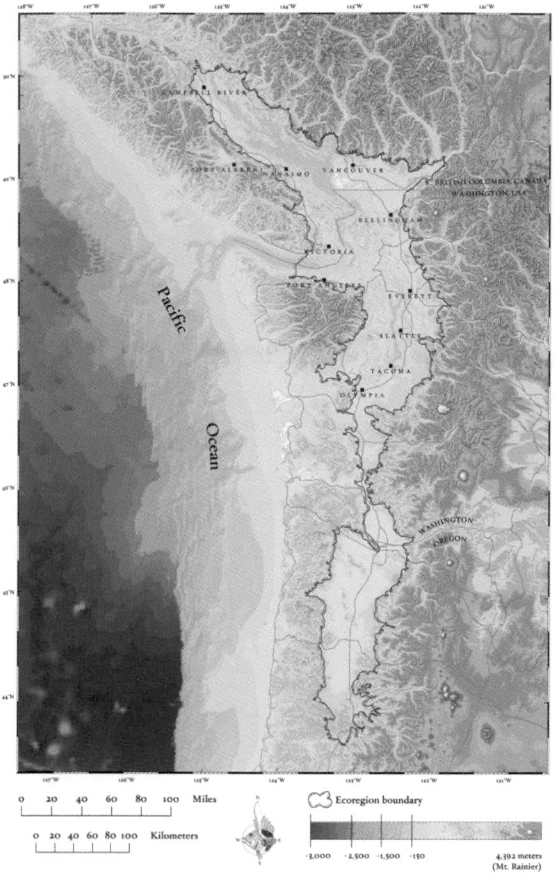

Figure 8.1. Willamette Valley-Puget Trough-Georgia Basin ecoregion of the Pacific Northwest

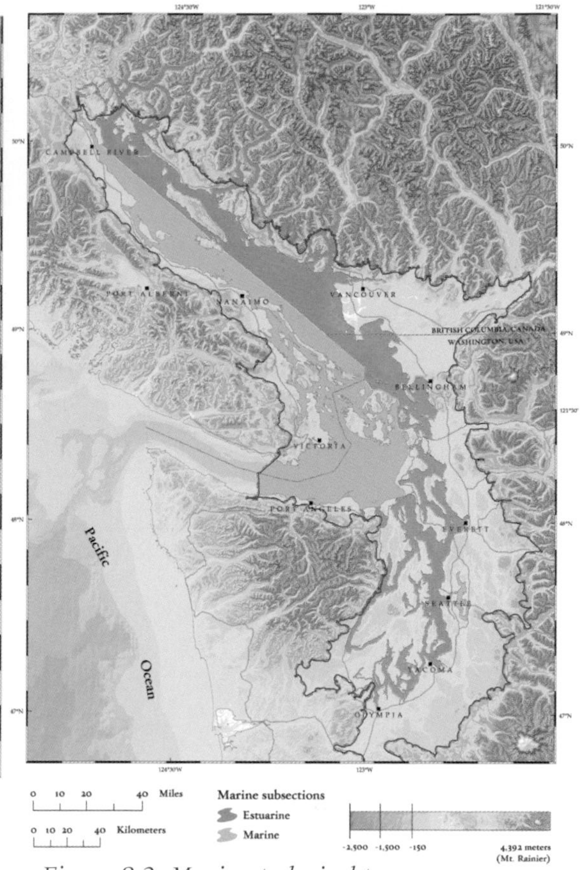

Figure 8.2. Marine technical teams separated the waters into subsections based on the dominance of freshwater outflow ("estuarine") and tidal conditions ("marine")

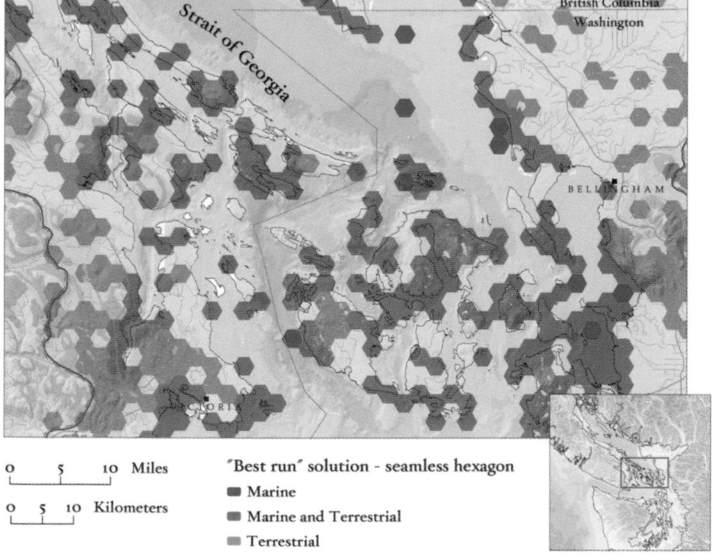

Figure 8.3. The analysis of marine and terrestrial conservation targets run within a single spatial planning unit. Transboundary region between the San Juan Islands in Washington and the Southern Gulf Islands in British Columbia

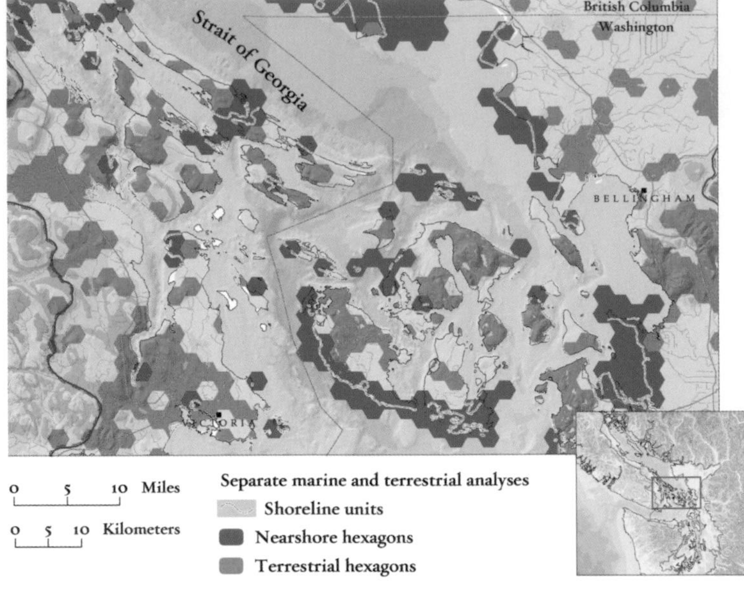

Figure 8.4. Marine and terrestrial conservation targets analyzed within three separate spatial planning units. Transboundary waters between Washington and British Columbia

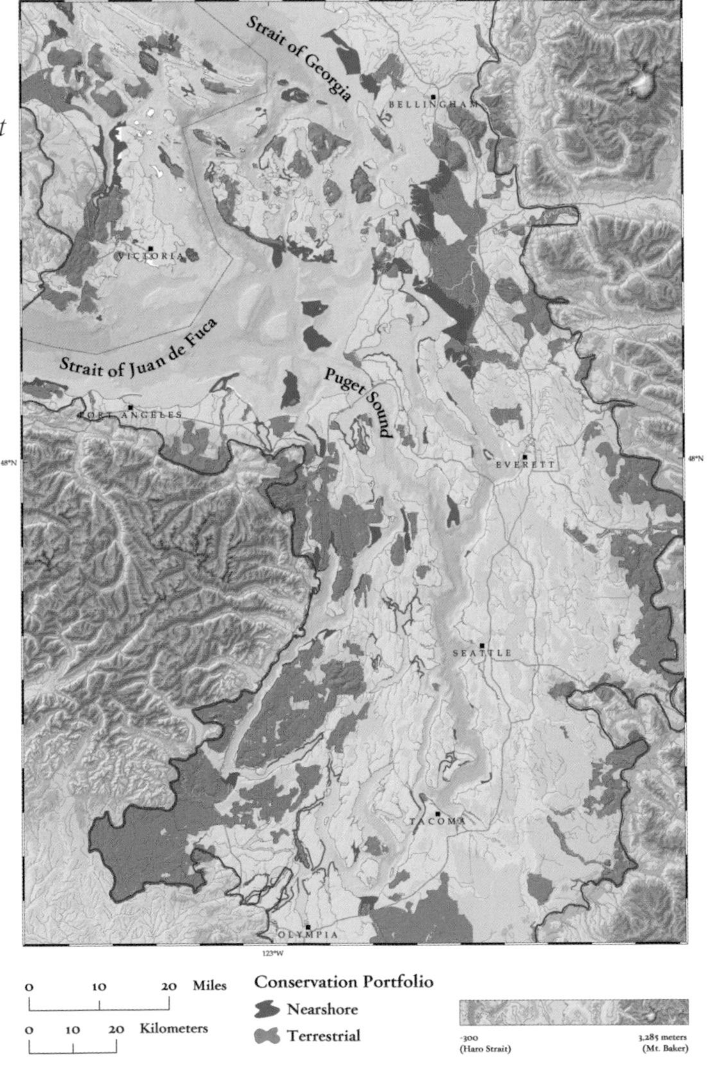

Figure 8.5. Final analysis and site delineation of nearshore marine and terrestrial high priority conservation areas. Puget Sound, Washington

Figure 9.1.Overlay of multibeam bathymetry (color, NOAA GOASEX) and ETOPO2 bathymetry (gray scale, Smith and Sandwell, 1997), demonstrating offset between the two data sources.

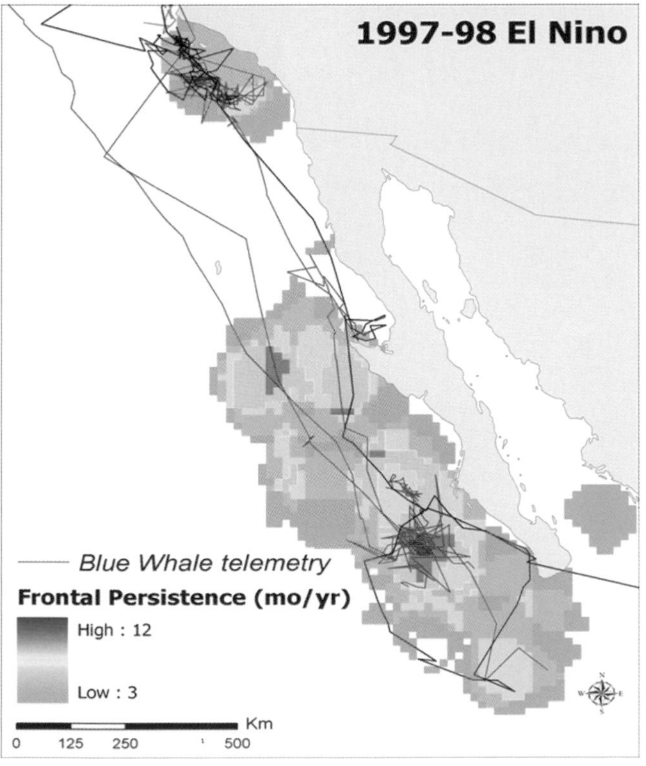

Figure 9.2. Results of frontal density analysis (1998) overlaid with blue whale tracks. Whales congregate in areas of high persistence of sharp SST discontinuities.

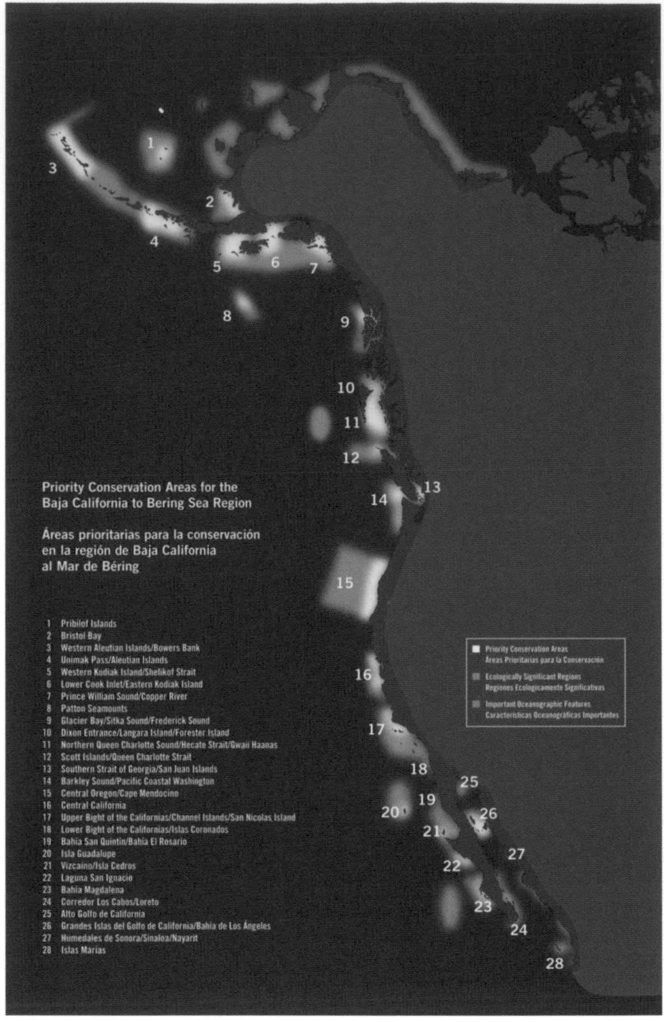

Figure 9.3. Priority Conservation Areas (gold), Ecologically Significant Regions (yellow) and Important Oceanographic Features (light blue) for the Baja California to Bering Sea Region.

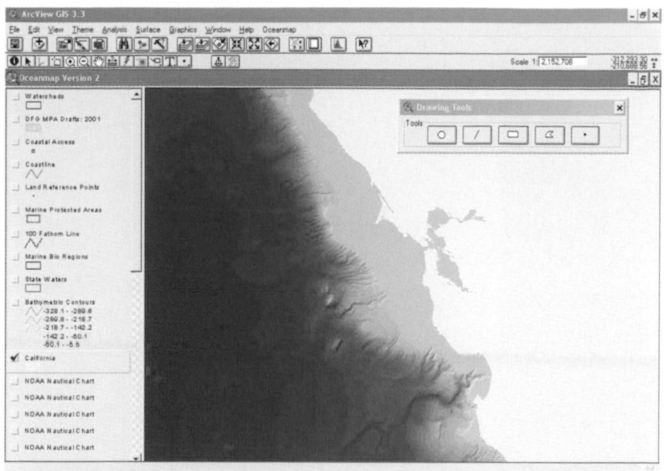

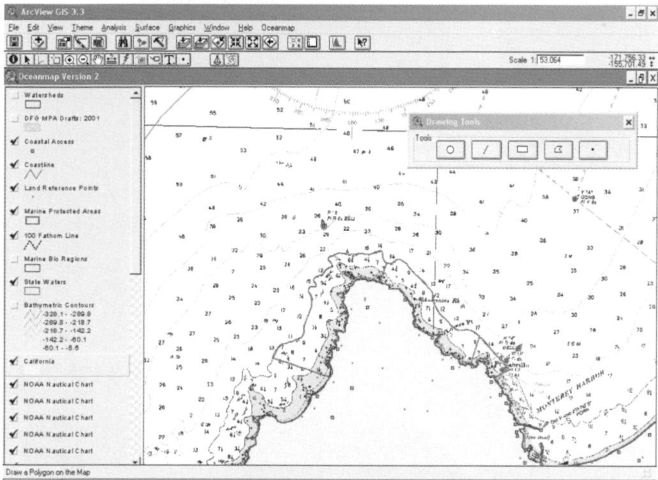

Figure 10.1. (a) GIS-based OceanMap full view. (b) GIS-based OceanMap enlarged view.

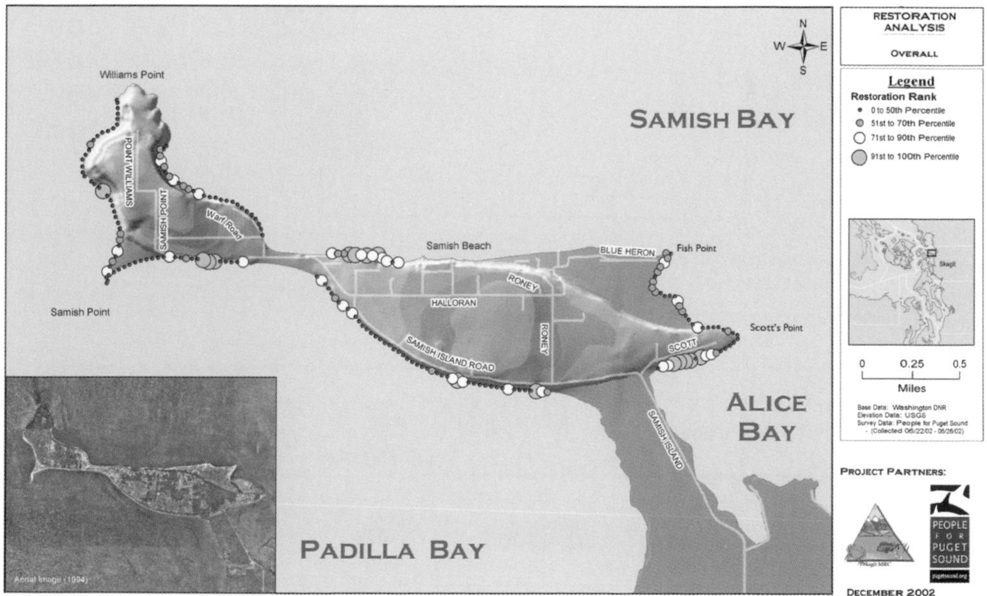

Figure 11.3. Example output for overall Restoration score, combined for all five models. Restoration ranks are relative.

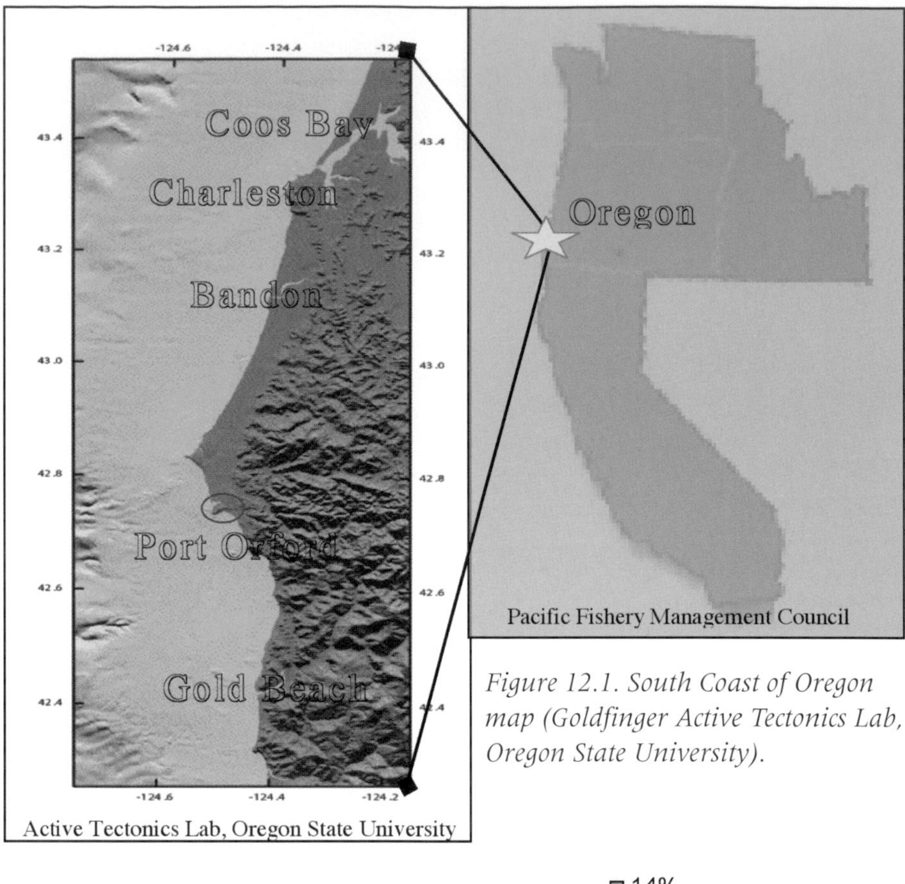

Figure 12.1. South Coast of Oregon map (Goldfinger Active Tectonics Lab, Oregon State University).

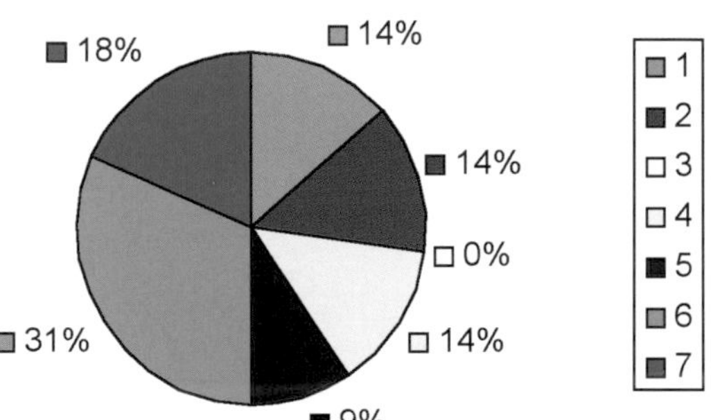

Figure 12.3. Percent of LKI participants and the number of target fisheries in their portfolio.

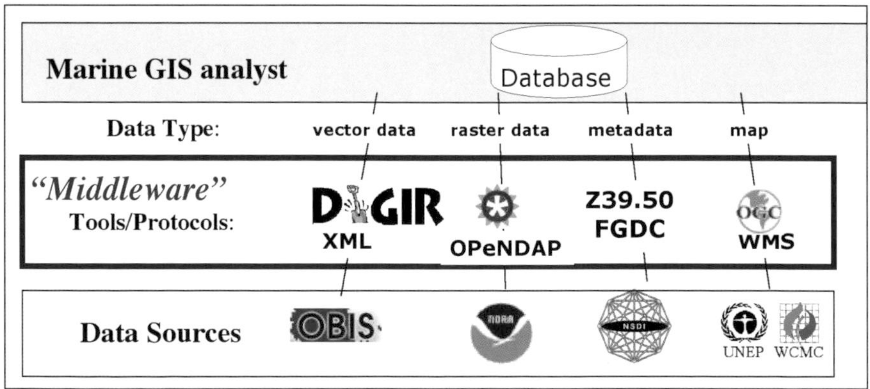

Figure 13.4. Common data discovery, data transport and Internet mapping tools, protocols and standards common to marine GIS operations (from Halpin, 2004).

Chapter 10

Using GIS to Elicit and Apply Local Knowledge to Ocean Conservation

Kate Bonzon, Rod Fujita, and Peter Black

Abstract

This chapter describes a protocol for using geospatial analysis tools based on the importance of eliciting and incorporating local expert knowledge and socioeconomic concerns into marine resource management decision-making processes. The GIS tool ("OceanMap") and rapid socioeconomic protocol described in this chapter grew out of past conflicts amongst stakeholders, specifically fishermen, regarding marine protected area planning processes in California, and builds on a pilot project conducted by the Pacific Coast Federation of Fishermen's Associations and Environmental Defense. The pilot project elucidated areas for improvement in the original protocol and GIS tool, resulting in changes and additions described here. Today, Environmental Defense continues to expand and improve the tool to address stakeholder needs, streamline the process of collecting and analyzing information, incorporate additional stakeholders, and accommodate other marine resource management efforts.[1]

Introduction

Many coastal communities are economically and culturally tied to the use of marine resources, both directly (as in fishing) and indirectly (as in recreational diving or nature-watching). Ocean resources are vital to the livelihood of these communities and policy decisions can directly affect individuals' lifestyles and economic well-being. Sound ocean resource management decisions (particularly if they are to be durable) often depend on acceptance by resource users, which in turn depends on perceived and actual social and economic impacts of policy decisions. Yet traditionally, marine conservation management has focused on the biophysical aspect of management first, while considering the socioeconomic aspect second, or not at all. Lack of detailed

Kate Bonzon, Rod Fujita, and **Peter Black**
Environmental Defense, 5655 College Avenue, Oakland, CA 94618, phone 510-658-8008, fax 510-658-0639
Corresponding author email: kbonzon@environmentaldefense.org.

socioeconomic information often hampers ocean conservation efforts both in the policy-making stage and in the implementation and enforcement phases. Recent efforts to implement marine protected areas in California highlight the benefit of integrating socioeconomic aspects of proposed management measures early in the decision-making process, and pilot studies have shown the effectiveness of utilizing geographic information systems (GIS) tools to achieve these goals.

Fully protected marine reserves (areas of the ocean that are off-limits to fishing and other extractive uses) are a relatively new tool for marine resource managers and have attracted both scientific support and political controversy. Much of the controversy is due to the perceived distribution of potential costs and benefits. For example, immediate costs of implementing reserves tend to be borne by the consumptive users of an area, i.e., commercial and recreational fishermen, in the form of restrictions that may adversely affect their income. Conversely, the benefits are often delayed and initially accrue to non-consumptive users (National Research Council, 2001; Carter, 2003), for example, by increasing the appeal of tourism. Marine management measures may also have social consequences by changing the profile and distribution of participation in marine recreational or commercial activities in an area. While federal (National Environmental Policy Act; Magnuson-Stevens Fisheries Conservation and Management Act) and state (California Environmental Quality Act) laws require consideration of socioeconomic costs and benefits of a management decision, such assessments are usually not comprehensive and are often conducted after the planning and public consultation process. Techniques to include local stakeholders in the planning process are needed to ensure consideration of socioeconomic impacts of marine protected areas and reduce conflicts among stakeholders.

Division and conflict between fisheries managers and fishing communities are already apparent (St. Martin, 2001), especially in the context of recent fishery declines (Gilden and Conway, 2002a; Gilden and Conway, 2002b) on the West Coast. Many user groups feel as if their concerns are ignored and their knowledge is underutilized. In order to halt this growing schism, agencies must understand the concerns of affected user groups regarding the costs of management measures. Additionally, incorporating local knowledge could fill important data gaps and be a more appropriate way to address socioeconomic impacts, especially considering the local nature of fisheries on the West Coast; and California's experience with MPA planning and implementation has elucidated the importance of socioeconomic analysis to address these concerns.

Recent projects in California show how GIS offer effective platforms for socioeconomic analysis and incorporation of local knowledge into policy processes. Recognizing the need for local participation in policy planning and implementation, and the ability of geospatial analytical tools to empower user groups, Environmental Defense (an environmental advocacy organization) has developed a GIS-based tool for California's coastal waters, OceanMap, to aid in California marine resource management (Fig. 10.1a; see page 203). OceanMap comprises numerous data layers that can be added or taken away from view, including geographical information, existing marine protected areas, habitat information, bathymetry, and nautical charts (Fig. 10.1b; see page 203). OceanMap was specifically designed to facilitate implementation of the California Marine Life Protection Act (MLPA) and grew out of the experience of creating a network of fully protected marine reserves within the state and federal waters of the Channel Islands National Marine Sanctuary (CINMS) off the southern coast of California.

In 2002, Environmental Defense collaborated with the Institute for Fisheries Resources (IFR; the research arm of the Pacific Federation of Fishermen's Associations) on the Local Knowledge Project, a pilot project to develop and test a participatory socioeconomic analysis protocol in the context of the MLPA. The project was designed to elicit fishermen's knowledge, test ways of incorporating their knowledge into the decision-making process, and to test spatially explicit methods for rapid socioeconomic assessments for MPA planning. The project was successful in achieving its goals and in elucidating areas for improvement of OceanMap and the protocol for rapid socioeconomic assessment. This chapter outlines the pilot project, important lessons learned, and the subsequent improvements in the tool. Environmental Defense will continue to use OceanMap and the improved data collection protocol to support MPA planning processes. Additionally, there is growing interest in expanding OceanMap to support MPA planning in other states, as well as other marine resource management efforts.

Including Local Knowledge in the California MPA Processes

There is a growing body of literature documenting the benefits of incorporating local ecological knowledge (LEK) and socioeconomic concerns into decision-making processes (McCay and Acheson, 1987; Feeny et al., 1990; Ostrom, 1990; Russ and Alcala, 1999; Berkes et al., 2000). LEK refers to the body of knowledge held by a specific group of people about their local ecosystems. The information is often site-specific, and can be a mixture of practical and scientific knowledge (Olsson and Folke, 2001). The Local Knowledge Project was predicated on the benefits of utilizing LEK and also grew out of practical experience with two MPA planning processes in California: the CINMS experience and the first attempt to implement the statewide Marine Life Protection Act in 2002. The MLPA requires the California Department of Fish and Game to implement a network of MPAs in state waters with an improved marine reserve (defined as no-take areas) component.

MPAs in California

Both the CINMS and the first attempt to implement the MLPA were highly contentious processes, with stakeholder groups including commercial and recreational fishing groups, environmental organizations, and resource managers often on opposite sides of the issue. In the CINMS case, a working group comprising scientists, fishermen, environmentalists, and other stakeholders were directed to design a network of marine reserves. Although the working group included a panel on socioeconomics and employed a team of consultants and academics to collect anecdotal and socioeconomic information from fishermen, many stakeholders were dissatisfied. Ultimately, the CINMS did not achieve consensus on one design alternative. Instead, agency staff drafted a number of design alternatives that attempted to meet scientific criteria while minimizing socioeconomic impacts. The Fish and Game Commission adopted the "preferred alternative" for state waters, in which 25% of the CINMS management area will be set aside in marine reserves (Department of Commerce, 2003), and a federal regulatory process will determine the outcome in federal waters.

The initial attempt to implement the MLPA in 2001 was also rife with controversy. The California Department of Fish and Game (CDFG) formed a Master Plan team to develop Initial Draft Concepts of potential marine reserve sites for public review. The Master Plan Team identified draft MPA candidate sites without stakeholder consultation and presented the maps to the public in a series of meetings along the coast in the summer and fall of 2001. The meetings were extremely charged with intense upheaval among numerous stakeholders, especially fishermen. In both the CINMS and MLPA instances, stakeholders felt as if they were inadequately consulted, that insufficient data were available for comprehensive socioeconomic assessment, and that the fishermen's local ecological knowledge was not appropriately incorporated into the process. The process polarized many fishermen and environmentalists, as the debate focused on trade-offs between conservation goals and economic concerns.

Due to the dissatisfaction and immense distrust created during the initial attempt to implement the MLPA, the Director of the CDFG disbanded the original process and started over. The new process, designed to be more participatory in nature, convened seven Regional Working Groups of representatives from the fishing, diving, scientific, and environmental communities. These stakeholders were charged with proposing sites for marine reserves that would be assessed for ecological benefits by the Master Plan Team and reviewed for socioeconomic impacts. The Master Plan Team was to then synthesize this stakeholder input, scientific analysis, and socioeconomic assessment into a Master Plan for a MPA network for the state, subject to the approval of the California Fish and Game Commission. In January 2004 (two years after its launch), the Regional Working Group process was abandoned due to California's budget crisis. Non-governmental organizations (NGOs) and community groups continued to develop petitions for individual marine reserves and began to construct a "civil society" process (sanctioned by the state, but implemented primarily by stakeholder groups and funded by foundations and other private-sector funders) to synthesize such petitions, assess them against scientific criteria, encourage petitions for reserves that would fill critical gaps in a coherent system of marine reserves, conduct scientific and socioeconomic analyses of individual sites and the proposed system, and present a master plan for approval by the Fish and Game Commission. In the Fall of 2004, the State committed $500,000 to re-start the MLPA process, and appointed a Blue Ribbon Task Force to design and oversee a process to implement a network of reserves, create a pilot project along the Central Coast, and develop a strategy for long-term MLPA funding. The panel has elicited comments from various stakeholders regarding the draft framework for the process and identified a Master Plan Science Advisory Team from a pool of nominated candidates. The Task Force is charged with implementing a pilot project in Central California by Fall 2006, and a statewide network of marine reserves by 2011.

The MLPA requires the inclusion of socioeconomic information as laid out in Section 2855 (c) of the Act: "(T)he department and team in carrying out this chapter, shall take into account relevant information from local communities, and shall solicit comments and advice for the master plan from interested parties on issues including [...] (2) socioeconomic and environmental impacts of various alternatives" (California Bill Number AB 993, 1999). MLPA implementation offers an opportunity to capitalize on local knowledge and assess the socioeconomic effects of management decisions, and the Task Force recognizes the need for active involvement of stakeholders and the general public. Restarting the MLPA implementation process creates an opportunity to incorporate local knowledge and socioeconomic analysis early in the planning process. While the Department convened an expert workshop on socioeconomics in the fall of 2002, it is still uncertain how to include such information in MLPA implementation. The Local Knowledge Project, described in the next section, was intended to contribute to the assessment of these complex effects in a participatory way, and to provide a protocol for inclusion of socioeconomic information. Lessons learned from the Local Knowledge Project have further improved the protocol that can be implemented in future versions.

The Local Knowledge Project

The Local Knowledge Project was a pilot project to test a protocol for rapid socioeconomic assessment for use in MPA planning processes. It was based on the need for socioeconomic information in the MLPA implementation process. The project employed GIS with the primary goal of developing methods and data for rapidly assessing potential socioeconomic impacts related to individual marine reserve sites in California state waters. Other design considerations included budgetary and time limitations, and the need for spatially explicit local knowledge data that can easily integrate with scientific information. While the goal of the project was to test a protocol for socioeconomic assessment rather than MPA siting alternatives, many fishermen were interested in using the maps of fishing activity, acceptable closure areas, and critically important economic areas as a platform for further discussion of MPA alternatives.

Methods

The pilot project focused on commercial fishermen and recreational charter boat captains in the north-central region of California (Fig. 10.2), from Point Año Nuevo to Point Arena. Five main ports and port groups were identified within this area: Mendocino County (Fort Bragg, Point Arena, Albion), Bodega Bay, Bolinas, San Francisco, and Half Moon Bay.

Fishermen were integral to the study, both during the design phase and as participants. Study design included several meetings between project staff and fishermen representatives, "port gatekeepers" associated with the Pacific Coast Federation of Fishermen's Associations (PCFFA). The group collectively developed the research questions to be asked of the participating fishermen and the gatekeepers provided names of about 10 initial fishermen who would be willing to participate in an interview. The rest of the participants (total of 30) were identified through the "snowball sampling" (Huck, 2000) method of having interviewees recommend other fishermen to be interviewed. All participants were recommended based on the length of their fishing career, their depth of knowledge, and their willingness to be interviewed.

Semi-structured, one-on-one interviews with fishermen comprised the core of the project. During a two-month period, field assistants trained in social science interview techniques interviewed 30 fishermen. Typically, the assistants contacted the participants via telephone to explain the project and ask if they would be willing to participate in an interview. All participants were informed that the decision to grant an interview was voluntary and they could relinquish information at their discretion. Response was overwhelmingly positive, with only two fishermen unwilling to grant an interview, citing scheduling conflicts or lack of time. Most fishermen saw this as a rare opportunity to share their knowledge, concerns, and opinions regarding marine reserve siting and MLPA implementation. Once a fisherman agreed to be interviewed, the interviewers traveled to the port and met at a time and location convenient for the fisherman, generally the fisherman's boat or a nearby restaurant.

Guided by a set of specific questions, each interview was a free-flowing conversation. Interviewers allowed the fishermen to discuss a multitude of subjects, but kept the process focused with a set of core questions. This allowed for a comfortable conversation while also achieving quantifiable results that could be recovered and coded in the data entry process for comparison across interviews. The participating

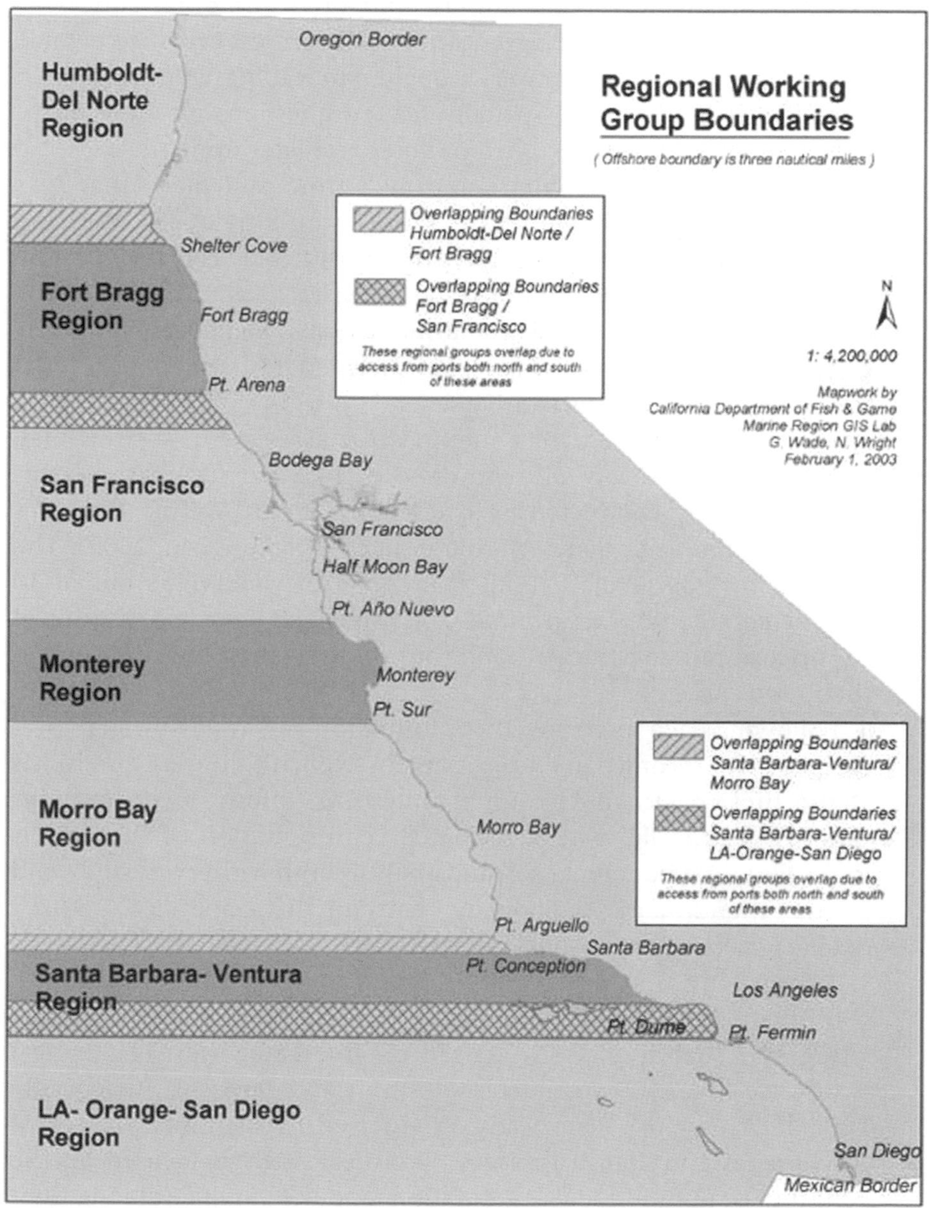

Figure 10.2. Map of MLPA regional boundaries

fishermen were asked a series of questions in four analytical categories: demographics (home harbor, years fishing experience, species targeted, gear and techniques used); oceanographic information (prevailing local weather and current patterns, weather-dependent fishing locations, observations about fish distributions based on physical oceanography, critical anchorages and transit passages, effects of ocean regime shifts such as the El Niño Southern Oscillation or the Pacific Decadal Oscillation); biological information (historically productive or "fished out" areas, known spawning sites, non-threatened or healthy species, threatened species or observed declines, biologically diverse areas, health of the fishery: past and present); and management (opinion of stock assessments, fishery management and environmental concerns, opinions of the MLPA process, economically critical areas, acceptable closure candidates). Interviewers brought nautical charts and fishermen recorded their answers to spatially related questions on these maps. Other information was recorded in notes and later transcribed.

All interviews were conducted at a time and place that was convenient for the fishermen. One complaint about existing socioeconomic analysis processes is the considerable cost they impose upon fishermen. Public meetings held at central locations are often the only option for participation in marine resource management, posing high actual and opportunity costs to fishermen. Fishermen must often take time off from fishing to drive several hours to a public meeting, all at their own expense. Other methods for assessment, such as mail surveys, often have low return rates—e.g., 14.6% in the case of a cost-earning survey conducted by the Pacific States Marine Fisheries Commission (Pacific States Marine Fisheries Commission, 2002). The Local Knowledge Project resulted in 27 viable interviews out of 30 conducted, or a 90% return rate. The fishermen were appreciative of the opportunity to participate and contribute on their own time and in their own space.

Following each interview, the information was coded by analytical category and recorded in Excel. Spatially explicit data were entered as shape files in OceanMap, Environmental Defense's GIS tool for California's coastal waters. A spatial representation of the data is important for data comparison and analysis both within this study and across other studies. Additionally, many of the shape files captured detailed species and seasonal information.

Recognizing the context of the MLPA and the goal to maximize conservation goals while minimizing socioeconomic impacts in MPA siting, the spatial analysis focused on five categories: (1) Critical Economic Areas; (2) Acceptable Closure Candidates; (3) Biologically Diverse Areas; (4) Historically Productive Areas; and (5) Critical Anchorages and Transit Passages. Statistical analysis focused on the congruence of fishermen's information, and the variance among their answers. Additionally, the Critical Economic Areas and Acceptable

Closure Candidates identified by the fishermen were aggregated and compared to the Department's original draft MPA maps. This analysis helped elucidate potential factors in fishermen opposition to the draft maps, such as the proximity of many proposed reserves near fishing ports.

A vital contribution to the success of the Local Knowledge Project was its iterative nature. Following the data collection, entry, and analysis phases, interviewers went back to each port and conducted plenary sessions with all participants from that port to review the statistical and spatial analysis. Information was only shown in aggregate and remained anonymous, but this gave fishermen the opportunity to review the data and correct any mistakes made in transcribing their information. Additionally, the plenary sessions facilitated discussion and understanding amongst the fishermen and often revealed as much information as the initial interviews.

Also central to the success of the project was confidentiality and anonymity. Fishermen's information regarding their fishing sites is proprietary and fishermen are often concerned about revealing information to competitors. Therefore, information collected during the project can be shown in aggregate, but never in fine-grain detail. Furthermore, the plenary sessions were used to gain permission for use of the aggregate data in publications and presentations. Due to the sensitive nature of this material, measures have also been taken to ensure fishermen ownership over the information.

Project Performance and Benefits

Information collected during the interviews was extensive and revealed an intricate use pattern over the oceanscape. Fishermen shared years of experience and first-hand knowledge that is invaluable, including for example, observations on oceanographic conditions, biological phenomena, and the effect of management decisions on ocean resources and their livelihoods. For a complete discussion of the results, see Scholz et al. (2004).

Overall, the project successfully achieved many of the goals set forth at its commencement (i.e., protocol for rapid socioeconomic assessment, and spatially explicit database of fishermen's ecological knowledge and socioeconomic concerns) and included numerous benefits over traditional methods of collecting and analyzing socioeconomic information. Importantly, the project was well received by most participants, many of whom expressed appreciation for the opportunity to share their knowledge and opinion. The project was relatively quick and inexpensive to conduct, is easily replicated, and achieved a high rate of return.

Through the pilot project, we created a spatially explicit database of fishermen's ecological information and socioeconomic concerns that can be easily integrated with biological information and used to inform future decision-making, and most importantly, successfully developed a protocol for rapid socioeconomic assessment for MPA planning processes.

Lessons Learned

The pilot project confirmed many of our preconceptions regarding the need for better processes for socioeconomic analysis and the ability of GIS-based tools to achieve these results. Engaging fishermen in conversations about the marine environment and MPA planning was invaluable in numerous ways: The process revealed information about the ocean, uncovered possible areas of compromise and common ground, and the project highlighted ways to improve OceanMap and the use of GIS tools for socioeconomic analysis.

In the context of this project, once engaged, fishermen were eager to share their knowledge and viewpoints about the ocean environment, fisheries biology, and marine resource management. In many cases, their observations of the marine environment correlate well with the scientific literature. Building upon this shared understanding can be invaluable to corroborate data and create policy applications that are supported by all stakeholders.

Fishermen generally disagree with scientists regarding the need for fully protected marine reserves (henceforth, "reserves"), where no fishing is allowed. While many scientists see reserves as a vital tool to manage marine resources and cite numerous studies showing that biomass, diversity, and fecundity are greatly enhanced in reserves (Castilla and Bustamante, 1989; Russ and Alcala 1996; Wantiez et al., 1997; Russ and Alcala, 1998; Halpern, 2003), fishermen often claim there is no scientific proof that reserves will benefit the fishery. Often missed is the fact that scientists and fishermen are frequently talking at cross-purposes, with scientists discussing benefits within reserves and benefits to fisheries in the form of insurance against management errors; while fishermen focus on fisheries yield enhancement (for which there is little evidence to date, due to the paucity of studies and lack of reserves of sufficient size to enhance fishery yield). Fishermen explain that existing strict regulations already make it nearly impossible to earn a living and additional restrictions will further jeopardize their future. Analysis revealed fishermen's extensive use of the marine environment, and according to participants, virtually every portion of state waters is important for catching a specific species or during a specific fishing season. Understanding the use patterns in state waters illustrates the challenge of creating reserves while limiting socioeconomic impacts. However, we found that despite their vehement opposition to the need for reserves, in light of impending legislation, most fishermen are willing to engage in conversation about reserve policy implementation, and many feel that the socioeconomic analysis is the only vehicle for their opinions.

The Local Knowledge Project highlighted shortcomings in both OceanMap and the protocol for socioeconomic assessment. Some attributes were anticipated while others were surprising. The scope of the project was too limited to conduct rigorous analysis or draw specific conclusions, which was expected based on the goals of the project. For example, some user groups were under-represented or omitted. This was intentional in the case of other marine resource users such as consumptive and recreational divers and surfers, but in addition, some types of fishermen, including surf fishermen and live rock fishermen, were left out. This is one problem with the snowball sampling method—as is unequal representation of gear types and species fished—and will be corrected in future iterations by ensuring that port gatekeepers also represent all fishermen groups that use the study area. Inclusion of all user groups and affected stakeholders is essential for comprehensive socioeconomic analysis of management implementation and policy decisions, and will be implemented in future phases of the project. Furthermore, the ability to draw conclusions from collected data is positively correlated with increased sample size; so, if this approach is used to formally assess socioeconomic impacts, sample size must be much larger.

While fishermen's shape files, as entered in OceanMap, contain species-specific and season-specific information, the resulting data analysis lacked detail. A significant amount of information could not be easily transcribed from notes to shape files because of constraints and limitations of the data entry protocol and OceanMap. For example, the only way to include information such as fishermen's name, homeport, and targeted species was to embed it in the title of a shape file. Additionally, some portions of the fishermen's information was coded and transcribed in Excel, while other portions were entered into OceanMap. Querying the data became a difficult and time-consuming process that could benefit from streamlining. Researchers also learned the importance of eliciting and representing all data in the greatest detail possible, by, for example, asking fishermen to specify the species to which they are referring when answering questions and recording that in OceanMap.

Related to insufficient detail, the project also lacked sufficient quantitative data and analysis. The interviewing technique and questions elicited a wide range of answers with limited standardization, creating a challenge for coding and comparison. There was no weighting mechanism to capture the relative importance of fishermen's data. For example, a total of 10 shape files identified by one fisherman as Critical Economic Areas had the same importance as one shape file identified by a different fisherman. This discrepancy is particularly evident in analyzing the congruence of these areas. Additionally, analysis did not discriminate between, for example, crab fishermen and salmon fishermen referring to salmon habitat. Lack of weighting mechanisms and techniques to derive relative importance of various answers limited the usefulness of the collected information and analysis. It is essential that future projects using this methodology to collect information to support policy processes incorporate more quantitative data and analysis.

Other challenges included how the project was perceived. There were some negative misconceptions of the project goals and techniques by fishermen who were not included in the study. Misunderstandings between project staff and project participants also occurred. The proprietary nature of fishermen's information exacerbates any miscommunication and highlights the need for clear communication from the beginning through the end. While most participants supported the project, some interviewees and other outside fishermen would have been more comfortable conducting the socioeconomic analysis themselves. Follow-up discussions resulted in ideas that can help strengthen future analysis.

GIS Tool Improvements and Future Directions

Local knowledge gained through the project described above not only informed marine management, but has also aided the design of innovative GIS-based tools and methods for socioeconomic analysis. Taking into account lessons learned from the Local Knowledge Project, we have redesigned both OceanMap and the protocol for socioeconomic data collection and analysis. The result is a more powerful, user-friendly tool that has generated interest among numerous marine stakeholders, and plans for additional studies to directly support MPA planning efforts in California.

Improvement of OceanMap and Study Protocol

Experience with the pilot project revealed numerous areas for improvement of our protocol and use of OceanMap. Focusing on fishermen, the first step was to expand the list of questions, making the study more comprehensive and quantitative in design. The revised protocol focuses on understanding and documenting fishermen's use patterns of state waters in a spatially explicit manner. The ultimate goal will be to compare fishermen's uses of the marine environment with scientific and jurisdictional information to identify areas that meet conservation goals, while minimizing socioeconomic impacts. In addition, we developed interview questions and protocols for other marine users including: consumptive and recreational SCUBA and free divers; surfers; kayakers; sailors; and shop owners who provide services for such users. These protocols were tested in summer 2004 with another iteration of the Local Knowledge Project and will be revised based on the results of that study.

FISHERMEN PROTOCOL

Changes to the interview questions reflect the new goal to compare information and identify areas for conservation with a minimal socioeconomic impact. First, we devised questions to be more quantifiable in nature. For example, fishermen will be asked to estimate the percentage of take-home income earned from fishing, broken down by species, to act as one metric for determining the relative importance of fisherman-identified Critical Economic Areas when conducting data analysis. Additionally, fishermen will only be asked to identify Critical Economic Areas (CEAs) and not Acceptable Closure Candidates (ACCs), as was included in the pilot study. The decision to include ACCs as an interview question is politically sensitive and depends upon the audience and the stage of the decision-making process. Groups of stakeholders that are using OceanMap internally may find it useful to identify ACCs, whereas meetings conducted across stakeholder groups will likely find it contentious to include ACCs early in the discussion, but may benefit by identifying them in later stages of negotiation. In the context of the Local Knowledge Project, fishermen were often hesitant to suggest any area for closure. By only inquiring about CEAs in the next phase of the project, we will create a data layer of fishing patterns in state waters that can inform marine management decisions while still being politically sensitive. All CEAs will be coded by species and weighted for importance by each fisherman. Interviewers will explain that each fisherman receives 100 total "points" to assign to their CEAs, to indicate relative importance amongst the areas they identify. For instance, one fisherman may identify one area and assign all 100 points to that area; while a second fisherman may identify 20 areas assigning 5 points to each one; and a third may identify six areas assigning 50 points to one area and 10 points to the remaining five areas. These two measures, weighting within and across fishermen's answers, will be recorded with each shape file and will act as proxies for relative importance of quantitative data. Additionally, OceanMap now has the ability to capture the scale at which a shape file was drawn, providing a broad ranking system of information. Shape files drawn at a smaller scale are more specific and contain more information. This can be accounted for in the analysis phase by attributing a multiplier to the scale of the shape file. These three weighting mechanisms will allow more quantitative and accurate analysis of fishermen's information and can be built into the analysis phase to create a more comprehensive spatial representation of use patterns and their socioeconomic importance.

The revamped OceanMap has significantly streamlined data entry and can now act as a centralized database for most information collected during the interview. OceanMap functions have been programmed for commercial and recreational fishermen and include six categories of shape files: Critical Economic Areas, Biologically Diverse Areas, Historically Productive Areas, Spawning Areas, Critical Anchorages,

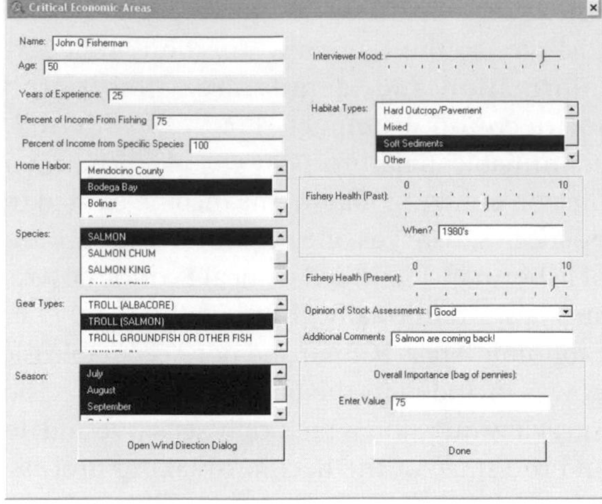

Figure 10.3. OceanMap pull-down menu screen for Critical Economic Areas

Table 10.1. OceanMap database of shape files and fishermen attribute information

	A	*B*	*C*	*D*	*E*	*F*
Fisherman Name	X	X	X	X	X	X
Home Harbor	X	X	X	X		
Species	X	X	X	X		X
Gear type	X					
Wind Direction	X		X	X		
Season	X	X		X	X	
Age	X					
Years Experience	X					
% income from fishing	X					
% income from species	X					
Fisherman mood	X					
Interviewer mood	X					
Habitat types	X					
Fishery Health (past)	X					
Fishery Health (present)	X					
Weighted importance of area	X					
Effect of regime shifts		X				
Pollution type						X
Pollution frequency						X
Pollution effects						X
Additional Comments	X	X	X	X	X	X

Column A = Critical Economic Areas; Column B = Biologically Diverse Areas; Column C = Historically Productive Areas; Column D = Spawning Areas; Column E = Critical Anchorages; Column F = Pollution and Polluted Areas. The user is guided through OceanMap with various pull-down menus, designed to build information into the program. All data are tied to shape files, categorized above, via customized pull-down menus (Fig. 10.3), designed specifically for that category of spatial information. For instance, inputting a fisherman-identified CEA requires choosing standardized options for numerous attributes including Home Harbor, Gear Type, Targeted Species, Wind Direction, Effects of El Niño, and Weighted Overall importance of area. The other categories of shape files have their own specified list of attributes displayed in the pull-down menus (Table 10.1). Associating this information with spatially-explicit shape files will ensure collection of complete information; create more detailed and useful data; and ease data organization, querying and analysis.

DIVER PROTOCOL

Recreational and consumptive SCUBA and free divers are another important group of marine resource users. We have also created interview protocols for these stakeholders and have built out the OceanMap program, in the same manner as for fishermen, to incorporate diver information. Customized pull-down menus have been created for both consumptive divers and recreational divers. While there are some differences between the protocol and pull-down menus for divers and fishermen, we intentionally designed the project to be comparable across user groups, and therefore the components for divers are largely the same as for fishermen.

For both recreational and consumptive divers, we programmed functions for six categories of shape files: Favored Dive Sites, Biologically Diverse Areas, Historically Productive Areas, Spawning Areas, Critical Anchorages, and Polluted Areas. Note that these are exactly the same as those for fishermen, except instead of Critical Economic Areas, divers have the category Favored Dive Sites, or user-identified areas that are most important for divers to have access for their activity and enjoyment. Identical to the fishermen shape files, users are guided through a series of pull-down menus where they specify different categories of information, which are tied directly to shape files by the OceanMap program. For a list of attributes that are displayed in the pull-down menus, see Table 10.2. As mentioned above, these are largely similar to the fishermen menus for comparison purposes, but include modifications such as average dive time, and size of dive area.

Importantly, Favored Dive Site also includes the weighting mechanism described in the previous section for the fishermen. Divers are asked to indicate the importance of each area by giving it a point value, with a total of 100 points for all of their areas combined. OceanMap also captures the scale at which the shape is drawn. The combination of these weighting mechanisms will be important for analyzing data within user groups, as well as across user groups. In

Table 10.2. OceanMap database of shape files and diver attribute information

	A	B	C	D	E	F	G
Diver Name	X	X	X	X	X	X	X
Home Harbor	X	X	X	X	X		X
Species Observed	X	X	X	X	X		X
Wind Direction	X	X	X	X	X		
Season	X	X	X	X	X	X	
Years Experience	X	X					
Species Targeted	X						
Harvesting Method	X						
% income from consumptive diving	X						
% income from specific species	X						
Interviewer mood	X	X					
Habitat types	X	X					
Species Health (past)	X	X					
Species Health (present)	X	X					
Average Dive time	X	X					
Average Size of Dive Area	X	X					
Access Method	X	X					
Weighted importance of area	X	X					
Effect of regime shifts			X				
Pollution type							X
Pollution frequency							X
Pollution effects							X
Additional Comments	X	X	X	X	X	X	X

Column A = Favored Dive Sites (Consumptive divers; Column B = Favored Dive Sites (Recreational Diver); Column C = Biologically Diverse Areas; Column D = Historically Productive Areas; Column E = Spawning Areas; Column F = Critical Anchorages; Column G = Pollution

particular, the scale feature will be helpful to compare fisherman and diver information, which are likely to be vastly different in size.

Marine Resource Stakeholders as Loci for Data Collection

The Local Knowledge Project has generated interest in socioeconomic analysis and marine reserve siting among numerous user groups, specifically in using the OceanMap tool. Many fishermen who participated in the pilot project requested copies of maps to use for fishermen-driven MPA planning efforts, and would like to see further data collection. Recreational diver representatives have seen demonstrations of OceanMap and have contributed their own data.

Environmental Defense has distributed numerous copies of OceanMap to interested parties for review and individual use.

OceanMap is designed to accommodate a variety of approaches for socioeconomic assessment. Environmental Defense can conduct research independently, partner with other organizations to collect data, or train interested stakeholder groups or organizations in the protocol and the use of OceanMap and let the groups collect data themselves.

OceanMap has additional built-in functions that allow stakeholder groups to work independently of GIS experts to easily perform initial data analysis. For instance, users can aggregate spatial information with the click of a button to quickly find congruence within the stakeholder data, thereby creating a map of their detailed spatial representation of their socioeconomic information. More advanced data analysis, such as comparing information across user groups, is also possible, but will likely occur after user groups have compiled their information into data layers. Environmental Defense can help groups synthesize and analyze OceanMap data initially, but will seek a "neutral party," perhaps an academic or government institution, to serve as the repository of data collected by OceanMap users. The flexibility and ease of OceanMap make it a powerful tool for reaching out to all stakeholders and conducting accurate and timely socioeconomic assessment that can easily be integrated with other types of information. Its application can facilitate the analysis and inclusion of socioeconomic data in implementing the Marine Life Protection Act and other marine management initiatives.

Over the summer of 2004, Environmental Defense, in partnership with Ecotrust and the Central California national marine sanctuaries, conducted the second iteration of the Local Knowledge Project, utilizing and testing the changes made in OceanMap. Participants included commercial and recreational fishermen, consumptive and non-consumptive SCUBA and free divers, kayakers, sailors, surfers, and shop owners. The results of the study will be published in a peer-reviewed journal, but have once again proven the effectiveness of OceanMap. We continue to refine interview protocols for each user group, expand the stakeholder-specific pull-down menus and methods for integrated data entry, and make other improvements to OceanMap.

Integration with other Spatially Explicit Information and Expansion Beyond California MPA Planning

OceanMap is a tool that allows spatial representation of disparate forms of marine-related data. One of the greatest benefits of the tool is its capacity to integrate numerous datasets, represented as individual layers of information, into one centralized location. Comparison and analysis of complex datasets is possible and relatively easy to accomplish. OceanMap already contains numerous data layers and will continue to assimilate information as it becomes available. The end result will be a rich database that can be used to derive MPA siting alternatives based on specified goals and objectives.

If coupled with simulated annealing software, OceanMap could serve as a decision-support tool, generating sets of reserve siting scenarios that all meet scientific criteria (e.g., for individual reserve size, total network size, habitat representivity, etc.). These siting scenarios could then be compared with socioeconomic data to choose scenarios that minimize costs to user groups.

OceanMap can also facilitate interaction between stakeholder groups. For instance, fishermen and scientists can view each other's information in a standardized format and identify areas of common agreement. Local knowledge is generally not subject to the same peer review as scientific information, but comparison of fishermen's information with other data sources would help validate it. Meetings between fishermen and scientists to achieve these objectives would be a logical extension of the project.

The protocol developed through the Local Knowledge Project and the capabilities of OceanMap can be generalized to other applications. These tools are flexible enough to support numerous marine policy implementation decisions and are not limited to MPA planning efforts. Groups working to derive marine decision support tools in other regions and states have expressed interest in the OceanMap model. The OceanMap design is replicable and, depending on available data, can be applied to any marine environment.

Conclusion

MPA planning is a complex and sometimes contentious process, with numerous stakeholders and data needs. Experience in California illustrates that early inclusion of socioeconomic information improves planning and implementation phases of the decision-making processes. Tools designed to facilitate inclusion of socioeconomic information in a spatially explicit manner and integration of different kinds of information are needed.

The Local Knowledge Project focused on developing a protocol for collecting local knowledge and standardizing it for integration into policy processes. The initial study yielded numerous products including: (1) a protocol for rapid socioeconomic assessment; (2) a database of fishermen's knowledge and information; and (3) a GIS of fishermen's ecological information and socioeconomic concerns for further use in the MLPA process (Scholz et al., 2004). Through this study, OceanMap was proven effective in representing spatial information and augmenting the MPA planning process.

Based on lessons learned during the pilot study and feedback from marine resource stakeholders, OceanMap has been further developed into a more comprehensive and powerful tool. The protocol for collecting socioeconomic information has also been improved for broader application, and with continued use and feedback, OceanMap development will persist. As this project has demonstrated, GIS-based tools are effective means of empowering user groups and representing information spatially. Such tools can greatly facilitate the collection and analysis of socioeconomic information, which is essential for sound policy making and durable marine management successes.

Notes

1. Parties interested in learning more about "OceanMap," or obtaining a copy should contact Peter Black at Environmental Defense. "OceanMap" currently runs on Microsoft 2000 or higher and with ESRI's Arcview 3.3.

References

Berkes, F., Colding, J., and Folke, C., 2000. Rediscovery of traditional ecological knowledge as adaptive management, *Ecological Applications*, 10: 1251-62.

California Bill Number AB 993, as introduced by Assembly Member Shelley, February 25, 1999.

California Department of Fish and Game, Marine Region. Marine Life Protection Act Initiative, http:// www.dfg.ca.gov/mrd/mlpa/ index.html. Last accessed November 18, 2004.

Carter, D. W., 2003. Protected areas in marine resource management: Another look at the economics and research issues, *Ocean & Coastal Management*, 46: 439-56.

Castilla, J. C., and Bustamante, R. H., 1989. Human exclusion from rocky intertidal of Las Cruces, central Chile: effects on *Durvillaea antarctica* (Phaeophyta, Durvilleales), *Marine Ecology Progress Series*, 50: 203-14.

Department of Commerce, 2003. Announcement of Intent to Initiate the Process to Consider Marine Reserves in the Channel Islands National Marine Sanctuary; Intent to Prepare a Draft Environmental Impact Statement, *Federal Register*, 68: 27989-90.

Feeney, D., Berkes, F., McCay, B. J., and Acheson, J. M., 1990. The tragedy of the commons—22 years later, *Human Ecology*, 18: 1-19.

Gilden, J., and Conway, F., 2002a. *An Investment in Trust: Communication in the Commercial Fishing and Fisheries Management Communities*, Corvallis, OR, Oregon Sea Grant Publication ORESU-G-01-004, 80 pp.

Gilden, J., and Conway, F., 2002b. *Fishing Community Attitudes Toward Socioeconomic Research and Data Collection by Fisheries Managers - Supplement to An Investment in Trust*, Corvallis, OR, Oregon Sea Grant, 35 pp.

Halpern, B. S., 2003. The impact of marine reserves: do reserves work and does reserve size matter? *Ecological Applications Supplement*, 13(1): 117-37.

Huck, S. W., 2000. *Reading Statistics and Research*, 3rd Edition, New York, Addison Wesley Longman, 126 pp.

McCay, B. M., and Acheson, J. M. (eds.), 1987. *The Question of the Commons: The Culture and Ecology of Communal Resources*, Tucson, AZ, University of Arizona Press, 439 pp.

National Research Council (NRC), 2001. *Marine Protected Areas: Tools for Sustaining Ocean Ecosystems*, Washington DC, National Academy Press, 288 pp.

Olsson, P., and Folke, C., 2001. Local ecological knowledge and institutional dynamics for ecosystem management: A study of Lake Racken watershed, Sweden, *Ecosystems*, 4: 85-104.

Ostrom, E., 1990. *Governing the Commons: The Evolution of Institutions for Collective Action*, Cambridge, Cambridge University Press, 280 pp.

Pacific States Marine Fisheries Commission, 2002. West Coast Catcher Boat Study: Summary 1997-1997, Seattle, WA, Pacific States Marine Fisheries Commission, http://www.psmfc.org/efin/docs/catcherboat.pdf. Last accessed August 30, 2004.

Russ, G., and Alcala, A., 1996. Marine reserves in fisheries management, *Aquatic Conservation: Marine and Freshwater Ecosystems*, 4: 233-54.

Russ, G., and Alcala, A., 1998. Natural fishing experiments in marine reserves 1983-93: community and trophic responses, *Coral Reefs*, 17: 383-97.

Russ, G., and Alcala, A., 1999. Management histories of Sumilon and Apo Marine Reserves, Philippines, and their influence on national marine resource policy, *Coral Reefs*, 18: 307-19.

Scholz, A., Bonzon, K., Fujita, R., Benjamin, N., Woodling, N., Black, P., and Steinback, C., 2004. Participatory socioeconomic analysis: Drawing on fishermen's knowledge for marine protected area planning in California, *Marine Policy*, 335-49.

St. Martin, K., 2001. Making space for community resource management in fisheries, *Annals of the Association of American Geographers*, 91(1): 122-42.

Wantiez, L., Thollot, P. and Kulbicki, M., 1997. Effects of marine reserves on coral reef fish communities from five islands in New Caledonia, *Coral Reefs*, 16: 215-24.

Chapter 11

Rapid Shoreline Inventory

A Citizen-based Approach to Identifying and Prioritizing Marine Shoreline Conservation and Restoration Projects

Philip L. Bloch, Jessemine Fung, Tom Dean, Lisa Younger, and Jacques White

Abstract

Development along marine and estuarine shorelines, coupled with the uncertain status of many species that use nearshore habitats, has generated a strong interest in both describing nearshore resources and understanding how to restore them. Advances in GIS technology and the wide availability of GIS software and inexpensive and powerful computers are improving our ability to manage complex natural and disturbed systems by helping us to understand the relationship between natural resources, habitat quality, and human uses. In this chapter, we discuss an innovative approach towards collecting, organizing, and analyzing shoreline data using volunteers known as the Rapid Shoreline Inventory. The Rapid Shoreline Inventory (RSI) is a project created with the assistance of regional experts by the Seattle, Washington based non-profit organization, People For Puget Sound. RSI links the information needs of resource managers and restoration practitioners to well-trained volunteer stewards who collect detailed data about marine and estuarine shorelines. The RSI consists of six key components: (1) the

Philip L. Bloch, Jessemine Fung, Tom Dean, Lisa Younger, and **Jacques White**[3]
People For Puget Sound, 911 Western Avenue, Suite 580; Seattle, WA 98104

Philip L. Bloch
Washington Department of Natural Resources, 1111 Washington St. SE, PO Box 47027; Olympia, WA 98504
Corresponding author: philip.bloch@wadnr.gov

Tom Dean
Vashon Island Land Trust, PO Box 2031,Vashon, WA 98070

Lisa Younger and **Jacques White**
The Nature Conservancy, 217 Pine St., Suite 1100, Seattle, WA 98101

identification of resource information needs; (2) permission to access shorelines from property owners; (3) careful classroom and field training of volunteers; (4) a rapid inventory of contiguous 45-m (150-ft.) sections of Puget Sound shoreline; (5) creation and maintenance of the associated GIS database and Web site; and (6) analysis and recommendations for specific conservation and restoration actions. RSI data are valuable both for describing the resources found on the shoreline and for developing predictive models of shoreline health. People For Puget Sound has developed a series of models that use RSI data to describe shoreline health and uses these models to identify restoration and conservation opportunities. Combining RSI data with aerial photos and broader scale nearshore characterizations can provide a comprehensive view of shorelines conditions, and provide data that can be used for a number of purposes. In addition to targeting restoration and conservation actions, RSI data can also be used to support several coastal and estuarine management activities, including identification of marine protected areas, regional and site land-use planning, research, monitoring, and oil-spill response.

Introduction

Environmental degradation attributable to anthropogenic sources has led to significant declines in a wide variety of coastal habitats and species (Vitousek et al., 1997). Development trends have been concentrated on coastal regions with 37% of the global population (more than two billion people) living within 100 km of the coast (Cohen et al., 1997). Coastal development has led to losses of wetlands, the straightening of rivers, and the armoring of shoreline banks. It is becoming increasingly clear that in addition to conservation of existing high-quality habitat, significant restoration actions necessary to ensure the continued survival of coastal marine wildlife and flora (Sinclair et al., 1995).

Many of these problems are caused by anthropogenic development and artificially imposed geographic stasis in an otherwise dynamic environment. Important ecosystem functions, including hydrology and sediment dynamics, are being more or less permanently disrupted, while historic development continues to impact habitat. Nearshore habitats are essential for prey resource production, refugia and reproduction of a variety of fish, shellfish and shoreline-dependent wildlife species (Nightingale and Simenstad, 2001). The impacts of historic habitat losses are compounded because people have reacted slowly or failed to recognize the severity of habitat and species losses (McMurray and Bailey, 1998). Restoration and rehabilitation of nearshore habitats and natural processes have been advanced as necessary for the continued viability of many coastal ecosystems (Restore America's Estuaries, 2002).

Most restoration assessments are designed to maintain the existing or near existing amount of habitat, species population level or ecosystem service (e.g., Fonseca et al., 2000), however in many urban and other severely impacted ecosystems, there is a need to restore habitat to a former, higher level in order to support sustainable populations or communities of desired species. For example, the urbanizing Puget Sound basin appears to have already lost more functional habitat and ecosystem processes than are required to maintain viable populations of many species, including forage fish, salmon and killer whales, all of which have been recently reviewed or listed under the Endangered Species Act in the region (Federal Register, 1999, 2001).

While there is a growing consensus that habitat protection and restoration are necessary strategies for preserving ecosystem integrity, the understanding of coastal systems is often regarded as insufficient to identify appropriate restoration sites and procedures. Ecological performance of coastal and estuarine restoration projects is not yet predictable with great certainty (Thom, 1997). Failures of restoration projects reflect our inability to accurately predict the trajectory of restoration processes in many instances, or perhaps, that we picked the wrong project in the wrong place.

One major factor limiting the development of successful conservation planning tools for nearshore areas in Washington State is the lack of cohesive, complete, and spatially detailed datasets focused on this ecological edge that functionally includes a combination of uplands, wetlands, and submerged habitats. Management of the nearshore area is fractured between several state, local, and tribal government agencies in the state, with no single entity having jurisdiction or the incentive to collect and maintain complete biological and physical datasets. For us to develop a tool to effectively prioritize preservation and conservation actions in the nearshore, it was first necessary to develop a tool to collect spatially explicit physical and biological information across a landscape that varied in space, type, ownership, and access.

Using new citizen-collected data describing a variety of shoreline characteristics, we developed site-specific indices to describe ecosystem preservation and restoration value for five targets. Targets include a mix of species guilds (forage fish, juvenile salmonids, and marine dependent wildlife), habitat (aquatic vegetation), and ecosystem processes (sediment transport). By combining these targets into a single index we have created a systematic mechanism for ranking sites based on their restoration or preservation value to the ecosystem. These indices prioritize sites that appear to be functioning, but are also measurably impacted by anthropogenic development. Restoration indices are based on characteristics intrinsic to a particular point along the shoreline, and do not incorporate characteristics of adjacent areas in the valuation process. However, viewing the spatial relationships of "scores" derived using these indices in GIS formats allows mangers to see groupings of high-priority sites. These indices have a variety of potential uses including: testing our understanding of species-habitat relationships, prioritizing sites for preservation or restoration, and creating monitoring sites for future research.

Methods

Study Area

This study focuses on a portion of the marine nearshore zone that includes the entire intertidal, parts of the shallow subtidal, and uplands immediately adjacent to shorelines. Indices were developed specifically for ecosystem components found in the inland marine waters of Washington State, including Puget Sound, Hood Canal, the Strait of Juan de Fuca, and the San Juan Archipelago. The inland marine waters are a series of interconnected, glacially scoured channels and have a total area of approximately 7,275 km^2 and a total of 3,973 km of shoreline (Washington Department of Natural Resources, 2001). In Whatcom, Skagit, San Juan, and King Counties, 40 km of shoreline have been inventoried using a high-resolution, volunteer-based inventory called the Rapid Shoreline Inventory (RSI).

Shoreline Inventory

Stretches of shoreline are selected for inventory in cooperation with local resource managers and citizens groups based on a review of wildlife and habitat distribution data. Once stretches of shoreline are identified, and landowner access permission is obtained, the shoreline is divided into 45-m linear sections that serve as the survey unit. Groups of trained volunteers work with "experts" to inventory shorelines during low tides. Volunteers receive 10 hours of training in the classroom and field before being allowed to collect data, and all data sheets are checked on-site by staff or volunteers that have received 40 or more hours of training before volunteers move to the next 45-m section. All data are collected during extreme low summer tide windows, with tide levels of 0 m Mean Lower Low or below. Data recorded for each 45-m shoreline section describe:

1. Beach location
2. Intertidal and backshore vegetation
3. Invasive species
4. Beach substrate
5. Bluff ecology
6. Streams, outfalls and signs of pollution
7. Shoreline structures
8. Adjacent land use
9. Wildlife sightings
10. Public access

Data are entered by volunteers and staff into a Microsoft Access database, and every 20th sheet is checked for accuracy by program

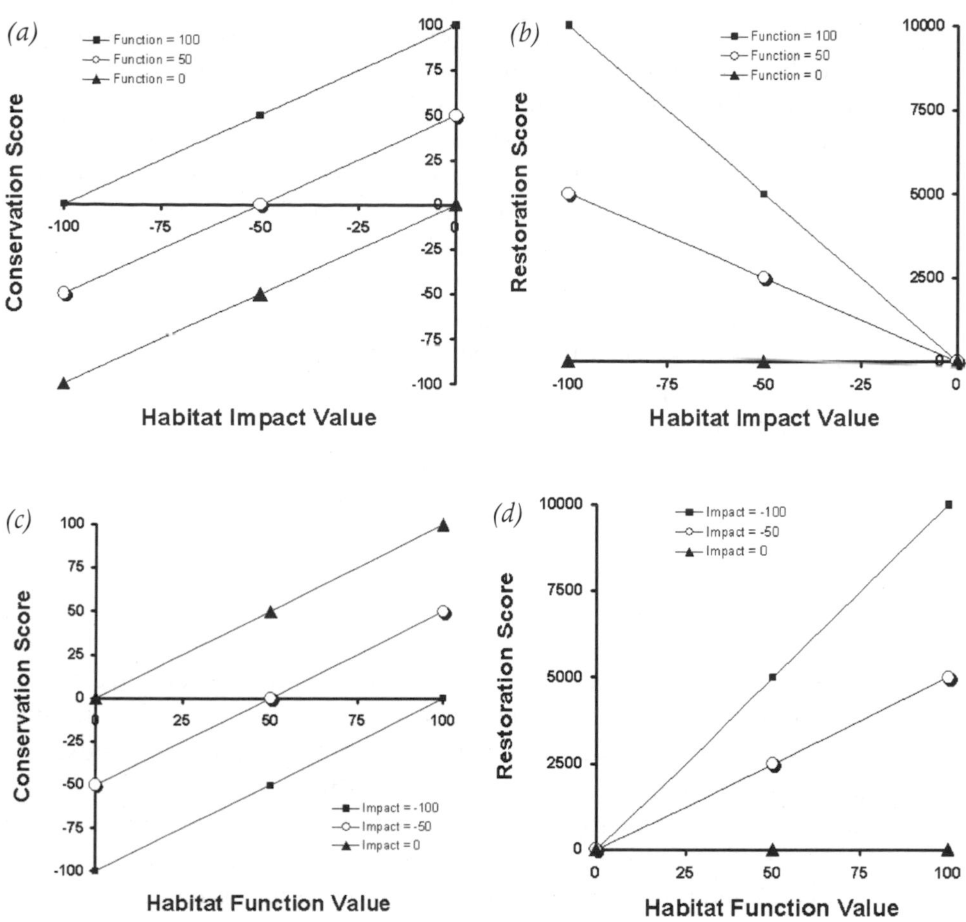

Figure 11.1 Relationship between conservation and restoration scores and habitat function and impact values (idealized for presentation, see Table 11.1). (a) Conservation score versus Habitat impact value, constant Habitat functions; (b) Restoration score versus Habitat impact value, constant Habitat functions; (c) Conservation score versus Habitat impact value, constant Habitat impacts; d) Restoration score versus Habitat impact value, constant Habitat impacts.

managers. Errors in data entry have characteristically been less than 0.05 %. Once data are collected, entered, and checked for accuracy, the geographic information system (GIS) is developed. Spatial coverages are created in ArcGIS using exported Access records and combined with spatial location data collected in the field with a Trimble GeoExplorer III GPS unit.

Index Development

Conservation and restoration planning begins with the development of clear goals. The three primary types of goals that have been identified relate to: (1) species; (2) ecosystem functions; and (3) ecosystem services (Ehrenfeld, 2000). Five ecosystem features of interest were identified that fit into these goal categories and appear to be tightly coupled to nearshore habitat conditions along sand and gravel beaches that dominate most of Puget Sound's shorelines. These features include: juvenile salmonids, forage fish, marine shoreline-dependent wildlife (especially birds), aquatic vegetation, and nearshore sediment supply and transport. For each of these targets, we developed semi-quantitative sub-indices using data collected during the RSI.

For conservation scores, each sub-index is the simple sum value of all positive attributes and all negative impacts. Conservation scores have a potential range of 0 to 100 with higher scores representing better conservation opportunities (Fig. 11.1a and c; Table 11.1). Restoration indices are calculated differently, reflecting the need to identify the "restorability" of a given site, balancing impacts with existing habitat value. For restoration, each sub-index is the absolute value of the product of all positive attributes multiplied by all negative impacts. Restoration scores have a potential range of 0 to 10,000 with higher scores representing better restoration opportunities (Fig.11.1 b and d; Table 11.1).

Since these conservation and restoration indices are semi-quantitative, the value assigned to a single site may have limited meaning relative to sites outside the study area, but in the context of a given study area, the relative value of a site is meaningful. The final conservation and restoration index value for each 45-m section is the average of the normalized rank order rankings for individual sub-indices. Therefore, conservation or restoration sites are ranked against others in the study area and have a score somewhere between the 1st and 99th percentile. Depending on regional conservation and restoration goals, it may be useful to examine different combinations of sub-indices rather than all five indices combined.

Table 11.1. Idealized habitat function and impact values for corresponding conservation and restoration scores (for demonstration purposes only, see Fig. 11.1a-d).

Function	*Impact*	*Conservation*	*Restoration*
100	-100	0	10000
100	-50	50	5000
100	0	100	0
50	-100	-50	5000
50	-50	0	2500
50	0	50	0
0	-100	-100	0
0	-50	-50	0
0	0	0	0

SUB-INDICES

Forage Fish. Forage fish, including populations of Pacific herring (*Clupea harengus*), surf smelt (*Hypomesus pretiosus*), and Pacific sand lance (*Ammodytes hexapterus*), are an essential component of the Puget Sound food web. Though phylogenetically unrelated, these three nearshore-dependent species comprise an essential trophic link within Puget Sound, and are a major component of the diet of many predatory species including salmonids (Bargmann, 1998). While relatively little is known about the adult life stages of forage fish, shoreline spawning preferences and requirements are generally understood. Our analysis is an important extension of existing surveys that identify actual forage fish spawning sites, because the model focuses on identifying all sites that have characteristics consistent with spawning needs, and therefore, identifies potential spawning habitat. While forage fish may use the same sites for spawning over long periods of time (Penttila, 1995), a site may be abandoned for no apparent reason only to become used again at some point in the future (Robards et al., 1999).

Shoreline surveys to identify spawning beaches have been conducted by the Washington State Department of Fish and Wildlife (formerly the Department of Fisheries) since 1972. Based on information obtained during these surveys, surf smelt and sand lance are thought to spawn selectively on shorelines that have deposits of either sand or pea-gravel sized sediment in the upper intertidal zone (Bargmann ,1998; Fig.11.2a). In addition to substrate preferences and requirements, forage fish eggs tend to have lower mortality when there is riparian vegetation adjacent to the shoreline that can shade the shoreline and moderate temperatures (Robards et al., 1999; Fig. 11.2a). Pacific herring vary slightly from smelt and sand lance in that herring spawn primarily in the lower intertidal and shallow subtidal zones, and therefore their habitat requirements are focused on vegetation such as eelgrass or algal turfs (Penttila, personal communication 2001; Fig. 11.2a).

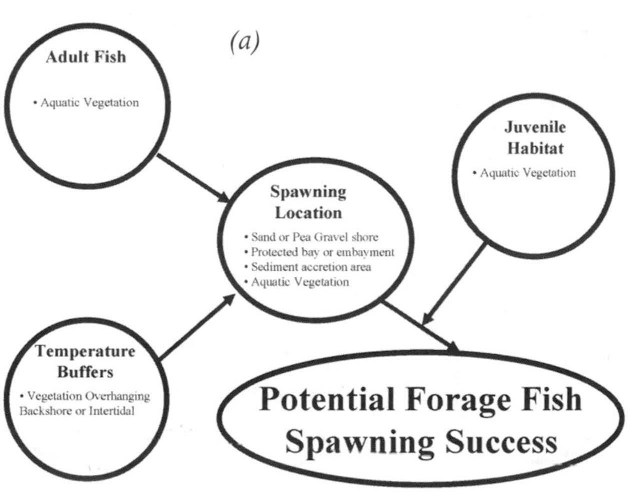

Figure 11.2 (a). Causal model describing the relationship between shoreline characteristics and ecosystem target indices for potential forage fish spawning success.

The forage fish analysis focuses on identifying those beaches with conditions that would seem to favor forage fish spawning and spawn survival (Table 11.2). Positive attributes for shorelines include appropriate sediment found in the upper intertidal, overhanging vegetation for shade, as well as aquatic vegetation that might be used for spawning.

Habitat impacts for this model are primarily those that interrupt or disturb potential spawning areas or the processes that form potential spawning areas. These include artificial outfalls that might supply excess nutrients or toxic chemicals to the shoreline, bulkheads that alter nearshore hydrography, or piers that shade subtidal vegetation .

Juvenile Salmon. The salmon habitat analysis relies on the assumption that nearshore habitats provide key functions for juvenile salmon development and survival (Fig. 11.2b). Nearshore marine habitat may serve as migration corridors, feeding areas, physiological transition zones, refuge from predators, or refuge from high-energy wave dynamics (Mason, 1970; MacDonald et al., 1987; Levings, 1994; Thom et al., 1994; Spence et al., 1996). All juvenile salmon utilize the shallow waters of estuaries and nearshore areas as migration corridors to move from their natal streams to the ocean (Willliams and Thom, 2001). Estuarine environments provide a gradual transition area for juvenile salmon to adjust physiologically to salt water (Simenstad et al., 1982). With declines in submerged aquatic vegetation that formerly served as feeding grounds and refugia for juvenile salmonids, it is likely that juvenile salmon have shifted their distributions and now utilize shallow water as an alternate refuge habitat (Ruiz et al., 1993).

This model focuses on valuing individual sites for their capacity to serve as feeding areas, refugia, or migration corridors (Table 11.2). Emergent vegetation (*Carex lyngbyei, Scirpus* spp., etc.) and riparian shrubs and trees have been identified as vital components that provide

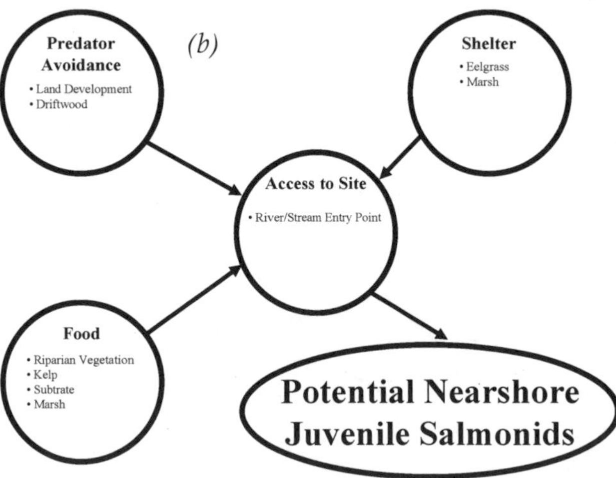

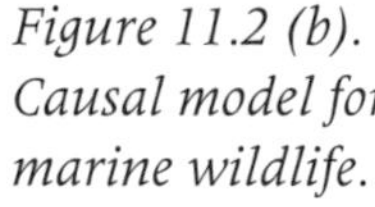
Figure 11.2 (b). Causal model for marine wildlife.

detritus and habitat for chinook food organisms (Levings et al., 1991; Tanner et al., 2002), and were therefore scored.

Habitat impacts are those features that are known to or believed to displace habitat or impede habitat forming processes. These include structures that reduce shallow, nearshore refuge habitat or adjacent land uses that may impact vegetation and upland food sources.

Marine Shoreline Wildlife. A variety of terrestrial animals will spend part or all of their lives within the nearshore environment and have a great impact on the composition and functioning of the nearshore ecosystem. An essential component of the nearshore ecosystem is marine birds. Marine birds are often the dominant predators along rocky as well as sandy beaches (Hori and Noda, 2001). In addition to being a dominant consumer of animals, most birds are omnivores and therefore play a critical role in structuring assemblages of animals as well as vegetation in the nearshore ecosystem.

This analysis focuses on habitat components that contribute to the feeding, rearing, and resting of shoreline dependent wildlife (Fig. 11.2c). This analysis looks at a variety of shoreline features that are beneficial for a variety of birds that depend on marine shorelines (Table 11.2). It awards points for fine sediments where shorebirds forage, niche habitats where rivers and creeks meet salt water, and dunes where some shorebirds nest. It awards points for a variety of vegetation directly beneficial to marine waterfowl (such as black brants) and indirectly beneficial to fish-eating birds (such as great blue herons and kingfishers). This model incorporates habitat impacts in the form of sources of disturbance such as trails accessing the shoreline as well as landuse patterns associated with human use.

Aquatic Vegetation. Primary production forms the base of any food web, and in Puget Sound, the primary producers are seaweeds, seagrasses, benthic microalage, kelps, marsh macrophytes, and phytoplankton. In Puget Sound, areas of increased algae and seagrass

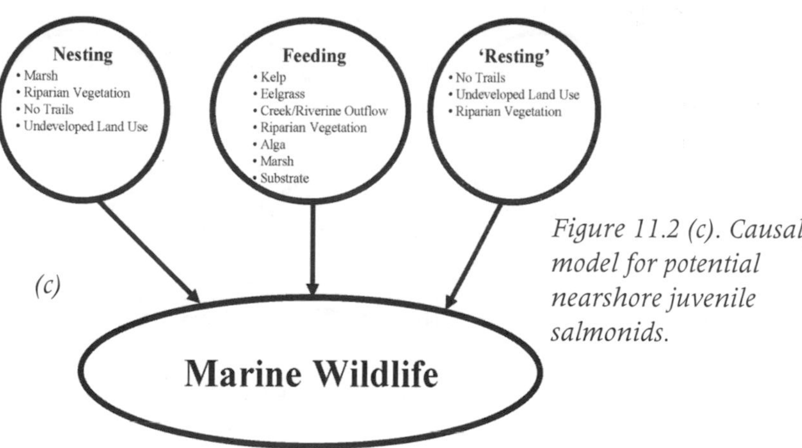

Figure 11.2 (c). Causal model for potential nearshore juvenile salmonids.

density, or biomass, contain more species and a greater abundance of epibenthic invertebrates than do areas of lower vegetative cover or structure (Cheney et al., 1994). With the exception of estuary marsh vegetation, which was formerly widespread in and around the major bays and deltas of Puget Sound (Bortelson, 1980), benthic primary production is limited to a relatively narrow band of habitat as a result of the steep fjord-like character of Puget Sound's nearshore habitat. Any attempt to determine the suitability of a certain area as habitat for aquatic vegetation must take into consideration light and parameters that modify light (epiphytes, total suspended solids, chlorophyll concentration, nutrients; Koch, 2001). Anthropogenic nitrogen loads to shallow coastal waters have been linked to shifts from seagrass to algae-dominated communities in many regions of the world (McClelland and Valiela, 1998). Propagules of most types of aquatic vegetation are generally found to be ubiquitous, so the absence of aquatic vegetation is generally a result of either inappropriate habitat for colonization and survival or displacement by another type of aquatic vegetation (Moore et al., 1996).

The focus of this analysis is on direct observations of aquatic vegetation with individual types of aquatic vegetation valued primarily for their ecological "services" (Fig. 11.2d). Implicit in the scoring of this model is the underlying assumption that each type of aquatic vegetation typically occupies a particular zone in the nearshore environment, from the subtidal to the upper intertidal. Species and multi-species assemblage scores are largely based on the ecological services they provide and the number of zones they occupy. Factors affecting light availability and nutrient loading as well as non-native competitors are assessed as habitat impacts in this model (Table 11.2).

Nearshore Sediment Supply and Transport. Puget Sound's shorelines are composed of hundreds of littoral cells that redistribute sediment along the shoreline. In the relatively protected waters of Puget Sound, the primary sources of sediment to the shoreline are alongshore

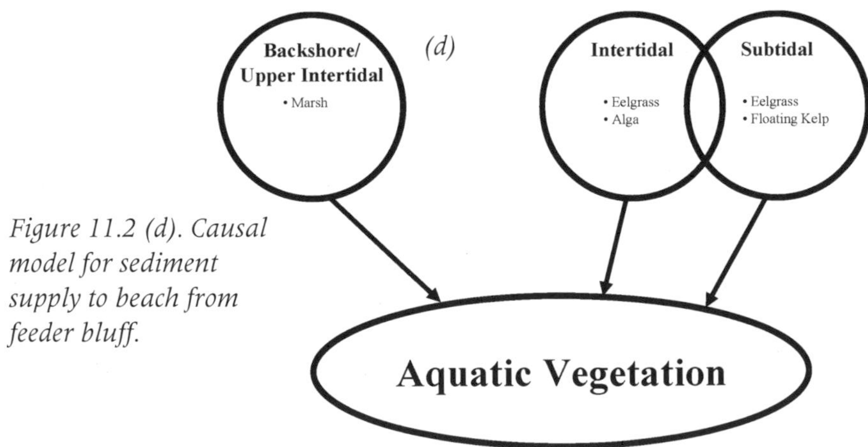

Figure 11.2 (d). Causal model for sediment supply to beach from feeder bluff.

Table 11.2. Description of model scores and justification for index targets.

Positive Attribute	*FF*	*JS*	*MW*	*AV*	*SS*
Geophysical Characteristics		-	-	-	-
Intertidal Substrate	-	10 to 15	10 to 20	10 to 15	-
Upper Intertidal Substrate	5	-	-	-	-
Sand/Pea Gravel Bed	20	-	-	-	-
Spit or Tombolo	10	-	-	-	10
Dune	-	-	15		-
Driftwood	-	5	-	5	-
Creek or River Mouth	-	5	5	5	10
Seep	5	-	-	-	
Bluff Height	5	-	-	-	10 to 50
Bluff Scars	-	-	-		10 to 15
Bluff Undercutting	-	-	-		10
Beach Energy	-	-	-		10
Vegetation Characteristics		-	-	-	-
Eelgrass (Z. marina)	10	15	-	15	-
Kelp and intertidal algae	10	5	-	5	-
Overhanging Vegetation	5 to 15	-	-	-	-
Riparian Vegetation	-	10 to 30	5 to 25	10 to 30	-
Marsh	5	15	10	15	-
Bluff//Bank Vegetation	-	3 to 5	3 to 5	3 to 5	-
Anthropomorphic Group		-	-	-	-
Undeveloped/Natural Landuse	5	5	5	5	-
No intertidal structures	10	-	-	-	-

and onshore transport, bluff erosion, and beach nourishment (Fig. 11.2e; Table 11.2). Sediment is lost from the beach as a result of erosion and longshore transport or deposition on spits (Downing, 1983). Shoreline development and armoring actively impact Puget Sound beaches by altering sediment supply and transport processes on shorelines and by directly modifying and occupying critical habitats (Shipman and Canning, 1993; Shipman, 1995).

In developing a causal model to assess the local functionality of the nearshore sediment budget, we adapted the results of other models that focus on the impacts of human activity on shoreline erosion (e.g., Lawrence, 1994). The focus of this analysis is on identifying signs that the sediment budget is being filled by looking for evidence of active erosion, in particular along bluff faces, and areas of deposition that are found at the end of drift cells such as tombolos and spits.

Negative Impact	*FF*	*JS*	*MW*	*AV*	*SS*
Shoreline Structures					
Intertidal Structures	-10 to -30	-30	-30	-20	
Shoreline Armoring		-10 to -30	-10 to -20	-10	-10 to -40
Boat Ramp	-20				
Adjacent Landuse					
Upland Land use	-10	-10 to -30	-10 to -30	-20	
Trails			-10 to -20		
Potentially Polluted Outfalls		-10	-10		-10
Invasive Plants					
Spartina				-30	
Purple Loosestrife				-20	
Sargassum				-10	

FF = Forage Fish; JS = Juvenile Salmonids; MW = Marine Wildlife; AV = Aquatic Vegetation; SS = Sediment Transpoty and Supply

Summary Data Presentations

Once the analyses are complete, potential nearshore conservation and restoration targets data are displayed on GIS-generated shoreline maps laid over topographic or orthophoto renderings of the adjacent water and land for perspective. Figure 11.3 (see page 203) shows the cumulative score for potential restoration targets on Samish Island in Skagit County, Washington, in northern Puget Sound. Notice that the complete island was not surveyed, a result of failure to obtain permission from all property owners to access their shoreline, and of time available during low summer tides. By looking at the map, it is possible to identify

Figure 11.2 (e). Causal model for aquatic vegetation.

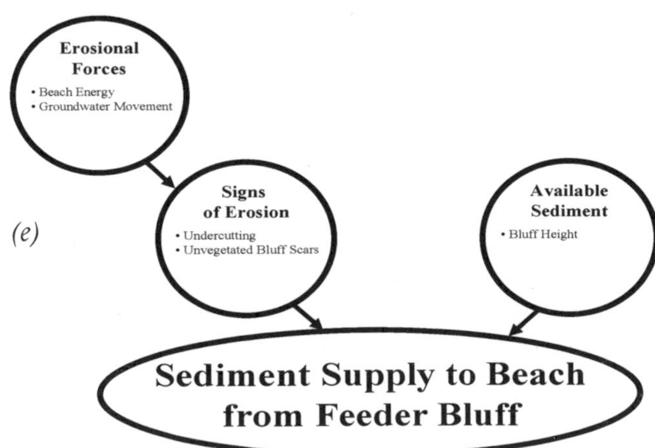

areas that have a higher density of high scores, and thus might be of particular interest for restoration actions.

Discussion

The models are designed to assess each site for both the current condition of the site (conservation opportunities) and for the potential condition of the site (restoration opportunities). Each model employs two series of "habitat attributes." One series of attributes is valued positively for perceived benefits or indication of benefits to habitat quality. The second series of "habitat impacts" is assigned negative values for impacts on habitat forming/maintaining processes, indications of physical disturbance, or direct impact on the model's focal species group.

For conservation opportunities, the models are used to rate individual 45-m sections of shoreline on a scale of -100 to 100, with higher scores reflecting higher quality habitat. Positive scores were assigned to positive attributes such as riparian vegetation or feeder bluffs. Negative scores were assigned to habitat impacts such as bulkheads or signs of pollution. The conservation score is then simply the net difference between the sums of positive and negative values accrued for each 45-m section.

This analysis is helpful for identifying areas of highly functional habitat as well as those places that are not being directly or indirectly impacted by habitat-altering processes related to invasive organisms or anthropogenic development. While scores vary linearly on this scale, it is important to recognize that this is a semi-quantitative model that provides a relative indication of site conservation value (sites scoring higher will generally be more favorable) for areas included in this study. The precise scores achieved may have little meaning taken outside the context of this specific cross-site analysis.

On the other hand, ranking sites for restoration potential is complex and must account for both existing habitat conditions and potential future conditions should the site be restored. Since no system currently exists for evaluating nearshore restoration potential, we were forced to create a novel scoring scheme. For our restoration-ranking scheme, the ultimate goal was to target high-value sites with restoration actions that produce the largest reduction in impacts. This scheme is designed to achieve the overall objective of identifying those sites with a high level of current ecosystem function or potential, and a significant degree of impairment.

We based our restoration analysis on the same scientific literature and data-driven, semi-quantitative rankings of site characteristics used in the conservation model. Our specific objective was to develop the most appropriate restoration model that would accentuate those sites scoring high in both the habitat attribute and habitat impact categories while giving relatively little value to sites that score low in either category. This objective was achieved by multiplying the habitat attribute score and the habitat impact score, and then taking the absolute value of the product of these two numbers. Thus, our restoration scores vary from zero—those sites that have either no current habitat function or no obvious habitat impacts, to 10,000—those sites that have both the maximum score in habitat attributes and impacts present. A site with high restoration potential might have multiple positive habitat attributes, such as pea gravel, a spit, eelgrass, and riparian vegetation, but also habitat impacts such as intertidal structures, a boat ramp, and several outfalls.

As with any model, the interpretation of scores requires care and consideration. We recommend that scores for this model be interpreted on a logarithmic scale. Since the model is semi-quantitative, the direction of scores (higher being more favorable than lower) is more important than the specific score or precise difference between scores.

One way to visualize our analyses is to plot conservation and restoration scores versus habitat function and impact values (the independent variables used to calculate the scores). Table 11.1 shows a series of idealized habitat function and impact values and the corresponding conservation and restoration scores. These values are plotted on Figures 11.1a-d. Notice that when conservation scores are plotted along lines of constant habitat function or habitat impact values, scores increase linearly with improvements in both habitat function and impact (i.e., less impact). The point of the conservation scoring system is to identify sites that have the greatest existing habitat value and the fewest negative impacts.

For the restoration analyses however, the scores increase along with increasing function and increasing intensity of impact (i.e., more impact equals a larger negative number). This result is because the impact and function values are multiplied instead of added. The implication of this model is that sites with very low habitat function or very low habitat impact, are not prime targets for restoration, whereas sites that still have substantial remaining or intrinsic habitat value, but also have significant impairment, represent the best opportunity to make significant gains for the ecosystem through restoration.

Overall, we believe this ranking system reveals those restoration opportunities that would provide the highest value to the living resources—not merely those that are the cheapest or most convenient. While sites identified using this tool are likely to provide ecosystem benefits if they are protected and restored, this ranking scheme should only serve as a guide and pre-ranking tool for further detailed site inspections and analysis of site-specific circumstances.

Because the specific interpretation of each individual scoring mechanism is uncertain, we believe it is best to compare sites within a given physical sampling area. In the specific examples presented here, we preferred to rank sites according to their scores and display those ranks rather than the raw scores. Those sites scoring in the highest decile (top 10%) are likely the most noteworthy sites and should be reviewed for potential conservation or restoration. Depending on the sampling area, sites in lower quantiles (the next 20%) may also be of interest for review. Overall conservation and restoration values were calculated by averaging the rank order (between 1 and 228 (the number of samples), with 228 being the highest scoring site) for the five models described here.

These conservation and restoration ranking schemes do not take into account the quality of immediately adjacent 45-m sections, or groups of adjacent sections. In this sense, the study and analysis does not explicitly account for habitat continuity along the shoreline. For example, multiple continuous sections of good to moderate quality habitat might be more important for conservation than one cell of excellent quality habitat in the middle of a larger area of very low-quality habitat. While scores for individual sections do not reflect this larger spatial context, viewing groupings of scores on the display maps can help identify important habitat "clusters," and at this point, the summary maps probably represent the appropriate tool for such integrative ranking of spatial relationships.

Finally, since this scoring system has only recently been developed, the model would benefit from further validation through: (1) taking conservation and restoration actions on sites identified by the model; (2) direct observations of target species and habitat processes at model sites; (3) further scientific inquiry into general habitat requirements of various species modeled here; and (4) review and exploration of the modeling method put forth here, incorporating newly collected information.

Results of the RSI data collection and analysis have provided support for several shoreline protection and restoration projects. While few of these projects originated as a result of these shoreline inventory efforts, the RSI provides project proponents with clear, objective information analyzing individual sites in a regional context. Restoration projects that have been supported include the restoration of a tidal channel on Samish Island, Washington, and the restoration of surf smelt spawning habitat along March Point, Washington. Protection efforts that have been supported include the acquisition of pristine shoreline habitats along Piner Point on Maury Island, Washington.

Conclusion

Restoration planning starts with the development of clear goals. The three primary types of goals that have been identified are: (1) restoration of species; (2) restoration of ecosystem functions and ecosystem management; and (3) restoration of ecosystem services (Ehrenfeld, 2000). An ecosystem approach requires an understanding of the fundamental linkages among ecosystem components, biological responses to physical and geochemical processes, rates and variability of these underlying processes, and the effects of disturbance and other modes of ecosystem change (Simenstad et al., 2000). The complexity of shoreline systems has delayed the development of assessment tools for either site specific or landscape level restoration planning.

While landscape metrics such as habitat connectivity may better describe some ecosystem components (Simenstad and Cordell, 2000), before these metrics can be applied suitable site-specific indices must be developed and examined. Therefore, the indices described here represent an early step in the development of restoration planning tools for marine shorelines.

References

Alsop, F. J., III, 2001. *Birds of North America*, New York, DK Publishing. 1024 p.

Bargman, G., 1998. *Forage Fish Management Plan*, Olympia, WA, Washington Department of Fish and Wildlife. 65 p.

Bortelson, G. C., Chrzastowski, M. J., and Helgerson, A. K., 1980. Historical Changes of

Shoreline and Wetland at Eleven Major Deltas in the Puget Sound region, Washington.

Atlas HQ-617. U.S.Geological Survey.

Cohen, J. E., Small, C., Mellinger, A., Gallup, J., and Sachs, H.,1997. Estimates of coastal populations, *Science* 278(5341): 1211-12.

Downing, J., 1983. The coast of Puget Sound: Its processes and development. University of Washington Press. 126 pp.

Ehrenfeld, J. G., 2000. Evaluating wetlands within an urban context. *Ecological Engineering* 15:253-65.

Federal Register,1999. Endangered and Threatened Wildlife and Plants; Listing of Nine Evoluntionarily Significant Units of Chinook Salmon, Chum Salmon, Sockeye Salmon and Steelhead, 64(147): 41835-39.

Federal Register, 2001. Endangered and Threatened Species; Puget Sound Populations of Copper Rockfish, Quillback Rockfish, Brown Rockfish and Pacific Herring, 66(64): 17659-68.

Federal Register, 2002. Endangered and Threatened Wildlife and Plants: 12-Month Finding for a Petition to List Southern Resident Killer Whales as Threatened or Endangered Under the Endangered Species Act (ESA), 67(126): 44133-8.

Fonseca, M. S., Julius, B. E., and Kenworthy, W. J., 2000. Integrating biology and economics in seagrass restoration: How much is enough and why? *Ecological Engineering*, 15: 227-37.

Hori, M., and Noda, T., 2001. Spatio-temporal variation of avian foraging in the rocky intertidal food web, *Journal of Animal Ecology*, 70(1): 122-37.

King County Department of Natural Resources. 2001. *State of the Nearshore Ecosystem*, Seattle, WA, King County Department of Natural Resources. 266 p.

Koch, E. M., 2001. Beyond light: Physical, geological, and geochemical parameters as possible submersed aquatic vegetation habitat requirements, *Estuaries* 24(1): 1-17.

Lawrence, P. L., 1994. Natural hazards of shoreline bluff erosion – A case-study of Horizon View, Lake Huron, *Geomorphology*, 10(1-4): 65-81.

Levings, C. D., Conlin, K., and Raymond, B., 1991. Intertidal habitats used by juvenile Chinook salmon (Oncorhynchus Tshawytsch) rearing in the north arm of the Fraser-River Estuary, *Marine Pollution Bulletin*, 22(1): 20-26.

Levings, C. D., 1994. Feeding behavior of juvenile salmon and significance of habitat during estuary and early sea phase, *Nordic Journal of Freshwater Research*, 69: 7-16.

MacDonald, J. S., Birtwell, I. K., and Kruzynski, G. M., 1987. Food and habitat utilization by juvenile salmonids in the Campbell River Estuary, *Canadian Journal of Fisheries and Aquatic Sciences* , 44: 1233-46.

Mason, J. C., 1970. Behavioral ecology of chum salmon fry (*Oncorhynchus keta*) in a small estuary, *Journal of Fisheries Research Board Canada*, 31: 83-92.

McClelland, J. W., and Valiela, I., 1998. Changes in food web structure under the influence of increased anthropogenic nitrogen inputs to estuaries, *Marine Ecology Progress Series*, 168: 259-71.

McMurray, G. R., and Bailey, R. J., 1998. Change in Pacific Northwest Coastal Ecosystems, NOAA Coastal Ocean Program Decision Analysis, Series No. 11, Silver Spring, MD, NOAA Office of Ocean and Coastal Resource Management. 342 pp.

Moore, K. A.,. Neckles, H. A., and Orth, R. J., 1996. *Zostera marina* (eelgrass) growth and survival along a gradient of nutrients and turbidity in the lower Chesapeake Bay, *Marine Ecology Progress Series*, 142: 247-59.

Nightingale, B., and Simenstad, C., 2001. White Paper: Dredging Activities, Marine Issues.

University of Washington, Wetland Ecosystem Team, School of Aquatic and Fisheries

Science, Seattle, Washington. 119 pp.

Penttila, D. E., 1995. Investigations of the spawning habitat of the Pacific sand lance *Ammodytes hexapterus* in Puget Sound, in Robichaud, E., (ed.), Puget Sound Research '93 Conference Proceedings, Olympia, WA, Puget Sound Water Quality Authority, p. 855-59.

Restore America's Estuaries, 2002. *A National Strategy to Restore Coastal and Estuarine Habitat*, Arlington, VA, Restore America's Estuaries, 156 pp.

Robards, M. D., Piatt, J. F., and Rose, G. A., 1999. Maturation, fecundity and intertidal spawning of Pacific sand lance in the northern Gulf of Alaska, *Journal of Fish Biology*, 54: 1050-68.

Ruiz, G. M., Hines, A. H., and Posey, M. H., 1993. Shallow-water as refuge habitat for fish and crustaceans in nonvegetated estuaries – An example from Chesapeake Bay, *Marine Ecology Progress Series*, 99: 1-16

Shipman, H., and Canning, D.,J., 1993. Cumulative environmental impacts of shoreline stabilization on Puget Sound, in Proceedings, Coastal Zone 1993, Eighth Symposium on Coastal and Ocean Management, New York, American Society of Civil Engineers, pp. 2233-42.

Shipman, H., 1995. The rate and character of shoreline erosion on Puget Sound, in Proceedings of Puget Sound Research, Olympia, WA, Puget Sound Water Quality Authority, pp. 77-83.

Simenstad, C. A., Fresh, K. L., and Salo, E. O., 1982. The role of Puget Sound and Washington coastal estuaries in the life history of Pacific salmon: An underappreciated function, in Kennedy, V.S. (ed.), *Estuarine Comparisons*, Toronto, Canada, Academic Press, pp. 343-65.

Simenstad, C. A., Brandt, S. B., Chalmers, A., Dame, R., Deegan, L. A., Hodson, R., and Houde, E. D., 2000. Habitat-biotic interactions, in Hobbie, J.E. (ed.), *Estuarine Science: A Synthetic Approach to Research and Practice*, Washington DC, Island Press, pp. 427-55.

Sinclair, A. R. E., Hik, D. S., Schmitz, O. J., Scudder, G. G. E., Turpin, D. H., and Larter, N. C., 1995. Biodiversity and the need for habitat renewal, *Ecological Applications*, 5: 579-87.

Spence, B. C, Lomnicky, G. A., Hughes, R. M., and Novitzki, R.P., 1996. An ecosystem approach to salmonid conservation. TR-4501-96-6057. ManTech Environmental Research Services Corp., Corvallis, OR. 356 pp.

Tanner, C. D., Cordell, J. R., Rubey, J., and Tear, L. M., 2002. Restoration of Freshwater Intertidal Habitat Functions at Specer Island, Everett, Washington. Restoration Ecology, 10(3): 564-76.

Thom, R. M., 1997. System-development matrix for adaptive management of coastal ecosystem restoration, *Ecological Engineering* , 8: 219-32.

Thom, R. M., Shreffler, D. K., and Macdonald, K., 1994. Shoreline Armoring Effects on Coastal Ecology and Biological Resources in Puget Sound, Washington. Olympia, WA.

Vitousek, P. M., Mooney, H. A., Lubchenco, J., and Melillo, J. M., 1997. Human domination of earth's ecosystems, *Science*, 277: 494-99.

Washington Department of Natural Resources, 2001. The Washington State ShoreZone Inventory user's manual. Olympia, WA. 23 p.

Williams, G. D., and Thom, R. M., 2001. *Marine and Estuarine Shoreline Modification Issues*. White Paper prepared by Battelle Marine Sciences Laboratory for Washington Department of Fish and Wildlife, Olympia, WA, Washington Department of Ecology and Washington Department of Transportation. 99 p.

Chapter 12

Port Orford Ocean Resources Team:

Partnering Local and Scientific Knowledge With GIS for Community-based Management in Southern Oregon

Victoria Wedell, David Revell, Laura Anderson, and Leesa Cobb

Abstract

The Port Orford Ocean Resource Team (POORT), a non-profit organization on the south coast of Oregon, combines scientific and local knowledge to address ocean resource and community management decisions. POORT is tackling community-based management on the scale of a small fishing community. The goal is to protect the long-term sustainability of the Port Orford fishery ecosystem and the economic and social systems dependant on it. To answer a series of scientific and management questions, POORT experimented with a process for documenting spatial information through local knowledge interviews (LKIs) with Port Orford community members. These interviews were conducted using acetate-covered base maps, which were then converted into digital GIS layers, aggregated, and incorporated into further GIS analysis. The LKI process and GIS analyses provided a needed biological and local economic baseline inventory, as well as valuable qualitative information on Port Orford's natural resource history and the changes occurring in this small-scale fishing community due to the changes in fisheries management on the West Coast of the United States. POORT is combining grassroots efforts and scientific

Victoria Wedell, Marine Resource Management Program, Oregon State University, OR 97331(now with NOAA Knauss Marine Policy Fellowship Program, National Sea Grant Office, Silver Spring, MD 20910)
Corresponding author: vicki.wedell@noaa.gov

David Revell, The Surfrider Foundation, Oregon Chapter (now at Earth Sciences Department, University of California, Santa Cruz, CA 95062)

Laura Anderson, Marine Resource Management Consultant, Newport, OR 97365

Leesa Cobb, Port Orford Ocean Resource Team, Port Orford, OR 97465

knowledge to demonstrate means to assess impacts to fishing communities, as is called for by the MagnusonStevens Fishery Conservation and Management Act.

Introduction

The Port Orford Ocean Resources Team (POORT) is using a geographic information system (GIS) as a tool to combine the best available science and local knowledge about the nearshore and coastal environment to support long-term planning for community-based resource management. POORT's overall goal is to engage the Port Orford fishermen and other community members in developing and implementing a strategic plan that enhances the sustainability of the Port Orford fishery ecosystem and social system dependant on it. Long-term planning objectives include: increased input into local fishery management decisions, diversification of economic opportunities, and ensuring that conservation strategies balance economic and ecological sustainability and social equity. Led by a community advisory board largely comprising commercial fishermen and supported by scientific advisors from agencies and academia, POORT developed a list of scientific, market, and management questions that are driving nearshore cooperative research and GIS development. A participatory GIS approach provided both the framework for capturing important information, and offered coastal citizens a process for active participation in management discussions about the marine environment.

This process provided a baseline inventory of the spatial ecology and economy experienced by Port Orford community members as well as an examination of the social and economic linkages important to this small-scale fishing community. Spatial questions generated by POORT included: "What are the abundance, distribution, and diversity of the flora and fauna of the Orford Reef area?" and "How are the recreational and commercial fishing effort and related socioeconomic value distributed?" Through semi-structured interviews with various community members, POORT documented the distribution and relative economic importance of areas targeted for commercial fishing activities as well as the distribution of recreational activities, ocean and coastal resources, and many species. The discussion was primarily limited to the area between Coos Bay and Gold Beach, Oregon, out to the edge of the continental slope. The Port Orford Ocean Resources Inventory validates a process for documenting local experiential knowledge and provides the first steps towards a more in-depth economic analysis to support community-based management in Port Orford.

The Port Orford Fishing Community

> *"Fifteen years ago, you could stand on this corner [Highway 101 and Harbor Drive, location of the POORT office] with your lunch pail, a pair of cork boots, and a pair of rubber boots and you could be certain to find work for the day. Port Orford used to be an easy place to work"* (Interview 007).

Established in 1851, Port Orford was the first European settlement on the Oregon coast and has depended on natural resource extraction throughout its history. Originally settled in hope of tapping into rich gold deposits, some pioneer families still own and execute mineral rights in nearby rivers (Interview 007). The original port dates back to 1856, with the port district being formed in 1911. Logging and milling supported the community for many years, with the timber industry peaking here during the 1930s, mainly with the shipment of Port Orford cedar. Shipping of lumber stopped shortly after the jetty was completed, in 1968, primarily due to market conditions and the decline of local timber (http://portorfordoregon.com/portofpo.html). Many current commercial fishermen were also loggers, or come from logging families. Port Orford also has a maritime history, with some families having third-generation fishermen with over 50 years of cumulative knowledge passed down through the generations.

Location

Located in Curry County, Port Orford is the most westerly incorporated city in the contiguous United States and is situated on an open bay, unlike most other Oregon ports, which are positioned along river channels (http://www.portorfordoregon.com/ relocate.html). Both by land and by sea, the town is physically and economically isolated compared to other Oregon ports. It is located about 50 mi. south of the nearest large population center of Coos Bay and 70 mi. north of the port of Brookings and the Oregon-California border (Fig. 12.1; see page 204). Large sand bars outside the nearest ports of Bandon and Gold Beach, both about 25 mi. away, can impose transportation barriers to the small fishing vessels of Port Orford. The Port Orford Lifeboat Station provided rescue services to the southern Oregon coast until 1970, when it was decommissioned (www.portorfordlifeboatstation.org/). Currently, the fishermen here must depend upon each other when trouble arises out at sea, risking their own safety and liability for others. The relative isolation of Port Orford may contribute to the town's true sense of community.

Cape Blanco, a prominent oceanographic feature in the California Current system located approximately 10 mi. northwest of Port Orford, separates two distinct oceanographic regions of the Northeast Pacific Ocean, as divided by the Global Ocean Ecosystem Project (Mackas et al., 2002). Generally, eastern boundary currents induce strong upwelling conditions in the nearshore and support diverse and abundant marine life, including fishes, invertebrates, marine birds, and marine mammals. The Orford and Blanco reefs together consist of about 7 mi. of rocky reef and bull kelp forest (*Nereocystis*) habitat. Several of these rocky islands breach the surface of the water, extending the three-mile limit of state jurisdiction to include most of the nearshore area.

Winds and rains are seasonal. Late fall, winter, and early spring account for 81% of the 72 in. of annual precipitation (http://www.wrcc.dri.edu/cgi-bin/cliGCStP.pl?orporf). January brings winter storms and gale force winds out of the southwest, from which there are no safe anchorages. "The only time it's calm in Port Orford is when the wind is blowing the same from both directions," a local resident only halfway joked during our interview (Interview 447; interview references throughout refer to the POORT project "Port Orford Ocean Resources Inventory and Local Knowledge Interviews," conducted August-December 2003). As the spring rains decrease, winds switch directions and come from the northwest throughout most of the summer. Moderated by the Pacific Ocean, temperatures range from 45 to 61° F. through the entire year (http://www.wrcc.dri.edu/cgi-bin/cliRECtM.pl?orporf).

Port Infrastructure

Port Orford has minimal fishing infrastructure: a pier and jetty, two commercial hoists, and one sport crane. The two buying stations at the port, Hallmark Fisheries and NorCal Seafood, purchase almost all of the fleet's seafood products; however, NorCal only buys Dungeness crab and live rockfish (Interview 138); Pacific Premium Seafood, the other buying station and processing plant, closed due in part to the decline in urchin harvesting and the increase in the live rockfish market, which reduced the demand for local processing capacity. There is cold storage and the port sells fuel (Scholz, 2003). The other marine businesses on the dock include Dock Tackle and Pac Nor West Charters.

Figure 12.2. The port of Port Orford Dock and hoists (photo: V. Wedell).

Dock Tackle is a unique combination of a tackle and gift store, nautical museum, and seasonal fresh fish market. Pac Nor West Charters runs recreational fishing and scuba trips off the dock. There is considerable financial pressure on the port to get returns on the decaying Premium Pacific Seafood building. Without a clear vision of the future of this port, precious space and infrastructure could be lost through insufficient land use planning.

A floating dock for recreational fishing boats on the side of the pier can be drawn up in bad weather. Sport fishing is less prominent here due to the large distance from nearby population centers and major airports, the adverse weather conditions, and the lack of nearby large rivers, although the Elk and Sixes rivers bring in some recreational anglers. Beachcombing, surfing and diving, kayaking, and whale and bird watching are other common recreational activities occurring in Port Orford.

The Port Orford Fleet

Port Orford vessels need to meet the weight and dimensional requirements of the commercial hoists that lift them in and out of the water every day, resulting in a homogeneous fleet as compared to other Oregon ports (Fig. 12.2). Vessels are restricted to a maximum length of 44 ft., maximum width of 15 ft. and no more than 44,000 lb. (http://discoverportorford.com/portofpo.php). Small vessel size restricts the range and duration of fishing activities, especially during adverse weather conditions, resulting in somewhat traditional fishing grounds. About 40 vessels homeport here, either secured to the dock on trailers or moored in the harbor during the summer.

The success of a small boat fleet depends upon the diversity of the fisheries they execute and the flexibility to move in and out of them as weather, ocean conditions, regulations, and market conditions permit. POORT interviewed people from approximately 50% of the vessels in the Port Orford fleet, with an average vessel length of 34 ft. Figure 12.3 (see page 204) shows that 72% of the 22 commercial fishermen interviewed currently participate in between four to seven fisheries over one year of fishing activities. Currently, the Port Orford fleet primarily targets salmon (86% of interviewees), Dungeness crab (82%), live rockfish (*Sebastes*) (73%), and sablefish (68%). However, albacore

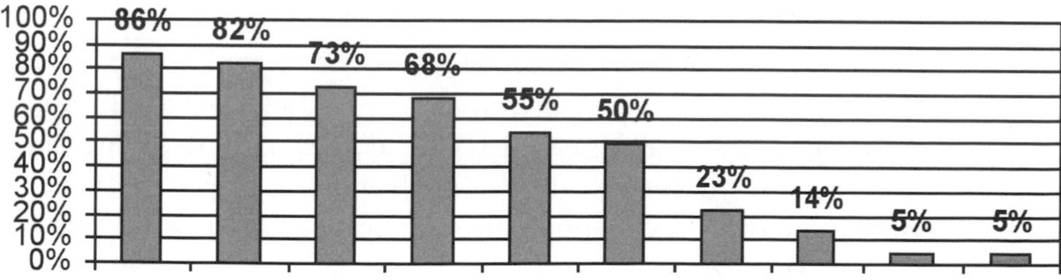

Figure 12.4. Percent of LKI participants targeting specific fisheries.

tuna (55%) and Pacific halibut (50%) are also important fisheries, as well as hagfish (23%) and urchins (14%) to a lesser degree (Fig. 12.4).

In years past, the deep-water shelf rockfishes would have been among the top five executed by Port Orford fishermen (73% of interviewees). Due to recent management measures, it is no longer economically viable for Port Orford fishermen to target these species. Historically, there was also a booming local urchin fishery, which went through its bust cycle in the mid-1990s. When a fishery closes, the traditional response by Port Orford fishermen is to shift effort to a new fishery.

Although some did shift their effort to the nearshore live fish fishery, these were primarily the displaced small boat urchin fleet, which found new opportunity with high-value live fish. Fishermen target nearshore groundfish species in this fishery because of increased fish mortality when fishing in depths greater than 27 fath. (Interview 193). The longline fishermen were reluctant to move into the live fish fishery, recognizing their added pressure would overfish the slow-growing, late-maturing rockfishes. "We stayed out of it, that was those guys only fishery and we had crab, blackcod (sablefish), and some salmon fishing" (Interview 207). The live fish fishery did not replace the income lost by the longline fishing businesses from the lack of opportunity to fish the shelf groundfish.

Social and Cultural Ties to Commercial Fishing

Whether we examined social and cultural linkages or economic indicators, Port Orford is a true fishing community supporting about 40 local fishing families (Andersion, 2001). The dock is the hub of social activity in the town. In addition to POORT, other industry-related organizations include the Port Orford Fishermen's Association and the Port Orford Women's Fishery Network. These fishing associations co-host the annual Salmon Bake and the men compete in the Dingy Race during the Port Orford Forth of July Jubilee, whose theme in 2003 was "Fishing the Wild Sea." The Blessing of the Fleet Ceremony, which takes place at the Fishermen's Memorial, honors those fishermen who have been lost at sea and prays for the continued safety of those who still make their living out on the ocean. The Port Orford Arts and Seafood Festival also celebrates Port Orford's fishing history. The cultural importance of the ocean and of commercial fishing is even evident in the many maritime murals and ocean-related names that adorn small businesses and schools in Port Orford.

Whereas other coastal towns are seeing much larger and more rapid increases in population, Port Orford almost refuses to grow. In 2000, the U.S. Census Bureau estimated 1,153 people currently living here, an increase of just over 11% in the last 30 years (http://bluebook.state.or.us/local/ populations/pop03.htm). In the next 15 years, the coastal zone is estimated to receive over half the nation's projected population growth, an additional 27 million people moving into coastal counties that cover only 17% of the land area of the United States (Beach, 2002). It is likely that Port Orford's population will grow at an increased rate. Common perceptions held by the community are that the immigrating people are predominately Californian retirees and that the limited living-wage jobs available in Port Orford are increasingly employing more of these people than local residents.

A relatively large proportion of the Port Orford population has jobs in the fishing industry. However, the estimates vary considerably, from as high as 30% to as low as 9%, as reported by Scholz (2003) and the U.S. Census (2000), respectively. Anderson (2001) reports that depending on the season, the community has between 100 and 150 people directly or indirectly involved in the day-to-day activities of commercial fishing, representing about 10-15% of the population. This relative proportion of fishing-related employment has statutory and management implications as set out by National Standard 8 of the 1996 reauthorization of the Magnuson-Stevens Fishery Conservation and Management Act, renamed the Sustainable Fisheries Act.

National Standard 8

National Standard 8 demonstrates the need for stronger emphasis on socioeconomic concerns in fisheries management, particularly a need for increased focus on communities. Specifically, it states that: "Conservation and management measures shall, consistent with the conservation requirements of this Act take into account the importance of fishery resources to fishing communities in order to (A) provide for the sustained participation of such communities, and (B) to the extent practicable, minimize adverse economic impacts on such communities" (National Marine Fisheries, Service, 2002).

Although Standard 8 requires that the impacts of fishing regulations to fishing communities be analyzed, it does not state how the boundaries of that place are drawn or its dependency measured. New incentives are needed to help groundfish management make the transition to new goals and objectives, such as National Standard 8 (Hanna, 2000).

Defining Fishing Communities and Fishing Dependency

Current research into the definition and practical application of the terms "fishing community" and "fishing dependency" intends to help management agencies determine what the differential economic impacts from fishing regulations and management measures are for communities. The Sustainable Fisheries Act defines a fishing community as: "... a community which is substantially dependent on or substantially engaged in the harvest or processing of fishery resources to meet social and economic needs, and includes fishing vessel owners, operators, and crew and United States fish processors that are based in such a community" (Hall-Arber et al., 2002).

Fishery management councils interpret the legislation to imply a place-based definition of a fishing community. Current research into identifying fishing-dependent communities determined that 15% fishing employment would qualify fishing-dependence, although the percentage is an arbitrary number (Jacob et al., 2001). However, limitations with this measurement include: the gross scale of census data (i.e., county-level) and the severe under-estimation of fishing employment using census methods (Jacob et al., 2002). Apparent dilution of dependency occurs with increases in the non-fishing proportion of populations, as with immigrating California retirees to Port Orford, or with the reduction in fishing employment, such as those retrained in the Groundfish Disaster Outreach Program.

Assessing Impacts to Fishing Communities

To better manage fisheries, you need to know the fishermen and the industry from their perspective and how perceptions, rationalities, and behavior change as a consequence to fisheries management (Jentoft, 1999). Socioeconomicimpacts among communities vary considerably and depend on fleet composition, infrastructure, specialization, social institutions and gentrification trends. The National Academy of Public Administration concludes that Fishery Management Plans do not have adequate social and economic goals and that social and economic data collected by NOAA Fisheries are inadequate for understanding the effects of past management on fishing communities or for predicting outcomes to these communities of management alternatives (Gade et al., 2002). The scale of economic data collection and the burden of turning that into useful information are insufficient to successfully assess impacts to specific communities.

Hall-Arber et al. (2002) not only looked at measures of fishing-related employment, but also traditional economic analysis, complexity of fishing infrastructure and degree of gentrification, and the port-profile approach, which looks at patterns of contracts, characteristics of community culture and institutions, and the local residents' views about their way of life and fisheries management. Because of the inherent complexity, a comprehensive analysis of the social and economic impacts of fishing regulations is impossible without new tools. The Groundfish Fleet Restructuring Information and Analysis Project is one such tool trying to resolve the gross scale of fish-ticket information through spatial GIS analysis to assess the impacts to ports of capacity reduction scenarios and area closures (Scholz, 2003).

The West Coast Groundfish Crisis

The National Marine Fisheries Service (NMFS) has responsibility to manage the nation's living marine resources within the exclusive economic zone. On the West Coast, the Pacific Fisheries Management Council (PFMC) has regional authority over the federally managed marine species in Washington, Idaho, Oregon, and California. PFMC is responsible for conservation and management of marine fish stocks, habitat, and fisheries in a sustainable manner while trying to equitably balance a variety of related human needs. However, the Council has taken on important fisheries problems on an individual issue basis: "The [PFMC] has responded to [economic hardship, uncertainty, polarization, low landing limits, over capacity] by trying to deal with individual issues on an ad-hoc basis. This short-term approach has been increasingly characterized by crisis management" (Pacific Fisheries Management Council, 2000).

The West Coast groundfish crisis is characterized by overfished groundfish species and reduced fishing opportunities occurring after a period of expansion and growth (Hanna, 2000). A 50% capacity reduction of the groundfish fleet (Pacific Fisheries Management Council, 2000). and subsequent depth-related shelf closure has impacted many fishing communities along the coast.

Port Orford has a long-established dependency on the groundfish fishery, with local longline vessels targeting lingcod, canary, yelloweye, and yellowtail. These deep-water shelf groundfishes were a valuable fishery for this small-scale fleet, as abundant fishing grounds are nearby. Quota reductions were placed on the fishery when it became evident to fishery managers that the trawl fleet was overfishing and discarding these species at a high rate. The restrictions were coast-wide and encompassed all gear types, effectively shutting down Port Orford's longline groundfish fishery. Many local fishermen lost a significant portion of their income, as much as 90% in one year for some fishermen (Interview 899).

Port Orford and The Groundfish Disaster Outreach Program

Port Orford fishermen received more than $800,000 in direct payments from the Groundfish Disaster Outreach Program (GDOP) to aid in transitioning people into new careers. Started in April 2000, with federal appropriation funds, the GDOP works with the Oregon Employment Department, which administers the Groundfish Transition Income, and the coastal Workforce Investment Agencies, which provides the actual training for new careers. In Port Orford, the target recipients for the transition income are: boat owners and their spouses, deckhands, and shore-side baiters and processor employees from Premium Pacific Seafood.

Unique to the Port Orford area are the displaced shore-side baiters. Traditionally, a longline vessel's crew does baiting of gear on the boat. However, Port Orford vessels have no room on board to do this. The shore-side baiting crews, who are often family members of fishermen, baited tubs seasonally from March through August until the groundfish cuts put them out of work, with no other employment opportunities available in Port Orford. Although, not everyone was successful in training for a new career, it is significant when examining the hurdles and challenges of stopping fishing and going on to another life. Barriers to career transition identified by the GDOP include: unwillingness to relocate, no GED or high school diploma, no driver's license, and drugs and alcohol. Of the 49 applicants, 29 have completed their transition successfully.

Communication in the Fisheries Management Process

The current fisheries management process does not allow for the meaningful participation of small-scale fishing representatives in management decisions. A documented communication problem already exists between the fishing industry and fisheries managers, characterized by blame, distrust, and stereotyping (Conway et al., 2002). Although the council process is designed to encourage and enable public participation, and to tailor management to local needs, customs, and interests, one problem often cited by Oregon's fishing communities is the industry's superficial involvement in the council process.

Potentially, the public can be involved at several levels of the council process: serving as council members, serving on advisory bodies, and providing public testimony at council meetings. However, fishery management and its regulatory process are complex, often contentious, and confusing to most people. For example, NMFS lists at least 61 different steps to develop and adopt fishery management plans and actions (Gade et al., 2002). The commercial fishing industry is distanced from NMFS and PFMC because of the lack of day-to-day communication and because they are often unsure when or to whom to approach with concerns. PFMC meetings are the primary venues for NMFS to interact with its constituents and partners, however the venue is often a considerable distance from the fishermen's homeport, making attendance difficult and expensive. For example, a PFMC meeting in Portland, Oregon, is about a six-hour drive from Port Orford.

PFMC meetings consist of a presentation of management options by the various committees followed by a public comment period, where citizens can either speak at a microphone or provide written comments. This communication process does not lend itself to the careful deliberation of citizen input and, consequently, experiential knowledge of long-time industry members is never considered in the decision process, often labeled as anecdotal and biased. Therefore, fishermen often perceive a lack of respect from scientists and managers. The formal nature of providing testimony in a public meeting is very different from the culture of fishing communities, where in-person, informal exchanges is preferred (Conway et al., 2002). Given the current process, the nearshore fishing industry is powerless to affect management decisions. In fact, most fishermen feel the decision has already been made before the meeting ever begins (Conway et al., 2002).

Dissatisfaction with current fisheries governance is evident by the increasing number of lawsuits challenging NMFS in federal courts. Since the mid-1990s, litigation against NMFS has grown 10-fold, an order of magnitude greater than in previous times and its record of defending management actions has dropped to less than 50% (Gade et al., 2002). Challenges come from both industry members and environmental organizations. This is a symptom of the management system's inability to reconcile its objectives of conserving fishery resources and maintaining optimum yields. Our fisheries management system has been slow to adopt its plans to accommodate the new national standards and the changes imposed by the 1996 Sustainable Fisheries Act.

Gade et al. (2002) report that, despite the frustrations, commercial fishermen believe that both sides could learn from each other and could successfully work on joint projects together. The respondents in this survey reflected that they would have a greater sense of ownership if they were involved in the design, implementation, and follow-through of research projects, beginning with collaboration in the design phase. Cooperative programs, with proper design and development, have achieved mutually agreed upon objectives between NMFS and the fishing industry. Some fishermen want a real voice in the decision-making because they have lost faith in the ability of the government's ability to solve management problems.

Cooperative and Community-based Fisheries Management as an Alternative

Community-based management provides an open framework for improving communication in fisheries management and encourages collaborative and innovative management strategies to address the unique environmental, economic, and social conditions at a manageable geographic scale. Fishing communities like Port Orford are looking for avenues to provide pertinent information to fisheries decision-makers in a way that is seen as valid and worthy of careful deliberation. Using GIS and rapid appraisal techniques provide an opportunity to collect local knowledge important to fisheries management at appropriate scales (Scholz et al., 2004). Mapping local knowledge is a process that also supports community-based fisheries management functions, which include: data gathering and analysis, logistical harvest decisions, habitat and water quality protection, and long-term planning (Pinkerton, 1989). It also allows the community to identify economic alternatives.

The Port Orford Ocean Resources Team

As a reaction to current management concerns, a non-profit organization has been formed to address science and management questions at the scale of a single fishing community. However, POORT realizes the importance of partners and has some at national, regional, and state levels and from academia, government, and conservation perspectives. POORT's Communications Coordinator is the gatekeeper of this whole project. POORT is seeking a balance that can maintain traditional fishing opportunities while diversifying its economic base.

Participatory GIS to Support Marine and Coastal Resource Management

Applying GIS science in a participatory setting for marine and coastal areas is an innovative approach for community-based management. Improvements in technology and increased availability of relevant data layers allow for a growing number of marine applications in a traditionally land-centric science. Over the last 40 years, increasing use of GIS by grassroots community organizations and participation in its use by ordinary citizens is possible because of the decreasing cost of hardware, improved user interfaces, and a trend towards a more human-centric vision of GIS (Craig et al., 2002). GIS offers communities a process for developing consensus about their environment and for engaging in long-term planning, and potentially, increased input into local management decisions. The state and federal agencies may be willing to share more power with groups they perceive as credible partners (Craig et al., 2002).

The methodology of the Port Orford Ocean Resources Inventory drew on the expertise of current research, including projects occurring in California (Scholz et al., 2004), and the Channel Islands National Marine Sanctuary Ethnographic Survey and in Canada (Macnab, 2002). Communities have combined fishermen's knowledge with GIS for marine protected area planning and for local area management.

Participatory GIS improves the communication of local knowledge about complex marine and coastal environments through using a common frame of reference, such as nautical charts, fath. contours, and local place names. Better information will help develop appropriate responses to management questions through spatial analyses. Participants in such processes may see opportunities to achieve individual goals through collective action, and also become empowered to do so. In this respect, participatory GIS may help bridge the competitive nature of fishing with collaborative learning approaches in community-based management.

Port Orford Community GIS Research Design and Methods

Introducing GIS: The First Step

POORT introduced GIS to the Port Orford community through two planning meetings in 2003. At the first meeting, the basic concepts of computer mapping and GIS were introduced, resulting in a valuable discussion of the different types of spatial data, and the uses and limitations of GIS. The proposed method of conducting local knowledge interviews was also introduced and a local commercial fisherman volunteered to participate in a pilot interview. At the second planning meeting, the full interview process and derived information layers were presented to the local advisory board. To further demonstrate how local knowledge is translated into a GIS layer, we also conducted a group exercise to document the navigation routes in and out of the dock. After some deliberation, general consensus was reached that for the community to have input into local fisheries management decisions, relevant information had to be collected and put into a format that would ultimately be digestible to managers. From that point forward, local knowledge interviews and GIS development became a priority project for POORT.

Developing the GIS Framework

Through ongoing POORT meetings, the local advisory board and the science advisory committee generated research and management questions. Listed here are some of the questions with a spatial component:

- What are the distribution, abundance and diversity of species near Port Orford?
- Are some species residents? What are their movement patterns?
- What are the distributions of recreational and commercial effort and related socioeconomic value?
- What is the catch per unit effort of different fishing gear types?
- What is the correlation between habitat and species distribution, abundance, and diversity?
- Are spatial management strategies appropriate for Port Orford? What kind and where might they be located?

From these questions, we developed our GIS framework, in which we determined what data is needed to answer some of the questions. We also established the database format to facilitate data storage, retrieval, and analysis. We then prioritized data for acquisition and creation (Table 12.1).

Conducting the Port Orford Ocean Resources Inventory

The overarching goals of the Port Orford Ocean Resources Inventory were: (1) to develop a foundation for local knowledge data collection and storage; and (2) to support the mission of the POORT charter of "using the best available science and local knowledge for the community to make local fishery management decisions." POORT agreed that GIS is a good tool to combine and analyze experiential and scientific information because focusing on location allows people to develop a common perspective of a shared marine environment. The POORT local advisory board determines information storage, access and use through a consensus process. Through an iterative participatory process with

Table 12.1. Spatial data list.

Data Layers	*Creator*	*Location*
NOAA Nautical Charts	NOAA Office of the Coast Survey	Valley Library, OSU
Side scan sonar of Orford Reef	Marine Program, ODFW	Hatfield Marine Science Center, Newport, OR
Urchin Surveys	Marine Program, ODFW	Hatfield Marine Science Center, Newport, OR
Fish Surveys	Marine Program, ODFW	Hatfield Marine Science Center, Newport, OR
Multi Beam Bathymetry off Oregon	Dr. Chris Goldfinger	Active Techtonics Lab, OSU
Geologic Substrate	Dr. Chris Goldfinger	Active Techtonics Lab, OSU
Oregon Shoreline	OR ORMTF	Oregon Coastal Atlas
Three-mile State Jurisdictional Boundary	OR ORMTF	Oregon Coastal Atlas
Digital Orthophotos Quarter Quadrangles	USGS	Oregon Coastal Atlas
Oregon State Boundary	BLM	Oregon Geospatial Data Clearinghouse
Oregon Counties	BLM	Oregon Geospatial Data Clearinghouse
Oregon Highways	ODOT	Oregon Geospatial Data Clearinghouse
Commercial Activities Distribution	POORT and Ecotrust	POORT & Ecotrust Offices
Recreational Activities Distribution	POORT and Ecotrust	POORT & Ecotrust Offices
Species and Resource Distributions	POORT and Ecotrust	POORT & Ecotrust Offices
Economic Importance of areas targeted for commercial fishing	POORT and Ecotrust	POORT & Ecotrust Offices

BLM = Bureau of Land Management; ODFW = Oregon Department of Fish and Wildlife; ODOT = Oregon Department of Transportation; ORMTF = Ocean Resources Management Task Force; OSU = Oregon State University; USGS = U.S. Geological Survey

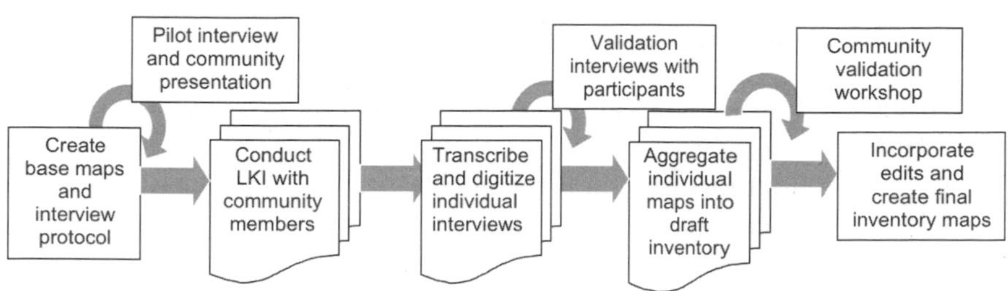

Figure 12.5. Flow chart of LKI process and GIS analyses.

Port Orford community members, we documented the local knowledge of the distribution of human uses and relative economic importance of areas targeted for commercial use, as well as the distribution of species and resources (Fig. 12.5).

IDENTIFYING THE STUDY AREA

Laminated base maps were the platform over which the interview took place. The smallest-scale base map (i.e., the one that covers the largest area) was 1:120,000 and displays the south coast of Oregon from Bandon in the north to Gold Beach in the south. However, through interviews, it was discovered that these maps did not show the complete extent of the Port Orford fleet's fishing activity. For logistical purposes, it was necessary to end somewhere and Bandon and Gold Beach were chosen because they are the nearest commercial fishing communities. Two of the base maps were larger scale, representing the marine area closer to the port of Port Orford, from north of Cape Blanco to Sisters Rock at Frankport, Oregon. The base maps displayed the relevant National Oceanic and Atmospheric Administration nautical charts, the latest bathymetric data from the Oregon State University (OSU) Active Tectonics and Seafloor Mapping Laboratory displayed as depth contours in fath. and some local place names. Using these maps to collect and display local knowledge was a natural outflow from the fishermen's experience in using nautical charts for navigation.

INTERVIEW PARTICIPANTS AND RECRUITMENT

The interview participants were not restricted by any gender or ethnic basis. The 33 interviews included 36 individuals from the Port Orford community who utilize the Port Orford marine environment for their occupation or leisure activities. The average age of participants was 51 years and the average experience in their primary activity was 20 years. Thirty-one men and five women were interviewed. Several people are not adequately represented by one category alone (Table 12.2). The 24 fishermen interviewed had a combined 524 years experience on the ocean and averaged 24 years of experience in commercial fishing. Port Orford commercial fishermen interviewed included 16 owner/captains and 8 deckhands. They work a combined total of over 2,000 days/year, averaging over 120 days/year out at sea (includes the four recently

Table 12.2. Local knowledge interview participants.

Number of People	*Participant Category*
24	Port Orford commercial fishermen
3	Recreational fishermen
6	Recreational users
3	Local fish buyers
1	Port of Port Orford staff
2	Oregon Department of Fish and Wildlife scientists

retired fishermen). Six out of 22 were second-generation fishermen, and one was third generation. The recreational users interviewed included a cross-section of user groups including recreational divers, surfers, kayakers, wild-life watchers, and beachcombers. Three representatives from the buying sector of the fishing industry participated. Although some of their knowledge is not direct at-sea observations, they possess a "common knowledge" of fishing locations and have great insights into the overall picture of the economic activities of this port.

A snowball sampling strategy started with volunteers from the local advisory board and identified potential participants through suggestions made by interviewees. This sampling method involved interview volunteers making unsolicited suggestions about other community members who are knowledgeable and might be interested in participating in the interview process. POORT recruited some participants at community meetings, through a flyer posted at the Port Orford dock and in the *Port Orford Today*! and informally through day-to-day communication. Volunteers contacted the POORT communications coordinator who then followed up to schedule a date and time for the interview.

CONFIDENTIALITY

Issues of access, representation, privacy, and confidentiality should not to be overlooked, as they contribute to the relative success or failure of participatory GIS processes. Interviewees signed a confidentiality agreement and informed consent document at the onset of the interview process to protect anonymity and the proprietary nature of information to the extent practicable under law. Contact information is taken only so that the interviewees could be contacted for the follow-up interview. A random identification number references an individual's local knowledge, appearing on the acetate overlays, interview notes, and in the computer records. Individual data were securely stored and were not be accessible to any person other than the interviewer and the people who input the information into the computer. After all information was collected and verified from all interviews, the acetate maps and written information were either returned to the participant or destroyed.

LOCAL KNOWLEDGE INTERVIEW MATERIALS AND METHODS

Trained POORT consultants and a graduate student conducted two- to three-hour, semi-structured interviews with commercial fishermen, recreational users, and other community members at the POORT office. The following materials were used.

- Confidentiality Agreement and Informed Consent Document
- Interview questions
- Checklist of activities/species/resources
- Interview response spreadsheet
- Approximately 36″ x 36″ base maps
  - 1:120,000 "Coos Bay to Gold Beach"
  - 1:40,000 "Port Orford Nautical Chart"
  - 1:24,000 "North of Port"
  - 1:24,000 "South of Port"
- Acetate overlays, scissors, and masking tape
- Colored wax pencils
- Tissue eraser
- Identification guides

Interviewees delineated areas of personal and observed human uses and locations of specific fish, invertebrate and plant communities on acetate-covered GIS base maps. Two interviewers provided a crosscheck for when transcribing the descriptive information into a database. The interviewers took handwritten notes on a standard response sheet. Participants verified the accuracy of the interview transcription and subsequent digitization during a one-hour follow-up interview. The only data used for community-based management purposes are the aggregate maps created from compiling individual information.

The interview process occurred mainly in five steps. First, demographic and vessel information was asked, then the locations of the interviewee's primary ocean-related activity. If the primary activity was commercial fishing, the participant was asked to assign a value to the relative economic importance of their areas targeted for particular fisheries. Referring to a list of human activities and species, including plants, invertebrates, marine birds and mammals, and fishes, the interviewers asked them to describe their personal observations of those species and activities in the Port Orford study area. The interviewees drew the location of their observations with wax pencils on clear plastic acetate overlaid on base maps. Some attribute information included: how the location was derived, the scale of base map (if applicable), trends over time as they relate to climate and weather, and habitat and/or depth associations. Identification guides were on-hand for reference. Lastly, the interviewee talked about anything else important to them about the ecology, or the economic and social conditions. This "open microphone" time helped identify the common themes of important issues to the community to help guide community-based management efforts.

INTERVIEW CONVERSION AND DATA AGGREGATION

The graduate student coded the acetate overlays and transcribed interview data into standardized Microsoft Excel databases, which would eventually become the polygon attribute tables. Location information was given to the interviewers in one of four ways: (1) directly drawn on the maps in wax pencils; (2) verbally referenced using a local place name; (3) using depth or distance associations; or (4) using another species association. Early in the process it was necessary to develop standardized polygons for those areas verbally referenced by a local place name. During an ad-hoc focus group with several fishermen, an OSU graduate student delineated the spatial extent of the local places referenced during interviews. Each polygon and otherwise-referenced area was given a unique identifier and its own record in the database. Attributes were populated from both interviewers' notes.

A team of consultants and the graduate student then digitized polygons from the individual interviews and joined them with the Excel databases to create individual digital map layers comprising all polygons assigned to the activity or species. Each interview generated anywhere from 10 to 50 data layers. One-hour follow-up interviews verified the content integrity following the conversion process. Participants specified any necessary edits to the polygons and associated attributes, which were then incorporated into the GIS databases.

POORT chose to aggregate all recreational activities, which included surfing, kayaking, diving, wind surfing, recreational fishing, shore fishing, whale watching, bird watching, and beachcombing, to produce a composite aggregate map of the intensity of recreational use in general. For commercially targeted areas, the spatial distribution and intensity of use of areas targeted for salmon, Dungeness crab, halibut, and sablefish (i.e., black cod) provided a good variety of economically important species, and were a less controversial subset of the whole Port Orford fishing portfolio.

Under contract with POORT, the non-profit environmental organization Ecotrust aggregated themes to create six draft inventory maps. Vector data was converted into 98-ft. grids. Each grid cell was assigned a value of 1 for poly presence and 0 for its absence. Then, cumulative totals for each grid cell were generated using an Arc Macro Language (AML) script. Nearest neighbor analysis with a 6-cell focal mean smoothed the data. The data were then classified using an equal area distribution of 7 classes and re-categorized in low, medium, and high usage. POORT then took the draft aggregation maps to a community workshop to solicit edits to further refine the community inventory (see Port Orford maps on the Web site accompanying this book).

COMMUNITY INVENTORY VALIDATION WORKSHOP

Conducted in January 2004, a community workshop provided validation of the process and resultant information and allowed the interview participants to suggest improvements to the aggregated map. This workshop provided forums for participants to propose edits in writing, individually, or in small groups. Propositions were then revised with the larger group and voted on. Suggestions for edits that were different between participants were dealt with by keeping it "as is." Votes of abstention were just as important as agreement and disagreement. The community concluded that the maps accurately represented the spatial extent and intensity of use of the study area. Few further edits will need to be incorporated.

Database Banking, Access and Immediate Use

Storage and accessibility of proprietary information concerned many fishermen and community members. The POORT office in Port Orford, Oregon, and disk space in the OSU Terra Cognita Spatial Analysis Laboratory, provided short-term secure storage for the data. Options for long-term storage of the GIS data include: the POORT office, the Oregon Coastal Atlas (a Web-based portal jointly managed by the Oregon Ocean-Coastal Management Program, OSU, and Ecotrust; Haddad et al., this volume), or Ecotrust's Inforain Server. A permanent solution for secure data storage will be decided through a series of community meetings, the first of which outlined the above options. Recommendations from the validation workshop suggest that Oregon Coastal Atlas would be the preferred alternative, although this must be voted on by the POORT local advisory board before adoption. Workshop participants also recommended immediate uses for the inventory.

Results

The community group achieved consensus on distribution of areas targeted by Port Orford for commercial and recreational activities in the draft maps. Through the local knowledge interviews, we have begun to directly answer some of the primary management questions that are guiding the POORT project. We are collectively gaining insight into the distribution and diversity of plants, fishes, and invertebrates around Port Orford, including some of the seasonality associated with each species and activity. The community is also beginning to relate the economic values of both recreational and commercial activities with the distribution of effort.

The spatial extent of effort for crabbing is determined by the movement of crabs in and offshore with the season and by competition with larger vessels that move in from other ports. For Dungeness crab, two areas of highest intensity are evident. Accessibility to these grounds is the primary reason for their higher level of use. However, north of Cape Blanco, weather can limit utilization of these grounds, as traveling above the Cape can be difficult.

Primary areas for targeting salmon are determined by the movement of forage species, oceanographic, and climatic conditions. The high-intensity area nearshore between Port Orford and Cape Blanco represents the North Beach season, a very important fishery because it is the only troll-caught salmon on the Oregon coast during part of the year. The area is very important economically due to its proximity and accessibility of the Port Orford fleet.

The halibut target area is primarily a large off-shore bank known locally as the High Spot. This highly utilized area is determined mainly by fisheries management, which sets a halibut opener (reduced from 72 to 10 hours in recent years). Although fishermen know halibut are located in other areas, when there are only 10 hours to fish, they go to the money spot. The community workshop did not suggest any changes be made to this draft map.

Rogue River canyon and the edge of the High Spot are the primary areas targeted for black cod. The large area extending from the edge of the High Spot represents areas accessible to trap gear, although not to longline gear. In the validation workshop, one small area off of the port of Bandon was determined to be an error and will be removed from the final map. The main suggestion for improvement was making high intensity use continuous from 100 to 300 fath. on the edge of the High Spot. This was the same suggested improvement for the draft map showing relative economic importance of black cod.

Discussion and Conclusions

To meet the requirements of the 1996 Sustainable Fisheries Act's National Standard 8, fisheries management must address issues at the community level, which means at a finer scale of data collection. For Port Orford, 450 ft. is appropriate. The process worked, as evidenced by so few changes being suggested at the validation workshop. A community-based GIS process, combined with scientific and local knowledge focused on a local area, allowed researchers for POORT the opportunity to collect this resolution of data.

Local knowledge interviews are a successful tool to understand a fishing community, its resources, and dependence on various areas. The open microphone time provided the participants an opportunity to drive the process and communicate in a more customary manner. Topics arose that shed light on potential market opportunities, rockfish spawning cycles, habitat-species associations, perceptions on and potential locations of marine parks, and the ecological changes and perceived drivers of these changes in the local area. Applicability to other communities depends on the relative homogeneity of the port and the degree of trust between those conducting the interviews and the industry participants.

Coupling scientific and local knowledge in GIS is an effective way to support community-based management objectives. Rapid rural appraisal techniques provide baseline information at a relevant scale and provide insight into the ecological nuances of a local area. Using the base maps provide a common frame of reference to improve communication about a shared marine and coastal environment. Using a consensus building-process to learn collaboratively, POORT is building social capacity for long-term planning and input into management decisions

Spatial representation of human uses, economic importance, and species distribution can guide area-based management strategies, including local area management and the selection of less impactful areas for marine protected areas. Addressing the community management question—"Is there an area in Port Orford suitable for marine protected areas?"—remains a significant challenge.

Developing a trust between industry, scientists, and managers is often difficult in community-based management approaches. Although many safeguards are in place to protect confidentiality, anonymity, and data access and use, some fishermen still worry that the data collected through these programs may be used against them in the end. This relates to the perception that fisheries managers place the most stringent regulations on the sectors that have the most scientific data collected about them. Therefore, a positive experience with cooperative research, such as participatory GIS, can be the first step towards an improved relationship of mutual ocean stewardship.

Future Work

This interview process did not answer each scientific, market, or management question; so additional data has been prioritized for acquisition. POORT received an $110,000 cooperative research grant from NMFS to help answer some of these questions. Potential next steps include examining linkages between the geologic substrate data provided by benthic habitat maps of the OSU Active Tectonics and Seafloor Mapping Laboratory, and species distributions as supplied by the collective experience of local fisherman and other local experts. Three separate biological projects will be undertaken in 2004 in addition to the GIS work: fish biological sampling, visual ecological surveys using a remotely-operated vehicle (ROV), and project design for a subsequent fish tagging and gear selectivity study. POORT will also conduct more in-depth economic surveys and spatial analysis. With the information and resources provided through the GIS and the cooperative research, POORT's long-term goal is to make management recommendations to state and federal regulatory entities charged with management of the marine and nearshore environment.

Acknowledgements

Special thanks to Charles Steinback of Ecotrust for assistance with GIS analysis and map preparation. For the lead author, this project constituted part of a Master of Science program in Marine Resource Management at Oregon State University (OSU), under the direction of Professors Jim Good and Dawn Wright of OSU, and Dr. Astrid Scholz of Ecotrust.

References

Anderson, L., 2001. *Fisheries Management and Marine Reserves in Oregon: A Question of Scale,* Oakland, CA, Environmental Defense, 17 pp.

Beach, D., 2002. *Coastal Sprawl: The Effects of Urban Design on Aquatic Ecosystems in the United States.* Arlington, VA, Pew Oceans Commission, 33 pp.

Conway, F. D. L., Gilden, J., et al., 2002. Changing communication and roles: Innovations in Oregon's fishing families, communities, and management, *Fisheries,* 27(10): 20-29.

Craig, W. J., Harris, T. M., and Weiner, D. (eds.), 2002. *Community Participation and Geographic Information Systems,* New York, Taylor and Francis. 370pp.

Gade, M. A., Garcia, T. D., Howes, J. B., Schad, T. M., and Shipman, S., 2002. *Courts, Congress, and Constituencies: Managing Fisheries by Default,* Washington, D.C., National Academy of Public Administration, 160 pp.

Haddad, T., Wright, D. J., Dailey, M., Klarin, P., Dana, R., Marra, J., and Revell, D., Klarin, P., Dana, R., Marra, J., and Revell, D., this volume. The tools of the Oregon Coastal Atlas, in Wright, D. J., and Scholz, A. J. (eds.), *Place Matters: Geospatial Tools for Marine Science, Conservation, and Management in the Pacific Northwest,* Corvallis, OR, Oregon State University Press.

Hall-Arber, M., Dyer, C., Poggle, J., McNally, J., and Gagne, R., 2002. *New England's Fishing Communities,* Cambridge, MA, Massachusetts Institute of Technology Sea Grant College Program, 417 pp.

Hanna, S., 2000. *Setting the Fisheries Management Stage: Evolution of the West Coast Groundfish Management,* Corvallis, OR, International Institute of Fisheries Economics and Trade.

Jacob, S., Farmer, F. L., Jepson, M., and Adams, C., 2001. Landing a definition of fishing dependent communities: Potential social science contributions to meeting National Standard 8, *Fisheries,* 26(10): 16-22.

Jacob, S., Jepson, M., Pomerory, C., Mulkey, D., Adams, C., and Smith, S., 2002. *Identifying Fishing Dependent Communities: Development and Confirmation of a Protocol,* MARFIN Project and Report to NMFS Southeast Fisheries Science Center, 214 pp.

Jentoft, S., 1999. Healthy fishing communities: An important component of healthy fish stocks, *Fisheries,* 24(5): 28-29.

Mackas, D., Strub, P. T., and Hunter, J., 2002. Eastern Boundary Current-California Current System Working Group Reports, Leonardtown, MD, U.S. Global Ocean Ecosytem Dynamics, http://www.usglobec.org/reports/ebcccs/ebcccs.contents.html. Last accessed August 30, 2004.

Macnab, P., 2002. There must be a catch: Participatory GIS in a Newfoundland fishing community, in Craig, W., T. Harris, and D. Weiner (eds.), *Community Participation and Geographic Information Systems*, New York, Taylor and Francis: 173-91.

National Marine Fisheries Service, 2002. Community Impact Analysis, http://www.st.nmfs.gov/st1/econ/impact.html. Last accessed March 7, 2004.

Pacific Fisheries Management Council, 2000. *Groundfish Fishery Strategic Plan: Transition to Sustainability*, Portland, OR, 66 pp.

Pinkerton, E., 1989. *Cooperative Management of Local Fisheries: New Directions for Improved Management and Community Development*, Vancouver, Canada, University of British Columbia Press.

Scholz, A., 2003. *Groundfish Fleet Restructuring Information and Analysis: Final Report and Technical Documentation*, San Francisco, CA, Pacific Marine Conservation Council and Ecotrust, 63 pp.

Scholz, A., Bonzon, K., Fujita, R., Benjamin, N., Woodling, N., Black, P., and Steinback, C., 2004. Participatory socioeconomic analysis: drawing on fishermen's knowledge for marine protected area planning in California, *Marine Policy*, 28(4): 335-49.

Related Web Sites

City Populations (by rank), No. 91-137. Oregon Blue Book. http://bluebook.state.or.us/local/populations/pop03.htm. Last accessed Jan. 15, 2004

Port of Port Orford. Copyright 2003. Port Orford Area Chamber of Commerce, http://discoverportorford.com/portofpo.php. Last accessed Feb. 26, 2004

Port of Port Orford: Circa 1856. Copyright 2003. http://portorfordoregon.com/portofpo.html. Last accessed Feb. 15, 2004

Port Orford Life Boat Station. Copyright 2000-2003. Point Orford Heritage Society. http://www.portorfordlifeboatstation.org/. Last accessed Feb. 26, 2004.

Port Orford Oregon: Gateway to America's Wild Rivers Coast. Copyright 2003-2004. http://www.portorfordoregon.com/relocate.html. Last accessed Feb. 15, 2004.

Port Orford 2, Oregon: Period of Record General Climate Summary - Precipitation. Western Regional Climate Center - Desert Research Institute. http://www.wrcc.dri.edu/cgi-bin/cliGCStP.pl?orporf. Last accessed Feb. 26, 2004.

Port Orford 2, Oregon (356784): Period of Record Monthly Climate Summary. Western Regional Climate Center - Desert Research Institute. http://www.wrcc.dri.edu/cgi-bin/cliRECtM.pl?orporf. Last accessed Feb. 26, 2004.

Appendix 12.1.
Local Knowledge Interview Recruitment Flyer

Port Orford Ocean Resource Team

POORT is engaged in a community-based management effort and is conducting a local inventory of the ocean region important to the Port Orford community.

We want to talk to commercial and recreational fishermen, recreationalists (divers, kayakers, surfers, etc.), and other citizens who have personal knowledge about the resources, species, and human activities that occur in the Port Orford ocean area.

Local knowledge interviews will be conducted in the POORT office:

351 W 6th St

Get involved! We want to talk to YOU!!!
Sign-up in the POORT office or call us to be a part of this unique opportunity!

Appendix 12.2.

POORT Confidentiality Agreement

Individual data will not be accessible by any person other than the Interviewer and the person who will input the data into the computer using geographic information system (GIS) software. Raw interview data will be securely stored until such time that all data are entered and verified. At that time, the information will be returned to the Interviewee, destroyed, or stored at said location with the Interviewee's permission.

Interview information will be aggregated with data from other interviews to produce compilation maps, which will NOT display any one individual's information. Furthermore, the POORT will NEVER share any one person's information without express written consent of the Interviewee.

The unique identification number below will be used to identify this interview in the computer database. The only place where your name and ID number will appear together is on this form, which will be securely stored indefinitely. By signing here you agree to the conditions of this confidentiality agreement.

Date: ____________________

POORT Interviewers:

____________________________ ____________________________

Interviewee:
Identification number: _______________
Name (please print and
sign):____

Address: ____
____

Phone: ___________________________

Appendix 12.3

Institutional Review Board Informed Consent Document

PROJECT TITLE: **PORT ORFORD OCEAN RESOURCES INVENTORY**

PRINCIPAL INVESTIGATOR: **JIM GOOD, MARINE RESOURCES MANAGEMENT PROGRAM**

RESEARCH STAFF: **VICKI WEDELL, LAURA ANDERSON, LEESA COBB, DAVE REVELL**

Purpose

The purpose of this research study is to conduct an inventory of the local knowledge of species, resources, and activities that occur in the marine environment important to the community of Port Orford. Computer mapping is used to document and display the information shared in the interview process. The purpose of this consent form is to give you the information needed to help you decide whether to be in the study or not.

We are inviting you to participate in this research study because you utilize the Port Orford marine environment for your occupation or recreational activities. A snowball sampling approach will be used to get an estimated 40 people in this interview process. Volunteers from POORT Advisory Board will be recruited first, while other willing participants will be identified through suggestions made by interviewees or other POORT members.

Procedures

If you agree to participate, your involvement in the interview process will last for three hours total. A two-hour interview will be followed a few weeks later by a one-hour consultation to verify the accuracy of the maps created. A community workshop will allow another opportunity to make modifications to the composite community map.

The following procedures are involved in this study. At least two interviewers are present for each interview. A random identification number will be used to reference your local knowledge maps. Confidentiality agreements are offered and signed at the onset of the interview. Then, you refer to a list of potential species and human uses and describe your personal observations of those that occur in the Port Orford study area. Identification guides are on-hand for reference, if needed. You use wax pencils to draw the areas of your observations on clear plastic mylar, which is overlaid on base maps having fathom contours and the relevant nautical chart displayed. Information shared at the interview process is taken back and digitally documented in map form. The maps are brought back to you after a few weeks for a 1-hour consultation where any necessary modifications are identified and corrected. After all consultations are completed for all participants, species and use maps will be aggregated and presented as the Port Orford Ocean Resources Inventory.

Risks

There are no foreseeable risks associated with participating in this research project. Sensitive information is protected through random identification numbers.

Benefits

There may be no direct personal benefit for participating in this study. However, society may benefit from this study by learning about a participatory process for computer mapping of local ecological knowledge.

Costs and Compensation

You will not have any costs for participating in this research project. You will be compensated with a rockfish poster even if you withdraw early.

Confidentiality

Records of participation in this research project will be kept confidential to the extent permitted by law. Individual data is not accessible to any person other than the interviewer and the person who will input the information into the computer. Raw interview data is securely stored until such time that all the data are entered and verified. Then the data is returned to the interviewee or destroyed. Information is aggregated with data from other interviews and compilation maps generated for exclusive use by POORT. Maps and information are not shared with outside groups without express written consent of the POORT Advisory Board members.

Voluntary Participation

Taking part in this research study is voluntary. You may choose not to take part at all. If you agree to participate in this study, you may stop participating at any time. You are also free to skip any question in the interview that you prefer not to answer.

Questions

Questions are encouraged. If you have any questions about this research project, please contact: Vicki Wedell at 541-619-4699 or vwedell@coas.oregonstate.edu or Jim Good at 541-737-1339 or good@coas.oregonstate.edu. If you have questions about your rights as a participant, please contact the OSU Institutional Review Board (IRB) Human Protections Administrator, at (541) 737-3437 or by e-mail at IRB@oregonstate.edu.

Your signature indicates that this research study has been explained to you, that your questions have been answered, and that you agree to take part in this study. You will receive a copy of this form.

Participant's Name (printed):

____

____

(Signature of Participant)
(Date) ____________________

RESEARCHER STATEMENT

I have discussed the above points with the participant. It is my opinion that the participant understands the risks, benefits, and procedures involved with participation in this research study.

____

(Signature of Researcher)

APPENDIX 12.4

Local Knowledge Interview Questions

INTERVIEW QUESTIONS:

1. User profile
   a. Identification number
   b. Age
   c. Sex
   d. Profession/activity (owner, captain, deckhand)
   e. Duration
      1. Start/end year
      2. Number of days/year in area
      3. How many years have you maintained this level of activity?
   f. What generation fisherman are you?
   g. What vessel(s) do you fish from?
   1. What are its length and size of engine?
2. Where are your primary (fishing) zones?
   a. What are the primary fisheries in each zone?
   b. What are the primary gears in each zone?
   c. Rank each zone on a scale of 1-5 for:
      1. The amount of effort you spend there
      2. Its economic importance

Effort (% of time fishing for a year)
1 = (0-20%) 2 = (21-40%) 3 = (41-60%) 4 = (61-80%) 5 = (81-100%)

Economic importance (% of yearly income)
1 = (0-20%) 2 = (21-40%) 3 = (41-60%) 4 = (61-80%) 5 = (81-100%)

3. What "resources" do you use or have you observed in the study area? **(Use list)**
   a. Where do you use/observe resource X?
   b. What is the current status of the resource in the study area? (abundance)
   c. Has the location or status of this resource changed since you have been involved in your activity in the study area? If so, how and why?
   d. What are the seasonal changes of this resource in the study area? (spawning locations, nursery grounds)
   e. What other changes have occurred with respect to this resource? When and why did they occur?
4. Is there anything else about the ecology of this area that you want to tell us?
5. Are there any other economic, social, or cultural factors to consider?
6. Anything else?

Epilogue

Spatial Reasoning for Terra Incognita
Progress and Grand Challenges of Marine GIS

Dawn J. Wright and Patrick N. Halpin

Introduction

"Just as fish adapted to the terrestrial environment by evolving into amphibians, so GIS must adapt to the marine and coastal environment by evolution and adaptation." — Goodchild (2000)

"Applying GISs to marine and coastal environments presents taxing, but particularly satisfying, challenges to end users and system developers alike." — Bartlett (2000)

After many years of focus on terrestrial applications, an increased commercial, academic, and political interest in the oceans throughout the 1990s has spurred fundamental improvements in the toolbox of GIS and its methodological framework for this domain of applications. The adoption of GIS for ocean by agencies and institutes, such as the National Oceanic and Atmospheric Administration (NOAA) National Marine Sanctuary Program and National Ocean Service, the U.S. Geological Survey (USGS), portions of the Woods Hole Oceanographic Institution, the Monterey Bay Aquarium Research Institute, the Nature Conservancy, and many others, speaks to its growing utility not only for basic science and exploration, but also for ocean protection, preservation, and management (e.g., Convis, 2001; Breman, 2002; Wright 2002; Green and King, 2003a). Indeed, "marine GIS" has progressed from applications that merely collect and display data to complex simulation, modeling, and the development of new coastal and marine research methods and concepts (and the term marine GIS is used here to mean applications to the deep ocean, but also to the

Dawn J. Wright
104 Wilkinson Hall, Department of Geosciences, Oregon State University, Corvallis, OR 97331-5506
Corresponding author: dawn@dusk.geo.orst.edu, phone 541-737-1229, fax 541-737-1200

Patrick N. Halpin
Nicholas School of the Environment and Earth Sciences, Duke University, Durham, NC 27708
phalpin@duke.edu

coasts, estuaries, and marginal seas, and by scientists and practitioners working as academic, government or military oceanographers, coastal resource managers and consultants, marine technologists, nautical archaeologists, marine conservationists, marine and coastal geographers, fisheries managers and scientists, ocean explorers/mariners, and the like). Numerous innovations in remotely sensed data (both satellite based and in situ acoustic), ocean sensor arrays, telemetry tracking of marine animals, hydrodynamic models, and other emerging data collection techniques have been added to the information data streams now available to answer marine science questions. And the commercial GIS sector continues to pay heed to the needs of marine and coastal GIS users, with many of the leading vendors entering into research and development collaborations with marine scientists and conservationists.

The preceding chapters of this book highlight many more of the success stories of marine GIS. Common themes include new methodologies for data analysis and implementation of the science and policy underlying the siting and design of shoreline conservation and marine protected areas, improved synthesis of information for policy-makers (particularly in map form), ways of incorporating local ecological knowledge and socioeconomic concerns, and ways to more effectively communicate the complexity of the marine realm to the general public. A common language of practice is developing for marine conservation GIS at many geographic scales from ocean basins to local marine habitats, while at the same time some distinctions still present challenges (such as the definitions of "habitat" and the varying ways of representing and analyzing benthic terrain in this regard, from measures of "benthic complexity" to rugosity to position indices).

It is the purpose of this chapter, however, to briefly review some longstanding challenges, challenges that underpin the successes of many of these applications but continue to provide avenues for further study, especially for posing important questions about the representation of spatial and temporal information in the marine environment (a marine GIS research agenda of sorts). In one way, the commercialization of GIS as a black box tool in the 1980s had the long-standing, beneficial effect of making GIS accessible to users who did not need advanced training in computer programming. But from an information technology perspective, it may also have had the detrimental effect of limiting the research into the underlying data structures and algorithms. To wit, most papers at GIS conferences during this time dealt with research using GIS; far fewer dealt with research on the information system itself, the data structures and spatial analysis algorithms, and innovative approaches to the integration of data, models, and analysis for use in scientific hypothesis generation, prediction, and decision making.

In the 1990s, the advent of geographic information science (GISci), the "science behind the systems," and the organized leadership of groups

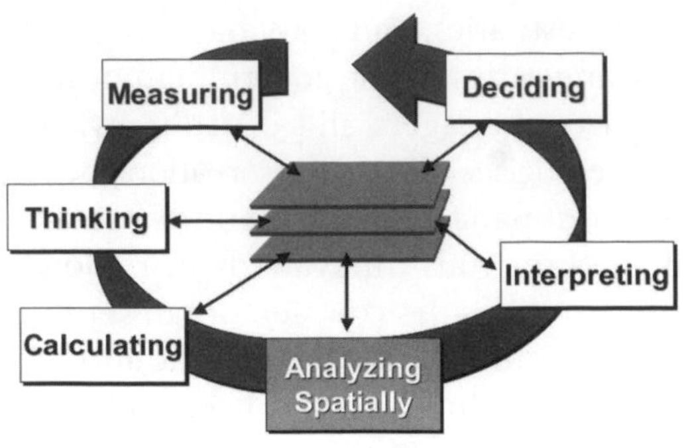

Figure 13.1. Illustration of the process of spatial reasoning, where the mechanics and issues surrounding the gathering and processing, and mapping of data in GIS lead the user to better understand and interact in the spatial "language" of GIS (rudimentary spatial analysis, spatial statistics, spatial process models, etc.). Modified from an Environmental Systems Research Institute (ESRI) GIS Day diagram.

such as the National Center for Geographic Information and Analysis (www.ncgia.ucsb.edu) and the University Consortium for Geographic Information Science (www.ucgis.org) changed this dramatically, where questions of spatial analysis (special statistical techniques variant under changes of location), spatial data structures, accuracy, error, meaning, cognition, visualization, and more came to the fore (for the most comprehensive treatment of GISci see Longley et al., 1999). Pursuant to GISci is the notion of "spatial reasoning," first defined by Berry (1995) as a situation where the process and procedures of manipulating maps transcend the mere mechanics of GIS software interaction (input, display and management), leading the user to think spatially using the "language" of spatial statistics, spatial process models, and spatial analysis functions in GIS (Fig. 13.1). This has been an important concept for the oceanographic community to embrace, as many have seen the utility of GIS only for data display and management (e.g., Wright, 2000).

For the coast and oceans, it is clear that the use of GIS is now crucial, but its use in this challenging environment can also help to advance the body of knowledge in general GIS design and architecture (Wright and Goodchild, 1997; Goodchild, 2000). The next section highlights a key motivation advancing the development of geospatial technologies: the need for more precision in marine resource science and management, followed by a brief review of current challenges of marine GIS in terms of: (1) data access and exchange; (2) spatial and temporal representation, and (3) the need for more temporally dynamic analytical models. These are discussed within the context of the benthic habitat, marine fisheries, and conservation focus of this book. Note that there is additional, detailed background on these challenges in Bartlett (1993a and b), Li and Saxena (1993), Lockwood and Li (1995), Wright and Goodchild (1997), Wright and Bartlett (2000), and Valavanis (2002).

Motivation: The Rapidly Increasing Demand for More Precision in the Management of Marine Resources

In direct parallel with developments in terrestrial natural resource management, managers and scientists are now being tasked with answering increasingly precise questions concerning physical, biological, and social resources of our coastal and marine environments. In the terrestrial realm, geospatial technologies (GIS, global positioning system, and remote sensing) have been widely and increasingly applied to assist in the "precision management" of agriculture, forestry, urban planning, business, and national defense issues. The application of geospatial technologies to terrestrial resource management has fueled a revolution in the process and practice of resource management. Farmers, foresters, urban planners, and business owners now regularly use geospatial technologies to optimize the management of their resources across a wide range of scales.

There is now emerging an equally strong demand for "precision management" of coastal and marine resources. The coastal and marine science and management community are challenged daily with increasing demands for more detailed analysis of the physical and biological processes. The coastal and marine community, however, faces additional challenges in the application of geospatial technologies. The three-dimensional nature of the marine domain, the temporal dynamics of marine processes, and the hierarchical interconnectedness of marine systems grossly increase the complexity of developing and applying geospatial solutions to marine management questions.

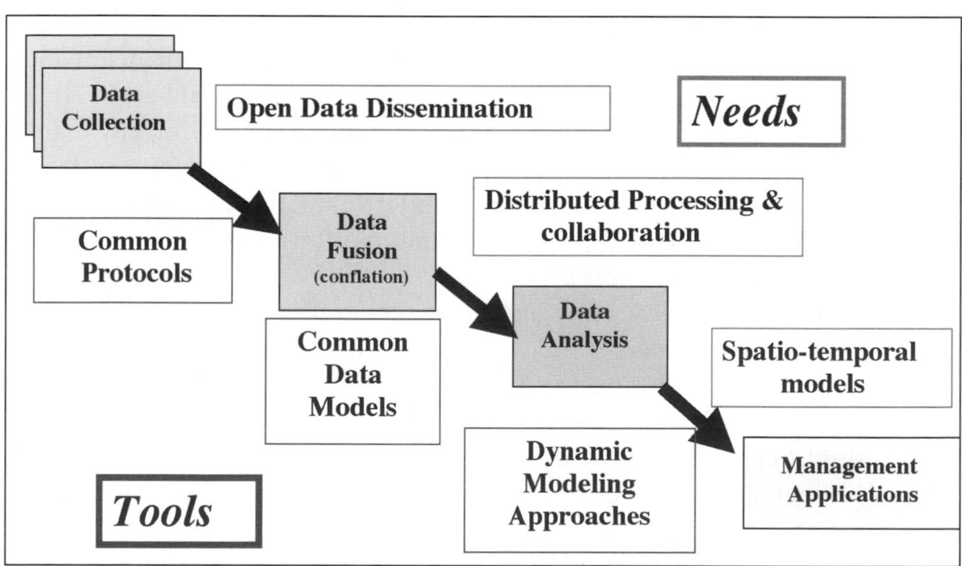

Figure 13.2. Marine spatial analysis needs and major areas of marine GIS tool development: common protocols, common data models and dynamic statistical and modeling approaches (from Halpin, 2004).

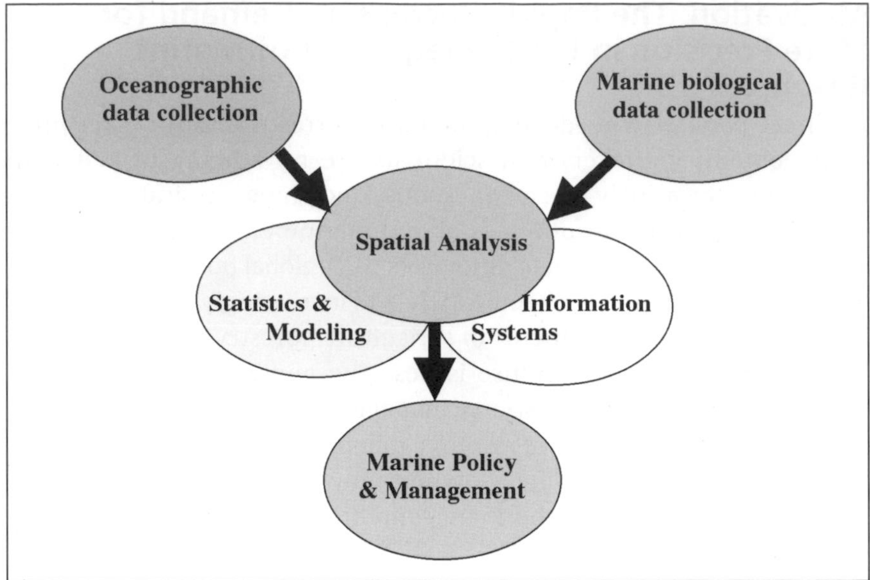

Figure 13.3. New advances in marine geospatial analysis will occur at the intersection of developments in spatial analysis, information systems and statistical modeling (from Halpin, 2004).

For example, the development of effective marine protected areas or time-area closures require scientists and managers to explicitly and precisely assess resource usage and potential conflicts in both space and time. The idealized goal of developing "win-win" management plans that optimize for both sustainable resource use and biological conservation will require an exceptionally high level of precision to ensure that economic and conservation resources can be separated in both space and time. Precision (as well as accuracy) in the delineation of the boundaries of these areas is a challenge (e.g., Treml et al., 2002), as they often transcend federal and state jurisdictions and may extend to the seafloor or into the subsurface. Descriptions of regulatory boundaries often are subject to misinterpretation (i.e., are imprecise), and if jurisdictional disputes arise, conservation and sustainability goals may be delayed or compromised.

In addition to the emerging challenges of precision management for marine practitioners is the vast quantity of data that are necessary for assessing, modeling, and monitoring our coastal and marine environments. A recent report assessing the geospatial data needs of the Integrated Ocean Observing System (IOOS; Hankin et al., 2003), estimated that the annual data flow of oceanographic data collected to support this effort will exceed ~2.9 terabytes per year.

In addition to the rapidly proliferating quantity of coastal and ocean data, much of the geospatial analyses that will be needed to be conducted in order to support scientific and management programs will require the fusion of multiple sources of physical, oceanographic, biological, fisheries, and management datasets together. In order to seamlessly merge data from disparate sources together, significant development will need to occur in the advancement of data dissemination tools, data standards, data transport protocols and Internet collaboration tools.

As can be observed in the general flow chart depicted in Figure 13.2, improvements in the geospatial analysis process will need to occur along the data collection, data fusion, data analysis and finally to management applications steps of the process. The three general areas of needs are: better data dissemination, better distributed processing and collaboration, and better spatio-temporal models. The specific areas for geospatial technology advancement to match these needs will come through the development of common protocols, common GIS data models, and more dynamic modeling approaches. All of these needs and tool development processes are highly interconnected. The most profound advancements in the field of marine geospatial analysis will likely not be tied to any single area of technological development, but instead will be found at the intersection of these new spatial analysis, information systems, and modeling disciplines (Fig. 13.3).

Grand Challenge: Data Access and Exchange

On one hand, there is still a comparative lack of data for marine GIS as compared to its terrestrial counterpart. The land abounds with accurate and unmoving geodetic control networks; satellite sensors can see the land through the atmosphere but not through water at all depths; aerial photographs aid us in delineating landforms, land ownership, cities, and the like at much larger cartographic scales than in the ocean, as does the Global Positioning System on land. As has been stated many times by various explorers and scientists with regard to the ocean floor, we have better maps of the moon, Venus, and even Mars, and we have sent more people to the moon than to the deepest parts of planet Earth (Challenger Deep in the Marianas Trench). Our mapping of the water column is extremely miniscule on a global scale, and the sensors that could provide detailed, three- and four-dimensional data about the dynamic marine environment generally do not exist, although enormous improvements in sensing technology have occurred in the past decade (Goodchild, 2000). Sampling or mapping may be rich in one-dimension (e.g., a vertical profile at a sampling station) but sparse horizontally, for which a great deal of interpolation must be relied upon in GIS (Wright and Goodchild, 1997; Schaefer and Schlueter, 2003).

On the other hand, there have indeed been tremendous advances in data collection techniques, that, as mentioned before, covering larger areas in two dimensions, add significantly to the information data streams now available to answer marine science questions. As such, we are faced with new challenges involving the synthesis, visualization, and analysis of these disparate data types to maximize the utility of past, present, and future marine data collection efforts. These challenges include critical needs for common data-sharing protocols and technologies, common marine data types, development of specialized analysis tools for temporally dynamic applications, and new statistical modeling frameworks for better forecasting. To meet these challenges, the marine science and management community will need to develop not only technological innovations, but also new priorities for the effective management and integration of marine science programs. The motivation for developing and accepting these new information systems approaches is found in the promise these approaches have for more accurately analyzing complex marine problems in a more objective and rigorous manner. The implementation of common data standards and protocols promises to allow for more efficient data sharing, higher quality analysis, and more direct linkage of spatial and temporal events in marine system.

There are three central areas of development that control our ability to effectively collaborate and exchange data: (1) data discovery and metadata standards; (2) data transport protocols; and (3) information system protocols. New developments in data discovery and metadata standards will provide the "card catalogue" for future marine scientists and managers to search Internet data warehouses and information system portal to discover and cross-reference data holdings. Because of the many spatial, temporal and trophic connections that may be inherent in any marine study, standards that control the way we locate relevant data are crucial. For example, a research project involving geospatial analysis to support a management question may need to identify appropriate ocean bathymetry data, ocean temperature, wind speeds, sea heights, ocean color, prey species, predator species, management conditions, and fisheries data, all for a specific period in time and spatial resolution.

The emerging tools being developed involve setting standards, authoritative information sources and common protocols. An example is the Ocean Biogeographic Information System – Spatial Ecological Analysis of Megavertebrate Animal Populations (OBIS-SEAMAP) program (http://obis.env.duke.edu/). A request for the name of a marine animal species is first sent to the IT IS (Integrated Taxonomic Information System) taxonomic service to validate the taxonomic naming conventions and then passed on to searches for other spatial data records within the OBIS network. These types of interlocking searches are possible through the use of common XML (Extensible Markup Language) protocols and the establishment of authoritative sources on the Internet.

Once data are discovered, common data transport protocols must be developed in order to allow researchers to exchange data uniformly between sites. An example of common data transport protocols is the development of the OPeNDAP (Open-source Project for a Network Data Access Protocol) developed by a consortium of ocean data development programs (http://www.opendap.org/). The OPeNDAP program and similar efforts allow for the transport of data from site to site in common exchange formats, allowing researchers to standardize processing tool development and expectations. Examples of the OPeNDAP applications can be found at Live Access Server (LAS) data sites (example LAS: http://las.pfeg.noaa.gov); as well as the National Virtual Ocean Data System (NVODS). Figure 13.4 (see page 204) depicts examples of the data discovery and data transport protocols and standards that marine GIS data users will regularly encounter when searching, retrieving, or publishing data over the Internet.

In addition to protocols specific to geospatial data and processes, the marine GIS community needs to be evolving their operations in compliance with new standards and protocols that affect the entire Internet computing environment.

The trend towards Internet-based, collaborative projects in the field of marine GIS also means that the roles of individual researchers and practitioners are changing. There are new categories of "data providers," "data aggregators," and "data users" emerging to define the role and specialization of different individuals and institutions in large marine GIS projects. The Gulf of Maine Biogeographic Information System GMBIS project provides explicit examples of these emerging roles (http://www.usm.maine.edu/gulfofmaine-census/Docs/Research/Gmbis2.htm). These emerging specializations define a departure from the role of the single researcher taking a project from data collection, geoprocessing, spatial analysis and cartographic production of final results, and highlight the move to a broader information systems approach in the field.

In addition to the need for common data protocols, there are different user communities that need to collaborate more closely in the future. The operational oceanography and the biogeographic informatics communities are making advances in large information systems programs, but tend to use mathematical scripting languages (e.g., MATLAB, IDL or Interactive Data Language, GMT or Generic Mapping Tools) to process spatial and temporal data. The "end user" marine management and conservation communities tend to use desktop commercial GIS packages. In order to bridge the gaps between these communities, efforts need to be made to develop more appropriate and interoperable software and data models for marine applications (e.g., Wright et al., 1998; Goldsmith, 2000).

As these varying communities interact, there will be a continuing need to formalize concepts and terms (i.e., ontologies) that will be used to aid the user in more effective searching and analysis of data and information (e.g., McGuinness, 2002). For example, in the search for data and resources one may use interoperable terms such as coastline vs. shoreline, seafloor vs. seabed, ecological resilience vs. robustness, scale vs. resolution, wetland buffering vs. GIS buffering. Here the development of ontology repositories for marine data will be important, along with "semantic integration and interoperability" (e.g., Goodchild et al., 1999; Egenhofer, 2002; Kuhn, 2003), to aid in fully describing the context in which data were collected for its proper use, or for appropriate legacy uses beyond the initial mission or target of the data collection (allowing the user to understand the details of data collection and purpose without being a science or policy expert in that particular field). Emerging also is the concept of grid computing, where not only the data are distributed but the computing power as well (e.g., data may be executed on one machine for a numerical model, sent on to another machine for GIS analysis, rendered in 3-D and 4-D on another, etc.). A very successful example is GEONGrid, a geosciences-oriented network of federated servers (i.e., "a cyber infrastructure") based on a common set of services for data integration, exchange, modeling and semantic interoperability (Allison et al., 2003; Baru, 2004; www.geongrid.org).

Grand Challenge: Representation of Marine Data and Common Data Models

One of the most powerful features of a GIS is the ability to combine data of various types simply by assigning coordinates and displaying these "layers" together. Of course, this representation runs into difficulty if the data are dynamic, with constant changes in location or attribute, and best viewed that way, when the data represent entities of different scales, or when its dimensionality is three, four, or greater. Marine applications, with tides, upwellings, ships, and vehicles moved by waves and currents, shorelines, and the like demonstrate all of these difficulties.

Shorelines are largely represented in GIS as fixed features, but the daily reality of tidal fluctuations leads to the question of a shoreline according to whom? States vary in their definition of the shoreline according to tidal datum, some using Mean High Water (MHW), while others use Mean Higher High Water (MHHW), or Mean Low Water (MLW). The depiction of a shoreline is fraught with uncertainty (where is the boundary for a rocky shore versus sandy shore versus tidal wetland?). There are significant differences between legal definitions and digital boundaries, as exemplified by a marine sanctuary boundary, where outer boundaries are explicitly described with coordinates, but inner boundaries follow a tidal datum such as mean high tide (Treml et al., 2002). When only half of the boundary specified is spatially explicit, then one is forced to make assumptions concerning scale, accuracy, and precision. And then there is the inimitable question: "How long is a shoreline?" (Mandelbrot,1967).

Much has been written about the importance of error and uncertainty in geographic analysis (e.g., from Chrisman, 1982 to Heuvelink, 1998), and with the challenge of gathering data in the dynamic marine environment from platforms that are constantly in motion in all directions (roll, pitch, yaw, heave), or in tracking fish, mammals, and birds at sea, the issue of uncertainty in position is certainly critical. We must accept that no representation in two-, three-, or four dimensions can be complete. And there are further uncertainties in what the data indicate about the marine environment, or what the user believes the data indicate about the environment.

As noted by Bartlett (2000), one of the most important lessons to be learned from collective experience in marine GIS is the importance of rigorous data modeling before attempting to implement a GIS database. Indeed, data models lie at the very heart of GIS, as they determine the ways in which real-world phenomena may best be represented in digital form. A data model for marine applications must undoubtedly be complex as modern marine datasets are generated by an extremely varied array of instruments and platforms, all with differing formats, resolutions, and sets of attributes. Not only do a wide variety of data sources need to be dealt with, but a myriad of data "structures" as well (e.g., tables of chemical concentration versus raster images of sea surface temperature versus gridded bathymetry versus four-dimensional data, etc.). It has become increasingly obvious that more comprehensive data models are needed to support a much wider range of marine objects and their dynamic behaviors.

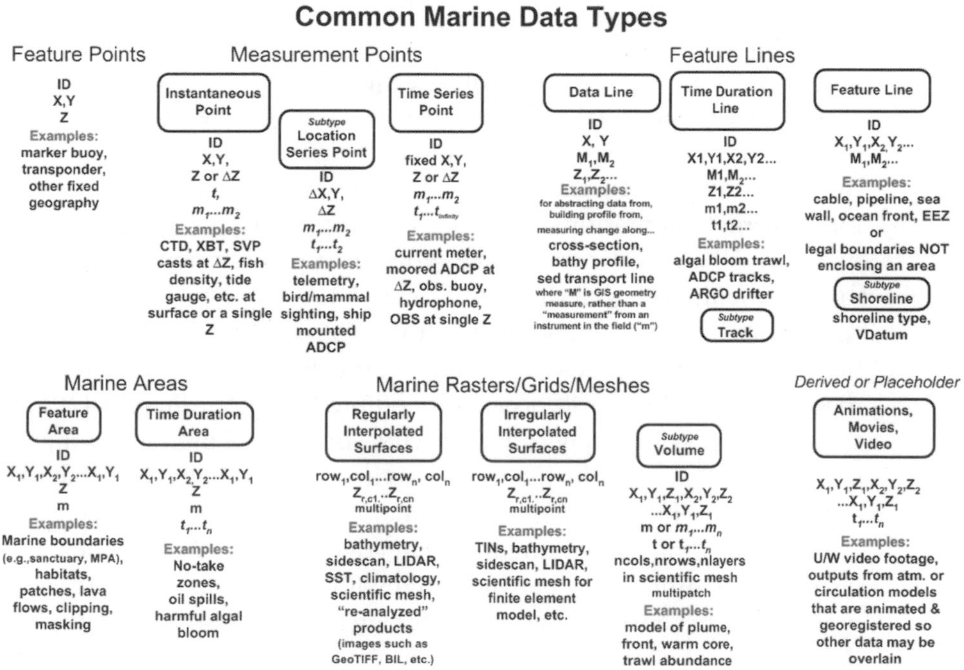

Figure 13.5. In order to develop more appropriate data structures to represent and relate coastal and marine GIS features, a draft set of common marine data types was developed as part of a fundamental conceptual framework for the ArcGIS Marine Data Model (http://dusk.geo.orst.edu/djl).

As an example, Figure 13.5 shows a summary of common marine data types that is part of the conceptual framework of the ArcGIS Marine Data Model, a software industry data model involving a collaboration of ESRI with Oregon State University, Duke University, the Danish Hydraulic Institute, and NOAA Coastal Services Center (http://dusk.geo.orst.edu/djl/arcgis; http://support.esri.com/datamodels). The common marine data types extend current GIS data structures (points, lines, polygons, and rasters) to include more temporally referenced data structures that will allow for better representation of spatially and temporally dynamic marine data. For example, an "instantaneous point" would provide for marine observations that are tied to a single moment in time, while a "time-duration line" feature would represent a ship track or other feature that moves along path in space and time. The "common marine data types" are intentionally generic, to provide the most basic spatial and temporal features and relationships needed to develop marine GIS application. Users involved in specific application areas would need to select and refine the core features they need to develop more detailed applications.

This ongoing project seeks to promote the interoperability of data and software for scientific and resource management users by providing the international marine GIS user community with a generic template to facilitate easier and faster input and conversion of data, better map creation, and most importantly, the means for conducting more complex spatial analyses by capturing the behavior of real-world objects in a geo-database.

Figure 13.5 focuses on the initial acquisition of marine data, and is thus concerned with the accurate sensing and collection of measurements from the marine environment, the dimensionality of these measurements, and their transformation from raw to processed for GIS implementation. Although it covers many of the data types used in all disciplines of oceanography and marine resource management, note that the 2-D, 3-D, and 4-D types are still classed as "placeholders" for the model (i.e., the GIS software is still unable to handle these data types satisfactorily and they are not available for many parts of the world ocean). As pointed out by Albrecht (2003), the development of application-specific conceptual models of objects and events, that include not only behaviors but also behaviors that can adapt to changing contexts, poses a major intellectual challenge.

In the end, how does one most effectively summarize, model, and visualize the differences between a digital representation and the real world? As the Earth's surface (water or land) is infinitely complex, decisions must be made about how to capture it, how to represent it in a digital system, how and where to sample it, and about what data format options to use in the GIS. This includes dealing with the inherent fuzziness of boundaries in the ocean, and addressing the multiple dimensionality and dynamism of oceanographic data, handling the temporal and dynamic properties of the seafloor, the water column, the sea surface, and the shoreline.

Grand Challenge: Dynamic Modeling in Space and Time

Probably the most interesting of the grand challenges facing marine GIS is the development of more dynamic models representing marine processes in space and time. The dynamic processes we are interested in may be geophysical, ecological, resource management or economic in nature, but all of them will require fundamental adaptations to the way we collect, process, analyze, and validate our data and our assumptions. It is still very difficult to imbed dynamic oceanographic models seamlessly into a GIS environment.

The questions that managers and policy-makers are asking are becoming increasingly specific. More than ever, geospatial analysts are now being asked to provide information to help forecast change over time. Parallel to the constraints we find representing a four-dimensional ocean environment with two-dimensional maps, our ability to forecast complex relationships at short time-intervals is constrained by statistical modeling approaches that were often originally developed for more static analyses. New developments in time-series and spatio-temporal modeling approaches are going to be crucial to completing the analytical framework of marine geospatial analysis. Many of these may be borrowed and adapted from the geocomputation, including diffusion modeling, time-series regression, cellular automata and network, extensions, differential equation modeling, and spatial evolutionary algorithms (e.g., Box, 2000; Yuan, 2000; Peuquet, 2002; Albrecht, 2003; Green and King, 2003b)

Conclusion

This chapter has reviewed the fundamental role of geospatial thinking and analysis to coastal and marine science and management, the current state of marine GIS and geospatial analysis, and some insights on longstanding challenges and future trends in data access and exchange, representation and modeling of marine data, and dynamic spatio-temporal modeling of processes (physical, ecological, and socio-economic). The demands on the marine GIS community for increased precision, accuracy, and more detailed analytical models have been increasing rapidly over the last several years and will continue to increase in the future. This, in turn, is forcing a rapidly increasing need for significantly more robust:

- data dissemination tools;
- spatio-temporal data standards & protocols;
- distributed processing & collaboration tools; and
- dynamic modeling & analysis tools.

As these demands for "precision management" and robust tools increase, it will be appropriate and timely to re-examine underlying data models in GIS and to develop new approaches particularly with regard to large-scale regional, interdisciplinary academic research projects. Such projects, within the new paradigm of "distributed" collaboration, will have an impact on both marine and terrestrial GIS. And marine GIS will continue to pose fundamental questions in the representation and analysis of spatial and temporal information, chief of which may be "how does one represent combinations of geometric objects and scalar fields, especially when the data are 'in flux'?" In order to take full advantage of new innovations in marine spatial analysis, end-users will need to keep up with emerging trends from the information systems, spatial analysis, and statistical analysis communities.

Future advances will take time. The archival nature of terrestrial GIS has meant that large GISs have been reticent to adopt new algorithms, much less new data models, as many users have needed a stable platform for their work. However, advocates of software component technology (e.g., Microsoft's Component Object Model, Sun's Java Beans, etc.) convincingly argue that the GIS of the future will not be monolithic, but will be composed of intercommunicating modules, once interfaces for geospatial information can be standardized and published. The Open Geospatial Consortium (http://www.opengeospatial.org) and others are pushing strongly in this direction. These efforts imply that prototypes that validate alternative representations or computational approaches, such as those posed by marine GIS, are especially valuable now, while standards are being considered and established. The increasing visibility of marine GIS and marine geospatial analysis as an essential tool for marine science and management is a testament to its growing usefulness across the field.

References

Albrecht, J., 2003. Dynamic Modeling, UCGIS Short-Term Priority White Paper, Washington, D.C., University Consortium for Geographic Information Science, http://www.ucgis.org.

Allison, M. L., Baru, C., and Jordon, T. H., 2003. Building the Geoinformatics systems: Coordination of the environmental cyberinfrastructure for the Earth science, Abstracts of the Geological Society of America Annual Meeting, Seattle, WA.

Bartlett, D. J., 1993a. *GIS and the Coastal Zone: An Annotated Bibliography*, Santa Barbara, CA, National Center for Geographic Information and Analysis.

Bartlett, D. J, 1993b. *Space, time, chaos, and coastal GIS*, International Cartographic Conference, Cologne, Germany.

Bartlett, D. J., 2000. Working on the frontiers of science: Applying GIS to the coastal zone. In Wright, D. J., and Bartlett, D. J., *Marine and Coastal Geographical Information Systems*, Taylor & Francis, London, 11-24.

Baru, C., 2004. GEON: *The GEON grid software architecture*, Proceedings of the 24th Annual ESRI User Conference, San Diego, CA.

Berry, J., 1995. Spatial Reasoning for Effective GIS, Fort Collins, CO, GIS World, Inc.

Box, P., 2000. Garage band science and dynamic spatial models, *Journal of Geographical Systems*, 2:49-54.

Breman, J. (ed.), 2002, *Marine Geography: GIS for the Oceans and Seas*, ESRI Press, Redlands, CA, 224 pp.

Chrisman, N., 1982. A theory of cartographic error and its measurement in digital databases. Proceedings, Fifth International Symposium on Computer-Assisted Cartography (Auto Carto 5), Falls Church, VA, American Society for Photogrammetry & Remote Sensing and American Congress on Surveying and Mapping, 159-168.

Convis, C. L., Jr. (ed.), 2001. *Conservation Geography: Case Studies in GIS, Computer Mapping, and Activism*, Redlands, CA, ESRI Press, 219 pp.

Egenhofer, M. J., 2002. Toward the semantic geospatial web, in Proceedings of the Tenth ACM International Symposium on Advances in Geographic Information Systems, McLean, VA.

Goldsmith, R., 2000. Some applications and challenges in extending GIS to oceanographic research, Proceedings of the 20th Annual ESRI User Conference, San Diego, CA, Paper 390.

Goodchild, M. F., 2000. Foreword. In Wright, D.J. and Bartlett, D.J., *Marine and Coastal Geographical Information Systems*, Taylor & Francis, London, xv.

Goodchild, M. F., Egenhofer, M. J., Fegeas, R., and Kottman, C. A. (eds.), 1999. *Interoperating Geographic Information Systems*, New York, Kluwer.

Green, D. R., and King, S. D. (eds.), 2003a. *Coastal and Marine Geo-Information Systems: Applying the Technology to the Environment*, Dordrecht, The Netherlands, Kluwer, 616 pp.

Green, D. R. and King, S. D., 2003b. Progress in geographical information systems and coastal modeling: An overview, in Lakhan, V. C. (Ed.), *Advances in Coastal Modeling*, Elsevier B.V., Amsterdam, 553-80.

Halpin, P. N., 2004. New innovations in marine spatial analysis, American Association for the Advancement of Science Annual Meeting, Symposium: New Approaches to Conserving Marine Animals in a Dynamic Ocean, Track: Living Oceans and Coastlines, Seattle, Washington, Session 10232.

Hankin, S., and the Data Management and Communications Steering Committee, 2003, The U.S Integrated Ocean Observing System (IOOS) Plan for Data Management and Communications (DMAC), Part I, Ocean.US, Arlington, VA, 12 pp., http://www.sccoos.ucsd.edu/docs/dmac_plan_exec_summ_hi.pdf. Last accessed March 13, 2004.

Heuvelink, G. B. M., 1998. *Error Propagation in Environmental Modelling with GIS*, Taylor & Francis, London.

Kuhn, W., 2003. Semantic reference systems, *International Journal of Geographical Information Science*, 17(5), 405-9.

Li, R., and Saxena, N. K.,1993. Development of an integrated marine geographic information system, *Marine Geodesy*, 16: 293-307.

Lockwood, M., and Li, R., 1995. Marine geographic information systems: What sets them apart?, *Marine Geodesy*, 18(3): 157-59.

Longley, P. A., Goodchild, M. F., Maguire, D., and Rhind, D. W., 1999. *Geographical Information Systems: Principles, Techniques, Applications, and Management*, 2nd edition, New York, Wiley, 1296 pp.

Mandelbrot, B. B., 1967. How long is the coast of Great Britain: Statistical self-similarity and fractional dimension, *Science*, 155, 636-38.

McGuinness, D. L., 2002. Ontologies come of age, in Fensel, D., Hendler, J., Lieberman, H., and Wahlster, W. (eds.), *Spinning the Semantic Web: Bringing the World Wide Web to Its Full Potential*, Cambridge, MA, MIT Press, http://www.ksl.stanford.edu/people/dlm/papers/ontologies-come-of-age-mit-press-(with-citation).htm. Last accessed August 31, 2004.

Peuquet, D., 2002. *Representations of Space and Time*, New York, Guilford.

Schaefer, A., and Schlueter, M., 2003. Marine GIS, http://www.awi-bremerhaven.de/GEO/Marine_GIS/, Alfred Wegener Institute Foundation for Polar and Marine Research. Last accessed February 29, 2004.

Treml, E., Smillie, H., Cohen, K., Fowler, C., and Neely, R. 2002. Spatial policy via the web: Georeferencing the legal and statutory framework for integrated regional ocean management, in Wright D.J. (ed.), *Undersea in GIS*, ESRI Press, Redlands, CA, pp. 211-30.

Valavanis, V. D., 2002. *Geographic Information Systems in Oceanography and Fisheries*, London, Taylor & Francis, 240 pp.

Wright, D. J., 2000. Spatial reasoning for marine geology and geophysics, in Wright, D. J., and Bartlett, D. J. (eds.), *Marine and Coastal Geographical Information Systems*, London: Taylor & Francis, 117-28.

Wright, D. J. (ed.), 2002. *Undersea with GIS*, Redlands, CA, ESRI Press, 253 pp.

Wright, D. J., and Bartlett, D. J. (eds.), 2000. *Marine and Coastal Geographical Information Systems*, London, Taylor & Francis, 320 pp.

Wright, D. J., and Goodchild, M. F., 1997. Data from the deep: Implications for the GIS community, *International Journal of Geographical Information Science*, 11(6): 523-28.

Wright, D. J., Wood, R., and Sylvander, B., 1998. ArcGMT: A suite of tools for conversion between Arc/INFO and Generic Mapping Tools (GMT), *Computers and Geosciences*, 24(8): 737-44.

Yuan, M., 2000. Modeling geographic information to support spatiotemporal queries, in Frank, A. U., Raper, J. and Cheyland, J. P. (eds.), *Life and Motion of Socio-Economic Units*, European Science Foundation (ESF) Series, London, Taylor and Francis.

Author Biographies

Satie Airame works on issues in marine science and policy. She was a science advisor to the Channel Islands National Marine Sanctuary during design, establishment, and initial monitoring of a network of marine protected areas. Currently, she is working on marine policy with an academic research consortium at the University of California, Santa Barbara.

Charles Alexander has been National Programs Branch Chief at NOAA's National Marine Sanctuary Program for the past five years. Earlier, Charles worked for over 15 years in NOAA's Strategic Assessments Division, conducting a series of natural resource management projects on a wide range of issues including coastal wetlands, estuarine nutrient enrichment, coastal data delivery, remote sensing, and shellfish growing waters. He completed an M.S. in Marine Science and a Masters in Public Administration at Louisiana State University in 1985. He was selected as a Thomas J. Watson Fellow in 1978, a National Sea Grant Fellow in 1983, and a Presidential Management Intern in 1985.

Laura Anderson has been a consultant to the non-profit Port Orford Ocean Resource Team since its inception in 2000. She has an intimate knowledge of the fishing industry, spending many summers commercial trolling and crabbing on her father's fishing boat. She worked for two years in coastal resource management for the U.S. Peace Corps in the Philippines, and subsequently earned a Master's in Marine Resource Management at Oregon State University. Her consulting work has involved social marketing and sustainability, coastal tourism management, groundfish fleet reduction, marine reserves communication, and watershed council outreach and strategic planning. In 2002 Laura co-founded an Oregon-based seafood company, Local Ocean Seafoods, with a focus on sustainable fisheries and fishing communities.

Jeff Ardron is recognized as a leading marine GIS specialist in British Columbia, and has been responsible for the development of several new analysis techniques and products. His work is mostly focussed on marine reserve design, though topics ranging from offshore oil and gas to salmon farms to sustainable fisheries have also crossed his desk. He is principal marine analyst for Living Oceans Society, a progressive non-governmental organization based in British Columbia. In addition to ecology, his passions include music, fine micro-brews, and sailing—though not necessarily in that order.

Joseph J. Bizzarro graduated from Dartmouth College in 1992 with a B.A. in Biology and will complete his M.S. degree in Ichthyology at Moss Landing Marine Laboratories (MLML) in the spring of 2005. He has worked extensively with Dr. Gary Greene at MLML's Center for Habitat Studies since 1997 and has been the Center's project manager since 2000. In addition to advanced video analysis and GIS skills, Joe has significant training and professional experience as an ichthyologist and fisheries biologist. He also maintains a position as staff scientist at MLML's Pacific Shark Research Center.

Peter C. Black is manager of geospatial information for Environmental Defense in Oakland, California, a national non-partisan environmental organization that uses science, economics, and law to create durable solutions to environmental problems. He specializes in creating custom GIS applications for environmental issues.

Philip Bloch provides scientific support for conservation and restoration initiatives on the more than 2.4 million acres of aquatic lands managed by Washington State Department of Natural Resources. He studied landscape ecology as a graduate student at Duke University and has published research on behavioral ecology, the Endangered Species Act, and marine conservation planning.

Kate Bonzon is Policy and Research Analyst for Environmental Defense, a national non-partisan environmental organization that uses science, economics, and law to create durable solutions to environmental problems. She researches sustainable financing for fisheries reform, and ways to incorporate socioeconomic concerns into policy decisions. She is co-founder of Independent Architecture Group, a company dedicated to designing sustainable development and redevelopment projects, and is an advisor to Pacific Marine Farms, a sustainable aquaculture company.

Villy Christensen is an Associate Professor at the Fisheries Centre, University of British Columbia. He is one of the main developers of the Ecopath with Ecosim ecosystem modeling approach, and is currently focused on estimating spatial biomass trends in marine ecosystems.

A long-term fishing community leader, **Leesa Cobb** lives on the southern Oregon coast in Port Orford, where she and her husband own and operate a commercial fishing business. Leesa has been working as communications coordinator for POORT for three years. She is a founding board member of Pacific Marine Conservation

Council, and previously served as president and treasurer. In June 2000 Leesa contracted with Oregon State University to work with the Groundfish Disaster Outreach Program, helping fishermen who want to transition out of fishing access existing programs and services. She has helped more than 50 members of the Port Orford fishing community begin their transition to new careers.

Randy Dana is GIS Coordinator for the Oregon Ocean-Coastal Management Program. He received a B.A. in General Sciences from Portland State University and completed coursework for an M.S. in Geography from the same institution.

Tom Dean is the Executive Director for the Vashon-Maury Island Land Trust and has been managing estuarine habitat restoration projects for more than eight years. He has also managed several restoration planning projects and collaborates with land trusts around Puget Sound on shoreline conservation efforts. He holds a B.A. in English from the University of Oregon.

Sylvia Earle, one of the world's most respected ocean scientists and explorers, is chair of Deep Ocean Exploration & Research, Inc., and an Explorer-in-Residence at the National Geographic Society. In addition, she serves as an honorary president for the Explorers Club, executive director for Global Marine Conservation for Conservation International, and program coordinator and advisorycouncil chair for the Harte Research Institute for Gulf of Mexico Studies. She is an adjunct scientist at the Monterey Bay Aquarium Research Institute, a director of Kerr-McGee Inc., a director for the Common Heritage Corporation, and serves on various boards, foundations, and committees relating to marine research, policy, and conservation. These include the World Resources Institute, World Environment Center, Woods Hole Oceanographic Institute, Duke University Marine Laboratory, Mote Marine Laboratory, Lindbergh Foundation, World Wildlife Fund, Natural Resource Defense Council, and the Ocean Conservancy. In 1998 she was named by *Time* magazine as one of its Heroes of the Planet. Sylvia is a Fellow of the American Association for the Advancement of Science, the Marine Technology Society, California Academy of Sciences, and World Academy of Arts and Sciences. She was one of the first scientists to research marine ecosystems, holds numerous diving records, and has spent more than 6,000 hours underwater. She is also the former chief scientist for NOAA and has written more than 100 scientific and popular publications and several books including *Wild Ocean: America's Parks Under the Sea* and *Dive: My Adventures in the Deep Frontier*.

Mercedes D. Erdey graduated from Saint Stephen University in Hungary with a degree in environmental protection and landscape management. Prior to matriculating at Moss Landing Marine Laboratories (MLML), she spent the academic year of 2000/2001 at the University of Gent in Belgium studying marine sciences, GIS techniques, and remote sensing. Mercedes currently works at MLML's Center For Habitat Studies on a variety of GIS, remote sensing, and habitat mapping projects.

Peter Etnoyer is a marine ecologist with a background in biogeography, octocoral systematics, and physical oceanography. He is president of Aquanautix, a consulting company that produces GIS analyses, investigations, and reports in marine protected area planning, endangered species, deepsea biology, and satellite oceanography.

Zach Ferdaña is a marine conservation planner for the Global Marine Initiative of The Nature Conservancy, with a focus on the Pacific Northwest coast. Using GIS technology and the Conservancy's ecoregional planning approach, Zach has focused the organization to adopt planning innovations in coastal, nearshore, and offshore environments. He received his degree in Environmental Studies at The Evergreen State College in Olympia, Washington, in 1994. He began his career in marine conservation with People for Puget Sound in 1997, constructing spatial databases and analytical techniques in marine and nearshore environments for their conservation and restoration projects. He joined The Nature Conservancy in 2000 and continues to work on marine planning efforts and GIS-related projects.

Rod Fujita is a senior scientist and marine ecologist at Environmental Defense, a leading non-governmental environmental research and advocacy organization. He has been working on climate change, fisheries management, and ocean wildlife protection since 1989, with a focus on crafting durable solutions that incorporate the human dimension. He is the author of numerous scientific papers, reports, and a recent book, *Heal the Ocean* (New Society Publishers), serves on a number of state, regional, and national advisory boards, and gives frequent public lectures and workshops.

Jessemine Fung is currently with CommEn Space, a non-profit spatial analysis center, working on conservation GIS projects throughout the Pacific Northwest and Northern Rockies region. Earlier, she managed People For Puget Sound's science and habitat program. She holds an M.S. in ecology and evolutionary biology.

Dr. H. Gary Greene received an M.S. (Geology/Geophysics) from San Jose State University/Moss Landing Marine Laboratories (MLML) in 1969, and a Ph.D. (Geology/Marine Geology) from Stanford University in 1977. Gary retired from the U.S. Geological Survey in 1994 after more than 28 years of service and took up the directorship of MLML. Presently, he is professor of Marine Geology and Director of the Center for Habitat Studies, which he founded in 1994. Gary is also a part-time senior scientist at the Monterey Bay Aquarium Research Institute. His expertise lies in the study of active plate margins. Presently, his research involves the characterization of marine benthic habitats and the study of underwater landslides. Much of his time is now spent in working to standardize the way the scientific community describes and maps marine benthic habitats.

Tanya Haddad is an information systems specialist with the State of Oregon Ocean-Coastal Management Program (OCMP) and coordinator of the Oregon Coastal Atlas. After obtaining degrees from Tufts University (B.S.) and Duke University (M.E.M.), she was nominated in 1998 by Connecticut Sea Grant for a NOAA Coastal Services Center Coastal Management Fellowship, and was placed with the OCMP to establish and expand a project known as the Dynamic Estuary Management Information System (DEMIS) to several Oregon estuaries and their watersheds. She has since incorporated DEMIS and several other OCMP data products into the Oregon Coastal Atlas.

Pat Halpin is an associate professor of the Practice and director of the Geospatial Analysis Program at the Nicholas School of the Environment and Earth Sciences at Duke University. Pat specializes in geospatial analysis for ecological and conservation applications in both marine and terrestrial environments. He is a principle investigator for the Ocean Biogeographic Information System (OBIS-SEAMAP) program providing geospatial data and analysis of marine mammals, sea turtles and seabirds of the world as well as numerous other marine GIS and marine spatial ecology programs.

Jamison L. Higgins is a physical scientist for the National Centers for Coastal Ocean Science, Center for Coastal Monitoring and Assessment, Biogeography Program. She works to develop information and analytical capabilities through research, monitoring, and assessment on the distribution and ecology of living marine resources and their associated habitats for improved ecosystem management.

Matt Kendall is a biologist with NOAA's Biogeography Team at the National Centers for Coastal Ocean Science, Center for Coastal Monitoring and Assessment. He is an expert in marine landscape ecology.

Paul Klarin is a senior policy analyst with the State of Oregon Ocean-Coastal Management Program (OCMP). His current and past projects with the OCMP include hazards program coordination, land use plan reviews, special area management plans, permit reviews, and special projects involving GIS as related to ocean shore resources. He holds degrees from the University of Washington (M.A. in Marine Affairs), and the University of California, Santa Barbara (B.A., Political Science).

Holly Lopez graduated from California State University at Monterey Bay with a B.S. degree in Earth Systems Science and Policy in 2001 and is currently a graduate student at Moss Landing Marine Laboratories (MLML) studying marine geology. She works at the Center for Habitat Studies at MLML, creating marine benthic habitat maps and interpreting geology from remote sensing data.

John Marra is a coastal hazards specialist at the NOAA Pacific Services Center in Honolulu, Hawaii. Prior to this post he was a hazards specialist with the Oregon Ocean-Coastal Management Program and principal of the Shoreland Solutions consulting firm, where he directed littoral cell management planning efforts underway in seven Oregon coast littoral cells. Other projects have included the refinement of models that can be used to predict extreme run-up and foredune retreat along dune-backed shorelines, as well a cooperative undertaking with the NOAA Coastal Service Center among others that, via a set of customized GIS tools, provides decision makers with the ability to rapidly project the total water level and the landward extent of foredune retreat under different storm event scenarios. John holds a Ph.D. in Geology from the University of Canterbury, New Zealand, and a B.A. in Geology from the University of Montana.

Mike Mertens has served as manager and senior GIS analyst of Ecotrust's GIS team for the last four years. He has over 10 years of experience in using GIS specific to various aspects of spatial data analysis and geographic information science, ranging from exploratory spatial data analysis to remote sensing, spatial statistics, and development of spatially explicit simulation tools.

Mark Monaco is a marine biologist who leads NOAA's Biogeography Program that focuses on the spatial ecology of marine, estuarine, and coral reef living marine resources. His research interests address defining and evaluating the efficacy of marine protected areas and the development of assessment approaches for place-based management. He serves as the chair of the NOAA working group for the mapping of coral reef ecosystems and co-chair of the US Coral Reef Task Force mapping and information synthesis work group.

Lance Morgan is chief scientist for Marine Conservation Biology Institute, a non-profit conservation organization dedicated to protecting ocean life through science and conservation advocacy. He is an affiliate professor at the University of Washington's School of Marine Affairs, and a marine advisor to the North American Commission for Environmental Cooperation.

Elliott A. Norse is president of Marine Conservation Biology Institute in Redmond, Washington. In 1980 he defined the concept of conserving biological diversity. His four books are *Conserving Biological Diversity in Our National Forests* (1986), *Ancient Forests of the Pacific Northwest* (1990), *Global Marine Biological Diversity* (1993) and *Marine Conservation Biology: The Science of Maintaining the Sea's Biodiversity* (2005).

Daniel Pauly is a professor at and director of the Fisheries Centre, University of British Columbia, and principal investigator of the Sea Around Us Project, devoted to studying, documenting, and mitigating the impact of fisheries on marine ecosystems. In 2001 he was awarded the Murray Newman Award for Excellence in Marine Conservation Research, sponsored by the Vancouver Aquarium, and the Oscar E. Sette Award of the Marine Fisheries Section, American Fisheries Society. He was named honorary professor at Kiel University, Germany in late 2002. In 2003 he was named one of UBC's distinguished university scholars and elected a fellow of the Royal Society of Canada (Academy of Science). In 2004, he received the Roger Revelle Medal from IOC/UNESCO, and the American Fisheries Society Award of Excellence.

David Revell is currently a doctoral candidate in Earth Sciences (Coastal Geology) at University of California, Santa Cruz. He has served on the staff of the Oregon Chapter of the Surfrider Foundation, and has been a NOAA coastal management fellow in coastal and ocean policy with the Oregon Ocean-Coastal Management Program. He holds a B.A. in Environmental Studies and Geography from the University of California, Santa Barbara, and an M.S. in Marine Resource Management from Oregon State University.

Charles Steinback is the lead fisheries GIS analyst and cartographer at Ecotrust. During the last four years his work has focused on developing the Ocean Communities "3E" ANalytical framework (OCEAN), a marine GIS platform for spatially integrating socio-economic and ecological data and analyses for community-based management. Charles also serves as the lead GIS analyst for the State of the Salmon Consortium, a joint venture between Ecotrust and the Wild Salmon Center that is dedicated to salmon stock monitoring and assessment across the Pacific Rim. He is a co-author of the consortium's first publication, *Atlas of Pacific Salmon, The First Map-Based Status Assessment of Salmon in the North Pacific*, to be published by the University of California Press.

Mitchell Tartt is a marine ecologist with the NOAA National Marine Sanctuary Program (NMSP). He is a member of the NMSP headquarter's science team, which focuses on NMSP characterization, research, and monitoring activities. Mitchell specializes in the integration of information in support of the management requirements of the NMSP.

Janet Tilden graduated from Eckerd College in 1999 with a B.S. in Marine Science and Moss Landing Marine Laboratories (MLML) in 2004 with an M.S. in geological oceanography . As a student in the Geological Oceanography Lab at MLML, she completed a multidisciplinary thesis involving a mix of biology, ecology, and geology. Janet currently works at the U.S. Geological Survey in Menlo Park, California.

Astrid J. Scholz is ecological economist and vice president for knowledge systems at Ecotrust, a Portland (Oregon) based conservation organization dedicated to strengthening economics and the environment from Alaska to California. She serves on the faculty of the Oregon State University College of Oceanic and Atmospheric Science, and oversees several research projects that integrate socioeconomic information into natural resource management.

Scott Wallace (Ph.D.) is a marine conservationist, researcher, naturalist, and educator presently involved with the assessment of marine fishes at risk, marine protected areas, and sustainable fisheries. He works independently through his company, Blue Planet Research and Education, and in this capacity consults to the marine conservation sector in British Columbia.

Reg Watson is a senior research associate at the Fisheries Centre, University of British Columbia. He combines expertise in database manipulation, GIS, modeling, and international fisheries to prepare detailed maps of global fisheries catch for the Sea Around Us Project.

Victoria A. Wedell is a 2004 graduate of the Marine Resource Management Program at Oregon State University. She is currently a John A. Knauss Marine Policy Fellow serving with the NOAA National Marine Sanctuary Program.

Jacques White recently joined The Nature Conservancy in Seattle to expand their fledgling Marine Conservation Program. He received a Ph.D. in marine, estuarine, and environmental science from University of Maryland and has done estuarine research along three major U.S. coastlines.

Dawn Wright is professor of Geography and Oceanography at Oregon State University, and the director of the Davey Jones' Locker Seafloor Mapping/Marine GIS Laboratory. Her research interests include geographic information science, marine geography, tectonics of mid ocean ridges, and the processing and interpretation of high-resolution bathymetric, video, and underwater photographic images. She has completed oceanographic fieldwork in some of the most geologically active regions of the planet, including volcanoes under the Pacific, Atlantic and Indian Oceans, and has had three dives in the Alvin submersible. Dawn serves on the editorial boards of the *International Journal of Geographical Information Science, Transactions in GIS*, and *Geospatial Solutions*, and was a member of the National Academy of Sciences' National Needs for Coastal Mapping and Charting Committee. Her other books include *Marine and Coastal Geographical Information Systems* (edited with D. Bartlett, Taylor & Francis, 2000), *Undersea with GIS* (ESRI Press, 2002), and *ArcMarine: A Geospatial Framework for Ocean and Coastal Analysis* (with M. Blongewicz, P. Halpin, and J. Breman, ESRI Press, 2006). She holds degrees from the University of California-Santa Barbara (Ph.D. in Physical Geography and Marine Geology), Texas A&M (M.S., Oceanography), and Wheaton College in Illinois (B.S., Geology).

Index

N

U

V

W

X

Z